Introduction to
Modern Virology

Introduction to Modern Virology

N. J. Dimmock
A. J. Easton
K. N. Leppard

All from:
Department of Biological Sciences
University of Warwick
Coventry

FIFTH EDITION

**Blackwell
Science**

© 1974, 1980, 1987, 1994, 2001 by
Blackwell Science Ltd
Editorial Offices:
Osney Mead, Oxford OX2 0EL
25 John Street, London WC1N
  2BS
23 Ainslie Place, Edinburgh EH3
  6AJ
350 Main Street, Malden
  MA 02148-5018, USA
54 University Street, Carlton
  Victoria 3053, Australia
10, rue Casimir Delavigne
  75006 Paris, France

Other Editorial Offices:
Blackwell Wissenschafts-Verlag
  GmbH
Kurfürstendamm 57
10707 Berlin, Germany

Blackwell Science KK
MG Kodenmacho Building
7–10 Kodenmacho Nihombashi
Chuo-ku, Tokyo 104, Japan

Iowa State University Press
A Blackwell Science Company
2121 S. State Avenue
Ames, Iowa 50014-8300, USA

First published 1974

Reprinted 1978
Second edition 1980
Third edition 1987
Fourth edition 1994
Reprinted 1995, 1996, 1998, 1999
Fifth edition 2001

Set by Best-set Typesetter Ltd.,
Hong Kong
Printed and bound in Great Britain
by T.J. International Ltd, Padstow,
Cornwall

A catalogue record for this title
is available from the British Library

ISBN 0-632-05509-X

Library of Congress
Cataloging-in-Publication Data
Dimmock, N. J.
  Introduction to modern virology /
  Nigel J. Dimmock,
  Andrew Easton,
  Keith Leppard.—5th ed.
    p.;   cm.
  Includes bibliographical
    references and index.
  ISBN 0-632-05509-X
  1. Virology.   2. Virus diseases.
  I. Easton, Andrew.   II. Leppard,
    Keith.   III. Title.
  [DNLM: 1. Virus Diseases.
  2. Viruses. QW 160 D582i 2001]
  QR360 .D56 2001
  616′.0194—dc21
                    00-069805

DISTRIBUTORS
  Marston Book Services Ltd
  PO Box 269
  Abingdon, Oxon OX14 4YN
  (*Orders*: Tel: 01235 465500
           Fax: 01235 465555)

The Americas
  Blackwell Publishing
  c/o AIDC
  PO Box 20
  50 Winter Sport Lane
  Williston, VT 05495-0020
  (*Orders*: Tel: 800 216 2522
           Fax: 802 864 7626)

Australia
  Blackwell Science Pty Ltd
  54 University Street
  Carlton, Victoria 3053
  (*Orders*: Tel: 3 9347 0300
           Fax: 3 9347 5001)

For further information on
Blackwell Science, visit our
website:
www.blackwell-science.com

The Blackwell Science logo is a
trade mark of Blackwell Science
Ltd, registered at the United
Kingdom Trade Marks Registry

Contents

Preface

Virology continues to move fast and, in response, another complete revision of this volume has been undertaken. With an augmented authorship, this edition has refined the balance of basic information with the latest advances in the field, which we hope will interest and stimulate the reader. The emphasis remains on animal viruses (where progress and human self-interest are greatest), but the major features of plant and bacterial viruses are maintained. Sadly space limits what we can include. Progress has necessitated major revisions in areas including HIV and AIDS, antiviral chemotherapy, viral carcinogenesis, evolution of influenza viruses and emerging infections. Gene expression and its regulation have now been expanded into two chapters, dealing with DNA and RNA viruses, respectively. We have used anglicized names for virus families, etc. throughout, which we hope will make for easier reading and remembering. In addition, although not caused by viruses, we have continued to cover prion diseases, essential reading for any student of infectious disease and, in acknowledgement of the increase in knowledge, this forms a new chapter. Finally, we should underline that, despite progress, much of virology remains to be elucidated, and we have attempted throughout the text to highlight deficiencies in understanding, particularly the development of safe, effective and affordable vaccines and therapeutic drugs. Virology is not just an academic subject, and scientific advances need to be made to protect and treat much of the world's population.

Nigel Dimmock, Andrew Easton, Keith Leppard

Chapter 1
Towards a definition of a virus

Towards the end of the last century, the germ theory of disease was formulated and pathologists were confident that for each infectious disease there would be found a micro-organism that could be seen with the aid of a microscope, cultivated on a nutrient medium and retained by filters. There were, admittedly, a few organisms that were so fastidious that they could not be cultivated *in vitro* (literally, in glass, meaning in the test tube), but they did satisfy the other two criteria. However, a few years later, in 1892, Iwanowski was able to show that the causal agent of tobacco mosaic, which manifests as a discoloration of the leaf, passed through a bacteria-proof filter, and could not be seen or cultivated. Iwanowski remained unimpressed by his discovery, but Beijerinck, on repeating the experiments in 1898, was convinced of the existence of a new form of infectious agent, which he termed '*contagium vivum fluidum*'. In the same year, Loeffler and Frosch came to the same conclusion regarding the cause of foot-and-mouth disease. Furthermore, because foot-and-mouth disease could be passed from animal to animal, with great dilution at each passage, the causative agent had to be reproducing and thus could not be a bacterial toxin. Viruses of other animals were soon discovered. Ellerman and Bang reported the cell-free transmission of chicken leukaemia in 1908, and in 1911 Rous discovered that solid tumours of chickens could be transmitted by cell-free filtrates. These were the first indications that viruses can cause cancer.

Although studies on bacterial viruses were later to prove an excellent system for investigating the virus–host relationship, they were the last to be discovered. In 1915, Twort published an account of a glassy transformation of micrococci. He had been trying to culture the smallpox agent on agar plates but the only growth obtained was that of some contaminating micrococci. Upon prolonged incubation, some of the colonies took on a glassy appearance and, once this occurred, no bacteria could be subcultured from the affected colonies. If some of the glassy material was added to normal colonies, they too took on a similar appearance, even if the glassy material was first passed through very fine filters. Among the suggestions that Twort put forward to explain the phenomenon was the existence of a bacterial virus or the secretion by the bacteria of an enzyme that could lyse the producing cells. This idea of self-destruction by secreted enzymes was to prove a controversial topic over the next decade. In 1917 d'Hérelle observed a similar phenomenon in dysentery bacilli. He observed clear spots on lawns of dysentery bacilli and, although realizing that this was not an original observation, resolved to find an explanation for them. On noting the lysis of broth cultures of pure dysentery bacilli by filtered emulsions of faeces, he immediately realized that he was dealing with a bacterial virus. As this virus was incapable of multiplying except at the expense of living bacteria, he called his virus a *bacteriophage* (bacterium-eater), or *phage* for short.

Thus, the first definition of these new agents, the viruses, was presented entirely in negative terms: they could not be seen, could not be cultivated and, most important of all, were not retained by bacteria-proof filters.

1.1 Assay of viruses

The observations of d'Hérelle, however, led to the introduction of two important techniques. The first of these was the preparation of stocks of bacterial viruses by lysis of bacteria in liquid cultures. This has proved invaluable in modern virus research, because bacteria can be grown in defined media to which radioactive precursors can be added to 'label' selected viral components. Many animal viruses can be similarly grown in cultures of the appropriate cell. Second, d'Hérelle's observations provided means of assaying these invisible agents. One method was to grow a large number of identical cultures of a susceptible bacterium species and to inoculate these with dilutions of the virus-containing sample. If the sample was diluted too far, none of the cultures would lyse. However, in the intermediate range of dilutions not all of the cultures lyse, because not all receive a virus particle, and the assay is based on this. For example, d'Hérelle noted that, in 10 test cultures inoculated with a volume corresponding to 10^{-11} ml, only three were lysed. Thus three cultures received one or more viable phage particles whereas the remaining seven received none, and it can be concluded that the sample contained between 10^{10} and 10^{11} viable phages per ml. It is possible to apply statistical methods to *end-point dilution* assays of this sort and obtain more precise estimates. The other method suggested was the *plaque assay method*, which is now the most widely used and most useful. D'Hérelle observed that the number of clear spots or plaques formed on a lawn of bacteria (Fig. 1.1a) was inversely proportional to the dilution of bacteriophage lysate added. Thus, the infectivity titre of a virus-containing solution can be readily determined in terms of *plaque-forming units* (PFUs) and, if each physical virus particle in the preparation gives rise to a plaque, then the *efficiency of plating* is unity.

Both these methods were later applied to the more difficult task of assaying the infectivity of plant and animal viruses. However, because of the labour, time, cost and ethical considerations, end-point dilution assays using animals are avoided where possible. For the assay of plant viruses, a variation of the plaque assay, the *local lesion assay*, was developed by Holmes in 1929. He observed that countable necrotic lesions were produced on leaves of the tobacco plant, particularly *Nicotiana glutinosa*, inoculated with tobacco mosaic virus and that the number of local lesions depended on the amount of virus in the inoculum. Unfortunately, individual plants, and even individual leaves of the same plant, produce different numbers of lesions with the same inoculum. However, the opposite halves of the same leaf give almost identical

Fig. 1.1 Plaques of viruses. (a) Plaques of a bacteriophage on *Escherichia coli*. (b) Local lesions on a leaf of *Nicotiana* sp. caused by tobacco mosaic virus. (Courtesy of National Vegetable Research Station.) (c) Plaques of influenza virus on chick embryo fibroblast cells.

numbers of lesions and it is possible to compare the same dilutions of two virus-containing samples by inoculating them on the opposite halves of the same leaf (Fig. 1.1b).

A major advance in animal virology came in 1952, when Dulbecco devised a plaque assay for animal viruses. In this case a suspension of susceptible cells, prepared by trypsinization of a suitable tissue, is placed in Petri dishes or other culture vessel. The cells attach and grow across the surface until a monolayer of cells is formed. The nutrient medium bathing the cells is then removed and a suitable dilution of the virus added. After a short period of incubation (~1 h) to allow the virus particles to attach to the cells, nutrient agar is placed over the cells. After a further period of incubation of around 3 days (but ranging from 24 h to 24 days), a dye is added to differentiate living cells from the unstained circular areas that form the plaques (Fig. 1.1c). These days plaque assays are conducted using permanent cell lines. Some tumour viruses are not cytopathic (i.e. do not kill cells) so they cannot be assayed by this means. An alternative, for those viruses that cause morphological transformation of cells (see

Chapter 17), is a focus-formation assay in which a single infectious particle forms a discrete colony of cells on the surface of the monolayer. However, not all viruses are cytopathic, and infected cells can also be recognized by the presence of virus protein or nucleic acid products that they contain.

1.2 Multiplication of viruses

Although methods of assaying viruses had been developed, there were still considerable doubts as to the nature of viruses. D'Hérelle believed that the infecting phage particle multiplied within the bacterium and that its progeny were liberated upon lysis of the host cell, whereas others believed that phage-induced dissolution of bacterial cultures was merely the consequence of a stimulation of lytic enzymes endogenous to the bacteria. Yet another school of thought was that phages could pass freely in and out of bacterial cells and that lysis of bacteria was a secondary phenomenon not necessarily concerned with the growth of a phage. It was Delbruck who ended the controversy by pointing out that two phenomena were involved, lysis from within and lysis from without. The type of lysis observed was dependent on the ratio of infecting phages to bacteria (*multiplicity of infection*). When the ratio of phages to bacteria is no greater than 1 : 1, i.e. low multiplicity of infection, then the phages infect the cells, multiply and lyse the cells from within. When the multiplicity of infection is high, i.e. hundreds of phages per bacterial cell, the cells are lysed and there is no increase in phage titre but, rather, a decrease. Lysis is the result of weakening of the cell wall when large numbers of phages are attached. Convincing support for d'Hérelle's hypothesis was provided by the one-step growth experiment of Ellis and Delbruck (1939). A phage preparation such as bacteriophage T4 was mixed with a suspension of the bacterium *Escherichia coli* and, after allowing a few minutes for the phage to attach, the culture was diluted to stop further attachment. Samples of cells and medium were then withdrawn at regular intervals and assayed for infectivity. The result obtained is shown by the dashed line in Fig. 1.2. After a latent period of 30 min in which no infectivity increase could be detected, there was a sudden rise in PFUs. This 'burst' size represents the average of many different bursts from individual cells. Before a burst an infected cell scores as one PFU regardless of how many phages it contains.

Although the growth curve in Fig. 1.2 demonstrated the nature and kinetics of the process by which bacterial viruses are formed, it gave no indication of the events taking place inside the cell. In 1952, Doermann infected cells of *E. coli* with bacteriophage T4 in the presence of cyanide to synchronize intracellular events. By diluting the culture in antibody that neutralized the infectivity of phages outside the cell, he ensured that no more phages could attach to cells. Antibody has no effect on phage inside the cell because it does not cross cell membranes. The dilution step also effectively removed the cyanide. The culture was then incubated and aliquots of cells were lysed at intervals by the experimenter, and assayed for phage. Initially, the number of T4 PFUs (solid line) present inside the cell decreased to undetectable levels in what is called the eclipse period (Fig. 1.2), whereas the number of infected cells remained constant until lysis. This was followed by an increase in PFUs until lysis occurs. Thus the intra-

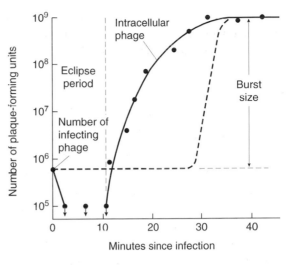

Fig. 1.2 A one-step growth curve of T4 bacteriophage after infection of susceptible bacteria. The dashed line shows the release of phages into the medium and the one-step growth curve. The solid line demonstrates that bacteriophages develop intracellularly (see text). Note that early samples contain fewer infectious phages than the original inoculum. In this period, virus is said to be eclipsed.

cellular nature of phage replication was demonstrated for the first time. It should also be noted that the kinetics of appearance of intracellular phage particles is *linear*, not exponential. This suggests that the particles are produced by assembly from component parts rather than by binary fission.

1.3 Viruses can be defined in chemical terms

The first virus was purified in 1933 by Schlessinger, using differential centrifugation. Chemical analysis of the purified bacteriophage showed that it consisted of approximately equal proportions of protein and deoxyribonucleic acid (DNA). A few years later, in 1935, Stanley isolated tobacco mosaic virus in paracrystalline form, and this crystallization of a biological material thought to be alive raised many philosophical questions about the nature of life. In 1937, Bawden and Pirie extensively purified tobacco mosaic virus and showed it to be nucleoprotein containing ribonucleic acid (RNA). Thus virus particles may contain either DNA or RNA, but not both. However, at this time it was not known that nucleic acid constituted genetic material.

The importance of viral nucleic acid

In 1949, Markham and Smith found that preparations of turnip yellow mosaic virus comprised two types of identically sized spherical particles, only one of which contained nucleic acid. Significantly, only the particles containing nucleic acid were infectious. A few years later, in 1952, Hershey and Chase demonstrated the independent functions of viral protein and nucleic acid. Bacteriophage T2 was grown in *E. coli* in the presence of ^{35}S (as sulphate) to label the protein moiety, or ^{32}P (as phosphate) to label the nucleic acid. Purified, labelled phages were allowed to attach to sensitive host cells and then given time for the infection to commence. The phages, still on the outside of the cell, were then subjected to the shearing forces of a Waring blender (Fig. 1.3).

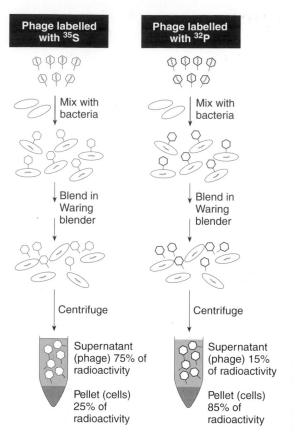

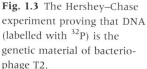

Fig. 1.3 The Hershey–Chase experiment proving that DNA (labelled with ^{32}P) is the genetic material of bacteriophage T2.

Treatment of the cells in this way removes any phage components attached to the outside of the cell, but does not affect viability. When the cells were removed from the medium, it was observed that 75% of the ^{35}S (i.e. phage protein) had been removed from the cells by blending, but only 15% of the ^{32}P (i.e. phage DNA). Thus, after infection, the bulk of the phage protein appears to have no further function and consequently it must be the DNA that is the carrier of viral heredity. The transfer of the phage DNA from its protein envelope to the bacterial cell upon infection also accounts for the existence of the eclipse period during the early stages of intracellular virus development, because the DNA on its own cannot normally infect a cell.

In another classic experiment, Fraenkel-Conrat and Singer (1957) were able to confirm by a different means the hereditary role of viral RNA. Their experiment was based on the earlier discovery that particles of tobacco mosaic virus can be dissociated into their protein and RNA components, and then reassembled to give particles that are morphologically mature and fully infectious (see Chapter 11). When particles of two different strains (differing in the symptoms produced in the host plant) were each disassociated and the RNA of one reassociated with the protein of the other, and vice versa, the type of virus that was propagated when the resulting 'hybrid' particles were used to infect host plants was always that from which the RNA was derived (Fig. 1.4).

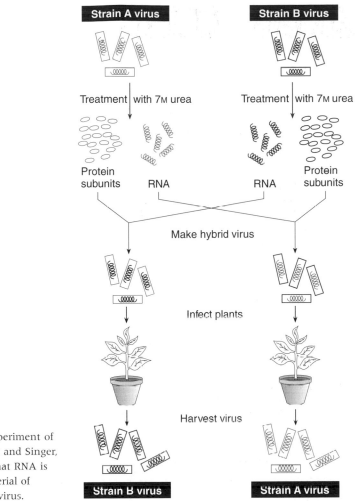

Fig. 1.4 The experiment of Fraenkel-Conrat and Singer, which proved that RNA is the genetic material of tobacco mosaic virus.

The ultimate proof that viral nucleic acid is the genetic material comes from numerous observations that under special circumstances purified viral nucleic acid is capable of initiating infection, albeit with a reduced efficiency. For example, in 1956, Gierer and Schramm and Fraenkel-Conrat independently showed that the purified RNA of tobacco mosaic virus can be infectious, provided that precautions are taken to protect it from inactivation by ribonuclease. In fact, the causative agent of potato spindle tuber disease completely lacks a protein component and consists solely of RNA. As these agents have no protein coat, they cannot be called viruses and are referred to as *viroids*.

Synthesis of macromolecules in infected cells

Knowing that nucleic acid is the carrier of genetic information, and that with bacteriophages only the nucleic acid enters the cell, it is pertinent to review the events occurring inside the cell. The discovery in 1953, by Wyatt and Cohen, that the DNA of the T-even bacteriophages T2, T4 and T6 contains hydroxymethylcytosine (HMC) instead of cytosine made it possible for Hershey, Dixon and Chase to examine infected bacteria for the presence of phage-specific DNA at various stages of intracellular growth. DNA was extracted from T2-infected *E. coli* at different times after the onset of phage growth, and analysed for its content of HMC. This provided an estimate of the number of phage equivalents of HMC-containing DNA present at any time, based on the total nucleic acid and relative HMC content of the intact T2 phage particle. The results showed that, with T2, synthesis of phage DNA commences about 6 min after infection and then rises sharply, so that by the time the first infectious particles begin to appear 6 min later there are 50–80 phage equivalents of HMC. Thereafter, the numbers of phage equivalents of DNA and of infectious particles increase linearly and at the same rate up until lysis, even if lysis is delayed beyond the normal burst time.

Hershey and his co-workers also studied the synthesis of phage protein, which can be distinguished from bacterial protein by its interaction with specific antibodies. During infection of *E. coli* by T2 phage, protein can be detected about 9 min after the onset of the latent period, i.e. after DNA synthesis begins, and by the time infectious particles begin to appear. A few minutes later there are approximately 30–40 phage equivalents inside the cell. Whereas the synthesis of viral protein starts about 9 min after the onset of the latent period, it was shown by means of pulse–chase experiments that the uptake of ^{35}S into intracellular protein is constant from the beginning. A small quantity (pulse) of ^{35}S (as sulphate) was added to the medium at different times after infection and was followed shortly by a vast excess of unlabelled sulphate (chase) to stop any further incorporation of label. When the pulse was made from the ninth minute onward, the label could be chased into material identifiable by its reaction with antibody (i.e. serologically) as phage coat protein. However, if the pulse was made early in infection, it could be chased into protein but, although this was non-bacterial, it did not react with antibodies to phage structural proteins. This early protein comprises mainly virus-specified enzymes concerned with its replication. The concept of early and late viral proteins is discussed in Chapters 9 and 10.

Being the genetic material, the nucleotide sequence in the viral nucleic acid has to be translated into proteins. Early work showed that protein synthesis takes place on ribosomes rather than on DNA, and other studies of phage-infected cells were to solve many of the mysteries of protein synthesis. Base ratios of the RNA from infected and uninfected cells were compared with those of the DNA from the infecting bacteriophage and from uninfected cells (Table 1.1). These and other data clearly showed that after infection only phage-specific RNA is synthesized, and with this information Brenner, Jacob and Meselson were able to show, in 1959, that this viral-specific RNA associates with pre-existing host cell ribosomes, and hence was responsible for directing protein synthesis.

1.4 Viruses can be manipulated genetically

One of the easiest ways to understand the steps involved in a particular reaction is to isolate mutants that are unable to carry out that reaction. Like all other organisms,

Table 1.1 Base ratios of DNA and RNA from uninfected and phage-infected cells.

Material	Ratio of (adenine + thymine/uracil)/ (guanine + cytosine)
DNA from uninfected cell	1.0
RNA from uninfected cell	0.85
DNA from phage	1.8
RNA from phage-infected cell	1.7

viruses sport mutants in the course of their growth, and these mutations can affect all properties, including the type of plaque formed, the range of hosts that the virus can infect and the physicochemical properties of the virus. One obvious restriction, however, is that many mutations will be lethal to the virus and remain undetected. This problem was neatly overcome in 1963 by Epstein and Edgar and their collaborators with the discovery of *conditional lethal mutants*. One class of these mutants, the *temperature-sensitive mutants*, was able to grow at some low temperature, the *permissive temperature*, but not at some higher, *restrictive* temperature at which normal virus could grow. Another class of conditional lethal mutants was the *amber* mutant. In such mutants a DNA lesion converts a codon into a triplet that terminates protein synthesis. These mutants can only grow on a *permissive* host cell, which has an amber-suppressor transfer RNA (tRNA) that can insert an amino acid at the mutation site during translation. The drawback to conditional lethal mutants is that mutation is random, but the advent of *recombinant DNA technology* has facilitated controlled mutagenesis, at least for those viruses for which infectious particles can be reconstituted from cloned genomic DNA or cDNA. What happens is that a piece of a cloned viral DNA or cDNA genome containing the target sequence is excised from the plasmid using two different restriction enzymes, so that it forms a unique restriction fragment which can eventually be reinserted in the correct orientation. The fragment is then modified by *oligonucleotide-* or *site-directed mutagenesis* via the polymerase chain reaction (PCR). Here, an oligonucleotide primer complementary to the region that you wish to mutate is synthesized chemically but with a single mismatched base. The other primer is unmutated. Thus all the resulting amplified fragments (amplicons) contain the mutation. The new fragment is then be reannealed into the original sequence in a plasmid, and can then be used to form a mutated virus particle. The PCR reaction is explained in Chapter 2.

1.5 Properties of viruses

Assuming that the features of virus growth just described for particular viruses are true of all viruses, we are now in a position to compare and contrast the properties of viruses with those of their host cells. Whereas host cells contain both types of nucleic acid, viruses contain only one type, as analyses of purified viruses have shown. However, just like their host cells, viruses have their genetic information encoded in their nucleic acid. Another difference is that the virus is reproduced solely from its genetic material, whereas the host cell is reproduced from the integrated sum of its components. Thus, the virus never arises directly from a pre-existing virus, whereas the cell always arises directly from a pre-existing cell. The experiments of Hershey and his collaborators showed quite clearly that the components of a virus are synthesized independently and then assembled into mature virus particles. In contrast, growth of the host cell consists of an increase in the amount of all its constituent parts, during which the individuality of the cell is continuously maintained. Finally, viruses are incapable of synthesizing ribosomes but, instead, depend on pre-existing host cell ribosomes for synthesis of viral proteins. These features clearly separate viruses from all other organisms, even *Chlamydia* species, that for many years were considered to be intermediate between bacteria and viruses.

1.6 Origin of viruses

This is a fascinating topic but, as so often happens when hard evidence is scarce, discussion can generate more heat than light. There are two theories: viruses are either degenerate cells or are vagrant genes. Just as fleas are descended from flies by loss of wings, viruses may be derived from pro- or eukaryotic cells that have dispensed with many of their cellular functions (*degeneracy*). Alternatively, some nucleic acid might have been transferred accidentally into a cell of a different species (e.g. through a wound or by sexual contact) and, instead of being degraded as would normally be the case, it might have survived and replicated (*escape*). Although half a century has elapsed since these two theories were first proposed, we still do not have any firm indications about the origin of viruses. Now, rapid sequencing of viral and cellular genomes is providing data for computer analysis. However, although such analyses may identify the progenitors of a virus, they cannot decide between degeneracy and nucleic acid escape.

It is unlikely that all currently known viruses have evolved from a single progenitor. Rather, viruses have probably arisen numerous times in the past by one or both of the mechanisms outlined above. However, once formed, viruses are subject to evolutionary pressures, as are all other organisms. Two processes that contribute significantly to virus evolution are recombination and mutation. Recombination takes place infrequently between the genomes of two related DNA or RNA viruses that find themselves in the same cell, and generates a novel combination of genes. Of far greater significance is the potential for genetic exchange between related viruses with segmented genomes. Here, we are dealing not with mutations, of which many will be non-viable, but with the reassortment of functional genes. The only restriction is the compatibility between the various individual segments making up a fully functional genome. Fortunately, this seems to be a real barrier to the unlimited creation of new viruses, although it is not invincible, because pandemic influenza A viruses (in 1918, 1957 and 1968) were created in this way (see Chapter 18). Mutation is of particular significance to the rapid evolution of RNA genomes which, in the absence of a molecular proofreading mechanism during RNA synthesis (unlike in DNA synthesis), accumulate mutations at a rate of approximately 3×10^{-4}/nucleotide per cycle of replication, compared with DNA at 10^{-9}–10^{-10}/nucleotide per cycle. In other words, an RNA virus can achieve, in one generation, the degree of genetic variation that would take an equivalent DNA genome between 300 000 and 3 000 000 generations to achieve. Once formed by reassortment, influenza A viruses evolve so rapidly that it takes only 4 years on average to mutate sufficiently to escape recognition by host defences and to reinfect that individual.

1.7 Plan of the book

In the preceding discussion, key experiments are described which led to our understanding of the nature of viruses. In addition, we have introduced some experimental techniques used by virologists. However, animal virologists usually work with animal cells in culture, and the special techniques that they use are outlined in Chapter 2.

It is clear that viruses consist of nucleic acid, the repository of the genetic information, surrounded by a protective protein coat; sometimes they are further enveloped in a lipid bilayer. In Chapter 3, consideration is given to the limited number of ways in which such protein shells can be constructed. Chapter 4 covers the structure of viral nucleic acid, which is very nicely rationalized by the Baltimore scheme. Next, attention is focused on mechanisms that viruses have evolved to enable them to infect susceptible cells (Chapter 5).

The replication of viral DNA and RNA occupies Chapters 6–8. Not all proteins synthesized in the infected cell are structural components of the virus particle, and the molecular events controlling their synthesis are discussed in Chapters 9 and 10. Finally, this analysis of the process of infection is concluded in Chapter 11, when the formation of mature virus particles from their different components is described.

The second half of the book is devoted to the biological interactions between viruses and their hosts, as distinct from the molecular aspects of replication. Chapter 12 describes lysogeny, the process whereby some bacteriophages are maintained in the cell in a state of 'suspended animation'. As well as being of intrinsic interest, lysogeny provided important clues about the interaction between viruses and animals. These interactions are outlined in the succeeding six chapters. Chapter 13 is a brief account of the interaction between viruses and cultured eukaryotic cells, and Chapter 15 is devoted to the interaction between animal viruses and their animal hosts. From the point of view of the virus, one significant difference between cells in culture and whole animals is that the latter possess a system of defence, the immune system, that modulates infection (Chapter 14). The way immunity can be stimulated by vaccines is covered in Chapter 16. Tumour viruses can cause cancer and Chapter 17 summarizes the latest information in this area. As for all living organisms, viruses continually evolve, and the effects of this evolution on the patterns of disease in humans are discussed in Chapter 18. The acquired immune deficiency syndrome (AIDS) pandemic is the best known and is the subject of Chapter 19. Although not caused by a virus, another recently emerged infection, bovine spongiform encephalopathy (BSE), and related transmissible spongiform encephalopathies are covered in Chapter 20.

Science itself is continuously evolving and virology is no exception. Chapter 21 contains an account of the latest trends. Finally, Chapter 22 contains a brief technical description of the major groups of viruses.

1.8 Further reading and references

Cann, A. (1997) *Principles of Molecular Virology* (2nd edn). London: Academic Press.

Fields, B. N., Knipe, D. M. & Howley, P. M. (eds) (1996) *Virology* (3rd edn), Vols 1 and 2. Hagerstown, MD: Lippincott-Raven.

Goodsell, D. S. (1991) Inside a living cell. *Trends in Biochemical Sciences* **16**, 203–207.

Granoff, A. & Webster, R. G. (eds) (1999) *Encyclopedia of Virology* (2nd edn), Vols 1–3. New York: Academic Press.

Harper, D. R. (1998) *Molecular Virology* (2nd edn). Oxford: BIOS Scientific Publishers.

Mahy, B. W. J. (1996) *A Dictionary of Virology* (2nd edn). London: Academic Press.

Matthews, R. E. F. (2001) *Plant Virology* (4th edn). New York: Academic Press.

Murphy, F. A., Gibbs, E. P. J., Horzinek, M. C. & Studdert, M. J. (eds) (1999). *Veterinary Virology* (3rd edn). New York: Academic Press.

Orgel, L. (1992) Molecular replication. *Nature (London)* **358**, 203–209.

Primrose, S. B., Twyman, R. & Old, R. (2002) *Principles of Gene Manipulation* (6th edn). Oxford: Blackwell Science.

White, D. O. & Fenner, F. J. (1994) *Medical Virology* (4th edn). New York: Academic Press.

Zuckerman, A. J., Banatvala, J. & Pattison, J. R. (eds) (1999) *Principles and Practice of Clinical Virology* (4th edn). Chichester: John Wiley & Sons Ltd.

Also check Chapter 22 for references specific to each family of viruses.

Chapter 2
Some methods for studying animal viruses

Viruses are too small to be seen except by electron microscopy (EM) and this requires concentrations in excess of 10^{11} particles per millilitre, or even higher if a virus has no distinctive morphology. Therefore viruses are usually detected by other indirect methods.

These indirect methods fall into three categories: (1) *multiplication* in a suitable culture system and detection of the virus by the effects that it causes; (2) *serology*, which makes use of the interaction between a virus and antibody directed specifically against it; and (3) detection of viral *nucleic acid*.

2.1 Selection of culture system

The culture system always consists of living cells, and the choice is outlined in Table 2.1. Which system is used depends on the aims of the experiment. These may be divided into the isolation of viruses, the biochemistry of multiplication, structural studies and the study of natural infections.

The investigation of any new virus starts with ways to cultivate it. There are still many that cannot be cultivated, particularly those occurring in the gut, but some of these occur in such high concentration that they were actually discovered by EM. Often a virus is suspected of causing a disease. By definition, disease can be studied only in an animal, preferably the natural host, although for humans this may be ruled out by ethical or safety considerations. Alternatively, organ cultures and cells can be used. Logically, these should be from the natural host and obtained from those sites where the virus multiplies in the whole animal. However, it may be that cells from unrelated animals are susceptible, e.g. human influenza viruses were first cultivated by inoculating a ferret intranasally and grow best in embryonated chicken eggs! Frequently, viruses grow poorly on initial isolation but adapt, as a result of the selection of mutants, on being passed from culture to culture; there is then the problem of knowing how similar the adapted virus is to the original primary isolate.

The usual way of detecting the presence of virus in an infected cell is by the pathology that it causes. This is known as a *cytopathic effect* or CPE. Often a virus or group of related viruses changes the morphology of the cell in a characteristic way, and this can be recognized by inspecting the cell culture through a microscope at low magnification. During the isolation of an unknown virus, such a CPE gives an excellent clue

Table 2.1 How to choose a culture system for animal viruses.

Culture system	Advantages	Limitations
Animal	Natural infection	Cost of upkeep is expensive. Variation between individuals even if inbred, so large numbers needed. Ethical considerations
Organ, e.g. pieces of brain, gut, trachea	Natural infection. Fewer animals needed. Less variation because one animal gives many organ cultures	Many cell types present. Unnatural because no longer subject to homeostatic processes and immune responses
Cell	Can be cloned therefore variation between individuals is minimal Best for biochemical studies as the environment can be controlled exactly and quickly	There are three types of cell culture: primary cells, cell lines and permanent cell lines. Primary cells are derived from an organ or tissue and remain differentiated but will survive only a few passages. Cell lines are dedifferentiated but diploid and survive a larger number (about 50) passages before they die. Continuous cell lines are immortal but dedifferentiated

as to which further, more specific, diagnostic tests to use. In the research laboratory, CPE provides a quick and easy check on the progress of the infection. An example of CPE is shown in Fig. 2.1.

Biochemical studies of virus infections require a cell system in which nearly every cell is infected. To achieve this, large numbers of infectious particles, and hence a system that will produce them, are required. Often, cells that are suitable for production of virus are different from those used for the study of virus multiplication. There is little logic in choosing a cell system, only pragmatism. Cells differ greatly and different properties make one cell the choice for a particular study and unsuitable for another. The ability to control the cell's environment is desirable, especially for labelling with radio-isotopes, because a chemically defined medium must be prepared that lacks the non-radioactive isotope. Otherwise, the specific activity of the radio-isotope would be reduced to an unusable level.

The investigation of natural infections and disease is best done in the natural host. However, these are frequently unsuitable and the nearest approximation is usually to use purpose-bred animals which, although usually not the natural host species, have a similar range of defence mechanisms and can be maintained in the laboratory. The mouse has been extensively studied; its genetics are well understood and inbred strains reduce genetic variability. Although the use of animals for studying viral diseases has been criticized by organizations concerned with animal rights, the student of virology

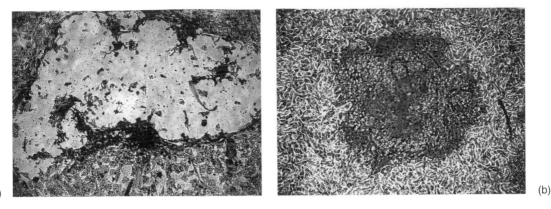

(a)

(b)

Fig. 2.1 Cytopathic effects caused by an influenza A virus and human respiratory syncytial virus (HRSV) in confluent cell monolayers. (a) Chick embryo cells infected by influenza A virus. In the clear central area infected cells have lysed; some cell debris remains, and cells in the process of rounding up can be seen on the edge of the lesion. There are healthy cells around the periphery of the photograph. (b) A monkey cell line infected with HRSV. HRSV does not lyse cells, but fuses them together to form syncytia. A collection of syncytia forms the dark area in the centre of the photograph. Individual cells are magnified to approximately 3 mm in length, and are packed close together. Note that the monkey cells have a slimmer morphology and are more regularly packed together than the chick cells.

will be aware, after reading Chapters 13–20, that there is, as yet, no alternative for studying the complex interactions of viruses with the responses of the host. Analysis of the processes involved would be so much easier if there were a test-tube system, but it is unlikely that any will appear in the foreseeable future.

Organ cultures

Organ cultures have the advantage of maintaining the differentiated state of the cell. However, there are technical difficulties in their large-scale use, and as a result they have not been widely used.

We consider here only organ cultures from the trachea, which have been used to grow a variety of respiratory viruses. Figure 2.2 shows the procedure used to prepare the cultures. Ciliated cells lining the trachea continue to beat in coordinated waves while the tissue remains healthy. Multiplication of some viruses causes the synchrony to be lost and eventually causes the ciliated cells to detach (Fig. 2.3). Virus is also released into fluids surrounding the tissue and can be measured if appropriate assays are available.

Cell cultures

Cells in culture are kept in an isotonic solution, consisting of a mixture of salts in their normal physiological proportions supplemented with serum (usually 5–10%), and in such a growth medium most cells rapidly adhere to the surface of suitable glass or

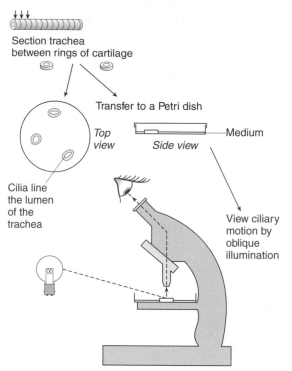

Section trachea
between rings of cartilage

Transfer to a Petri dish

Top
view

Side view

Medium

Cilia line
the lumen
of the
trachea

View ciliary
motion by
oblique
illumination

Fig. 2.2 Preparation of
tracheal organ cultures.

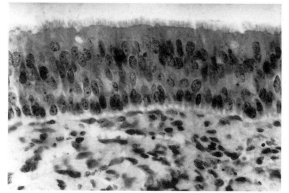

(a)

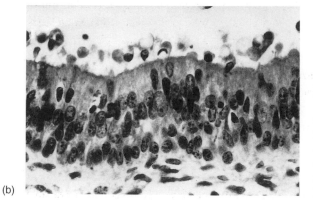

(b)

Fig. 2.3 Sections through tracheal organ cultures (a) uninfected and (b) infected with a
rhinovirus for 36 h. Note the
disorganization of the ciliated
cells (uppermost layer) after
infection. (Courtesy of B.
Hoorn.)

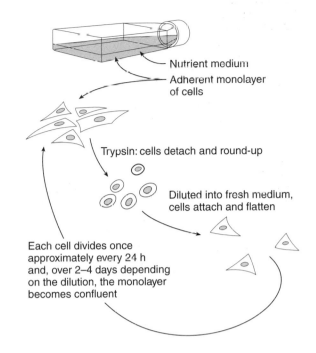

Nutrient medium

Adherent monolayer
of cells

Trypsin: cells detach and round-up

Diluted into fresh medium,
cells attach and flatten

Each cell divides once
approximately every 24 h
and, over 2–4 days depending
on the dilution, the monolayer
becomes confluent

Fig. 2.4 Cell culture.

plastic vessels. Serum is a complex mixture of proteins and other compounds, without
which mitosis does not occur. Synthetic substitutes are now available but these are
expensive and employed mainly for specialized purposes. All components used in cell
culture have to be sterile and handled under aseptic conditions to prevent the growth
of bacteria and fungi. Antibiotics have been invaluable in establishing cells in culture,
and routine cell culture dates from the 1950s when they first appeared on the market.
Figure 2.4 shows the principles of cell culture.

Cultured cells are either diploid or heteroploid (having more than the diploid
number of chromosomes but not a simple multiple of it). Diploid cell lines undergo a
finite number of divisions, from around 10 to 100, whereas the heteroploid cells will
divide for ever. The latter are known as *continuous cell lines* and they originate from
naturally occurring tumours or from some spontaneous event that alters the control
of division of a diploid cell. Diploid cell lines are most easily obtained from embryos
by reducing lungs, kidneys or the whole body to a suspension of single cells.

Modern methods of cell culture

The methodology described above is suited for research and clinical or diagnostic laboratories, but is difficult to scale up for commercial purposes, such as vaccine manufacture. There are now various solutions to the problem, all aimed at increasing cell
density. One of the earliest was to grow cells in suspension, and this has been refined,
using hybridoma cells (which are immortalized antibody-synthesizing or B cells) that
produce *monoclonal antibodies* (MAbs). However, many cells grow only when anchored
to a solid surface, so the technology has sought to increase the surface area available

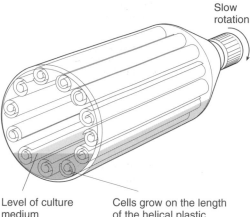

Slow
rotation

Level of culture
medium

Cells grow on the length
of the helical plastic
inserts

(a)

(b)

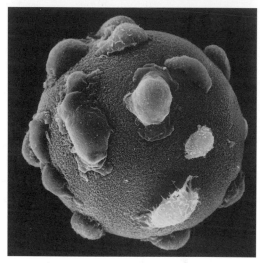

(c)

Fig. 2.5 (a) One way to increase cell density is by increasing the surface area to which cells can attach; this is a view from the end of a bottle lined with spiral plastic coils. The bottle is rotated slowly (at about 5 rev/h) so that a small volume of culture fluid can be used. (b, c) Cells growing on microcarriers. (b) Scanning electron micrograph of pig kidney cells. (Courtesy of G. Charlier.) (c) Removal of cells from a microcarrier bead by incubation with trypsin. The cells are rounding up and many have already been detached. Each bead is about 200 µm in diameter. The microcarriers shown are Cytodex (Pharmacia Ltd), and reproduced by the permission of the company.

by, for example, providing spiral inserts to fit into conventional culture bottles (Fig. 2.5a). Another method is to grow cells on 'microcarriers', tiny particles (about 200 µm diameter) on which cells attach and divide. The surface area afforded by 1 kg of microcarriers is about 2.5 m² and the space taken up (a prime consideration in commercial practice) is very economical. This method combines the ease of handling cell suspensions with a solid matrix for the cell to grow on (Fig. 2.5b,c).

2.2 Identification of viruses using antibodies (serology)

Antibodies are proteins produced by the immune system of higher vertebrates in response to foreign materials (antigens) that these cells encounter. The antibodies are

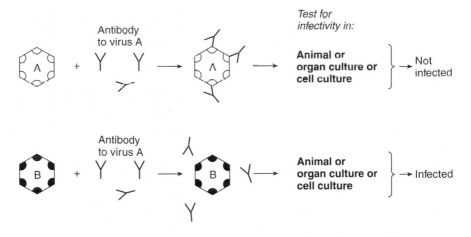

Fig. 2.6 A neutralization test. Virus A loses its infectivity after combining with A-specific antibody (it is neutralized). A-specific antibody does not bind to virus B, so its infectivity is unaffected. The complete test requires the reciprocal reactions.

secreted into the body fluids and are usually obtained from the fluid part of the blood (serum), which remains after clotting has removed cells and clotting proteins. This is then known as an *antiserum*.

The principle of identifying infectious virus by antibody is shown in Fig. 2.6, in which the interaction between virus and antibody is inferred by its effect on infectivity. However, any property of a virus that is affected by antibody can be measured, such as inhibition of haemagglutination. Some viruses attach to molecules on the surface of red blood cells (RBCs) and, at a certain virus:cell ratio, the RBCs are linked together by virus and the cells agglutinated. This has nothing to do with infectivity and, when the infectivity of a virus has been deliberately inactivated, that virus can agglutinate RBCs efficiently, provided that its surface properties are unimpaired. A quantitative haemagglutination test can be devised by making dilutions of virus in a suitable tray and then adding a standard amount of RBCs to each well (Fig. 2.7). The amount of virus present is estimated as the dilution at which the virus causes 50% agglutination. This test has the advantage of speed—it takes just 30 min, compared with an average of 3 days for a plaque assay. However, it is insensitive, e.g. approximately 10^6 plaque-forming units (PFUs) of influenza virus are needed to cause detectable agglutination.

In the haemagglutination-inhibition test, a small amount of virus is added to serial dilutions of antibody before the addition of RBCs (Fig. 2.8). Blocking of agglutination indicates that antibody has bound to a virus particle, and hence identifies it. It can be used with a known antibody to identify an unknown virus, or vice versa using a known virus to identify the presence of virus-specific antibody in a serum sample. Alternatively, virus that has been aggregated by reaction with a specific antibody can be directly visualized with the EM virus.

Virus dilution: 1/

| 2 | 4 | 8 | 16 | 32 | 64 | 128 | 256 | 512 | 1024 |

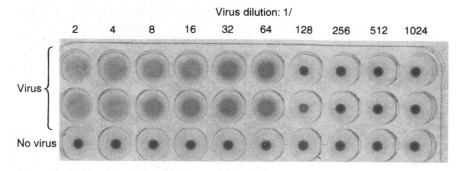

Virus

No virus

Fig. 2.7 Haemagglutination titration. Here an influenza virus is serially diluted (from left to right) in depressions in a plastic plate. Red blood cells (RBCs) 0.5% v/v are then added and mixed with each dilution of virus. Where there is little virus, cells settle to a button (from 1/128), indistinguishable from RBCs to which no virus had been added (row 3). Where sufficient virus is present (up to 1/64), cells agglutinate and settle in a diffuse pattern. (Photograph by A. S. Carver.)

Antiserum dilution: 1/

| 100 | 200 | 400 | 800 | 1600 | 3200 | 6400 | 12 800 | 25 600 | 51 200 |

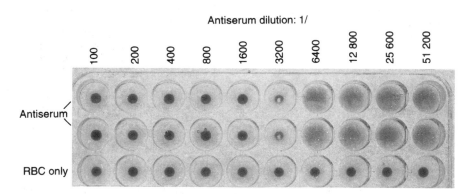

Antiserum

RBC only

Fig. 2.8 In the haemagglutination-inhibition test, antibody is diluted from left to right. Four haemagglutination units (HAUs) of an influenza virus are added to each well. The antibody–virus reaction goes to completion in 1 h at 20°C. Red blood cells are then added to detect virus that has not bound antibody. In this test, haemagglutination is inhibited up to an antibody dilution of 1/3200. (Photograph by A. S. Carver.)

Antibody can also be employed to detect viral antigens inside the infected cell. When the cell is alive, antibodies cannot cross the plasma membrane and will therefore react only with antigens exposed on the surface of the cell. This permeability barrier is destroyed by 'fixing' the cell in organic solvents such as acetone or methanol, which permeabilizes the plasma membrane and enables antibody to enter the cytoplasm and nucleus, and attach to antigens. Antibodies are 'tagged' before use with a marker substance and hence can be detected *in situ*. Tags such as fluorescent dyes can be detected microscopically when illuminated with ultraviolet (UV) light (Fig. 2.9); enzymes (e.g. peroxidase, phosphatase) that leave a coloured deposit on reaction

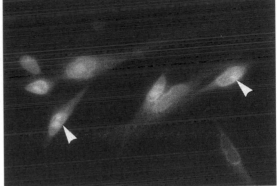

Fig. 2.9 Fluorescent antibody staining. An antibody covalently bound to a fluorescent dye has been used to detect an antigen present mainly in the nucleus (arrowed) of influenza virus-infected cells.

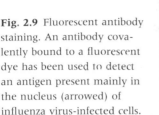
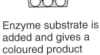

An unknown virus is attached to surface of microtitre well

Antibody of known specificity is bound and the remainder washed away

Enzyme-linked anti-immunoglobulin is added. Wash again

Enzyme substrate is added and gives a coloured product on reaction with enzyme that can be quantitated using a spectrophotometer

Fig. 2.10 Identification of an unknown virus by enzyme-linked immunosorbent assay (ELISA). Virus bound to the plate is used in parallel with cognate virus as a standard. The unknown virus is positively identified when the antibody reaction is identical to that using the control virus.

with substrate can be seen by light microscopy; radioactive substances can be detected by deposition of silver grains from a photographic emulsion; electron-dense molecules (e.g. colloidal gold particles or ferritin, an iron-containing protein) can be visualized by EM.

Antibodies covalently linked to a marker enzyme are now commonly used in a quantitative assay called the enzyme-linked immunosorbent assay (ELISA), as shown in Fig. 2.10. In the example shown, a panel of antibodies is being used to identify an unknown virus. Equally it can be used with a constant amount of virus bound to the assay tray and serial dilutions of antibody to measure antibody concentration. In this example the primary antibody is unlabelled, and it is the secondary antibody that is covalently linked to the enzyme. The secondary antibody binds to conserved epitopes on the primary antibody and, as several molecules of secondary antibody can bind to primary antibody, the test is made more sensitive. The coloured product that results from reaction of the enzyme with added substrate is proportional to the amount of primary antibody bound, and can be measured spectrophotometrically. ELISAs can easily be automated to deal with large numbers of routine samples.

2.3 Detection, identification and cloning of virus genomes using PCR and RT-PCR

Serological methods of virus detection are effective but not without limitation. Neutralization tests are simple but are confined to viruses that can be cultivated. They are slow to give a result because this depends on the time a virus takes to kill a detectable number of cells, and this can range from several days to several weeks. Such a situation is far from ideal, and the problem was solved by the discovery of a technique that makes many, many copies of a chosen part of the virus genome. This is the polymerase chain reaction (PCR), which synthesizes DNA from a DNA template that was devised in 1985 by Kari Mullis. If the virus of interest has an RNA genome, the region of interest is first converted into DNA using a primer (see below) and the retrovirus enzyme, reverse transcriptase (RT) (see Section 8.1). If a unique sequence is chosen and there is a positive result, the virus present is immediately identified. The system is highly

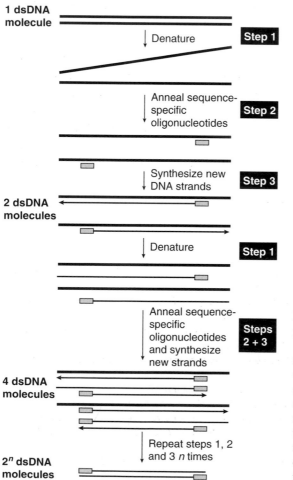

Fig. 2.11 An outline of the polymerase chain reaction (PCR). Step 1: denaturation; step 2: annealing of oligonucleotide primers; and step 3: synthesis of new DNA by added polymerase. This is repeated n times. Note that the end-product, the amplified DNA fragment or amplicon, is not formed until after the third annealing process.

sensitive and can detect one copy of a DNA genome or around 1000 copies of an RNA genome. Thus, it is a detection and identification system all in one. Normally a region of 100 bp (base-pairs) or so is amplified, but with care whole genomes of up to 15 000 bp can be copied, leading to infectious nucleic acids from which a whole virus can be recovered. PCR has the added advantage of detecting virus in primary tissue, so that mutations associated with adaptation to cell culture are avoided. It is no more expensive than a neutralization assay.

The only prerequisite for PCR is knowing the sequence of the genome, so that oligonucleotide primers that are complementary to a sequence on each strand of DNA can be made. PCR requires two primers, each of around 20–30 nucleotides in length, and these are chemically synthesized. The DNA is denatured by heating to around 90°C and the primers added in high molar excess (Fig. 2.11). On cooling, the primers anneal to their respective template strands and the template is copied by the enzyme. It is convenient to use a polymerase that is not inactivated at high temperatures, such as *Taq* polymerase from *Thermophilus aquaticus*, a bacterium that lives in natural hot springs; otherwise fresh polymerase would have to be added after each denaturation step. The mix is again denatured and cooled so that further primers can anneal, and the next round of DNA synthesis take place. The defined PCR product is now present, but many rounds of synthesis (around 30) are required before there is sufficient product to be analysed. To determine whether the PCR result is positive, DNA is extracted from the reaction mix and electrophoresed in agarose together with DNA size markers to detect an amplified fragment (amplicon) of the expected size.

This technique rapidly gained acceptance and has very many applications. It is as widely used for diagnostic clinical virology as for research purposes. For diagnostic purposes PCR may be carried out in two phases. In the first, primers are chosen to amplify regions of the genome that are common to a whole group of viruses known to occur, say, in the gut. On proving positive, primers to a region unique to each virus type can then be used for exact identification.

2.4 Further reading

Adolph, K. W. (ed.) (1994) Molecular virology techniques. In: *Methods in Molecular Genetics*, Vol. 4. New York: Academic Press.

Clementi, N. (2000) Quantitative molecular analysis of virus expression and replication. *Journal of Clinical Microbiology* **38**, 2030–2036.

Desselberger, U. & Flewett, T. H. (1993) Clinical and public health virology: a continuous task of changing pattern. *Progress in Medical Virology* **40**, 48–81.

Harlow, E. & Lane, D. (1988) *Antibodies: A laboratory manual*. Cold Spring Harbor, NY: Cold Spring Harbor Press.

Kurstak, E., Marusyk, E., Murphy, F. A. & van Regenmortel, M. H. V. (eds) (1994). *New Diagnostic Procedures* (Applied Virology Research, Vol. 3). New York: Plenum.

Lennette, E. H. (ed.) (1992). *Laboratory Diagnosis of Viral Infections*. New York: Marcel Dekker, Inc.

Mahy, B. W. J. & Kangro, M. O. (eds) (1995) *Virology Methods Manual*. New York: Academic Press.

Old, R. W. & Primrose, S. B. (1994) *Principles of Genetic Manipulation* (5th edn). Oxford: Blackwell Scientific Publications.

Payment, P. & Trudel, M. (1993). *Methods and Techniques in Virology*. New York: Marcel Dekker, Inc.

Tompkins, L. S. (1992) The use of molecular methods in infectious diseases. *New England Journal of Medicine* **327**, 1290–1297.

Wiedbrauk, D. L. & Farkas, D. H. (eds) (1995). *Molecular Methods for Virus Detection*. New York: Academic Press.

Also check Chapter 22 for references specific to each family of viruses.

Chapter 3
The structure of viruses

Purified viruses particles are composed of 50–90% protein. This protein has three main functions: it protects nucleic acids from shearing and nuclease degradation to which they are susceptible; it comprises cell identification and genome release systems which ensure that genetic information is transferred only into an appropriate target cell; and, for many viruses, it provides enzymes that are essential for virus infectivity. At first sight, it would appear that there is an enormous variety of ways in which the protein could be arranged round the nucleic acid. However, only a limited number of designs are observed and we discuss briefly the limiting factors.

3.1 Viruses are constructed from subunits

Before considering the architecture of viruses, it is worth remembering that, although proteins may have regular secondary structure elements in the form of an α helix and β structure, the tertiary structure of the protein is not symmetrical. This, of course, is a consequence of hydrogen bonding, disulphide bridges and the intrusion of proline in the secondary structure. Although we might naively think that the nucleic acid could be enveloped by a single, large protein molecule, this cannot be so because proteins are irregular in shape, whereas most virus particles have a regular morphology (Fig. 3.1). However, this can also be deduced solely from considerations of the coding potential of nucleic acid molecules. A coding triplet has a relative molecular mass (M_r) of approximately 1000, but specifies a single amino acid with an average M_r of about 100. Thus, a nucleic acid can at best specify only one-tenth of its weight of protein. As viruses frequently contain more than 50% protein by weight, it should be apparent that more than one identical protein must be present.

Obviously, less genetic material is required if the single protein molecule specified is to be used as a repeated subunit, but it is not essential that the coat be constructed from identical subunits, provided that the combined relative molecular masses of the different subunits are sufficiently small in relation to the nucleic acid molecule that they protect. There is a further advantage in constructing a virus from subunits—greater genetic stability, because reducing the size of the structural units lessens the chance of a disadvantageous mutation occurring in the gene specifying it. If, during assembly, a rejection mechanism operates such that faulty subunits are not included in the virus particle, then an error-free structure can be constructed with the minimum of wastage.

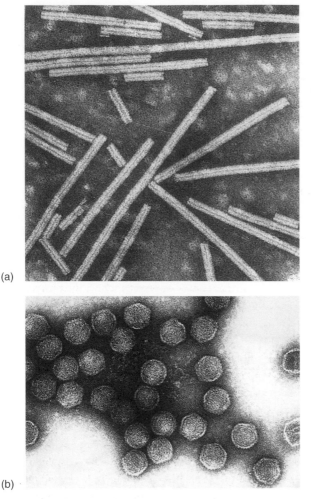

(a)

(b)

Fig. 3.1 Electron micrographs of viruses showing their regular shape. (a) Tobacco mosaic virus; (b) bacteriophage Si1.

The necessary physical condition for the stability of any structure is that it be in a state of minimum free energy, so we can assume that the maximum number of interactions are formed between the subunits of a virus particle. As the subunits themselves are non-symmetrical, for the maximum number of interactions they must be arranged symmetrically, and there are a limited number of ways in which this can be done. Shortly after their seminal work on the structure of DNA, Watson and Crick predicted on theoretical grounds that the only two ways in which asymmetrical subunits could be assembled to form virus particles would generate structures with either cubical or helical symmetry. To date, all virus structures determined have conformed to those predictions. (However, an important rider to the energy status of virus particles is that at least some are suspected of being metastable, and that their true minimum energy state is reached only after they have undergone uncoating during the process of infecting a cell (see Section 3.7).)

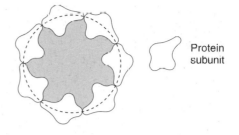

Protein
subunit

Fig. 3.2 Arrangement of identical asymmetrical components around the circumference of a circle to yield a symmetrical
arrangement.

3.2 The structure of filamentous viruses and nucleoproteins

One of the simplest ways of symmetrically arranging non-symmetrical components is
to place them round the circumference of a circle to form discs (Fig. 3.2). This gives
us a two-dimensional structure. If we stack a large number of discs on top of one
another, we obtain a 'stacked-disc' structure. Thus, we can generate a symmetrical
three-dimensional structure from a non-symmetrical component such as protein and
still leave room for nucleic acid. Examination of published electron micrographs of
viruses reveals that some of them have a tubular structure. One such virus is tobacco
mosaic virus (TMV) (see Fig. 3.1). However, close examination of TMV reveals that
the subunits are not arranged cylindrically, i.e. in rings, but helically. There is an
obvious explanation for this. A helical nucleic acid could not be equivalently bonded
in a stacked-disc structure. However, by arranging the subunits helically, the
maximum number of bonds can still be formed and each subunit equivalently bonded,
except, of course, for those at either end. All filamentous viruses examined so far are
helical rather than cylindrical, and the insertion of the nucleic acid may be the factor
governing this arrangement. A helical arrangement offers considerable stability
because of the subunits. This is greater than would be found with a cylinder, which
has no linking along the long axis (see Section 11.3). Many nucleoprotein structures
inside enveloped viruses (see Fig. 3.16) are constructed in the same way.

3.3 The structure of isometric viruses

A second way of constructing a symmetrical particle would be to arrange the smallest number of subunits possible around the vertices or faces of an object with cubic
symmetry, e.g. tetrahedron, cube, octahedron, dodecahedron (constructed from 12
regular pentagons) or icosahedron (constructed from 20 equilateral triangles). Figure
3.3 shows possible arrangements for objects with triangular and square faces. Multiplying the minimum number of subunits per face by the number of faces gives the
smallest number of subunits that can be arranged around such an object. The
minimum number of subunits is determined by the symmetry element of the face,
i.e. a square face will have four subunits, a triangular face will have three subunits,
etc. For a tetrahedron, the smallest number of subunits is 12, for a cube or octahedron it is 24 subunits, and for a dodecahedron or icosahedron it is 60 subunits.

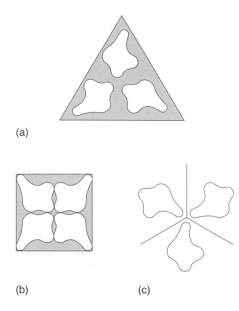

(a)

(b)   (c)

Fig. 3.3 Symmetrical arrangement of identical asymmetrical subunits by placing them on the faces of objects with cubic symmetry. (a) Asymmetrical subunits located at vertices of each triangular facet. (b) Asymmetrical subunits placed at vertices of each square facet. (c) Arrangement of asymmetrical subunits placed at each corner of a cube with faces as represented in (b).

Although it may not be immediately apparent, these represent the few ways in which an asymmetrical object (such as a protein molecule) can be placed symmetrically on the surface of an object resembling a sphere. (The reader can check by using a ball and sticking on bits of paper of the shape shown in Fig. 3.3.) Examination of electron micrographs reveals that many viruses appear spherical in outline, but actually have icosahedral symmetry rather than octahedral, tetrahedral or cuboidal symmetry. There are two possible reasons for the selection of icosahedral symmetry over the others. First, as it requires a larger number of subunits to provide a sphere of the same volume, the size of the repeating subunits can be smaller, thus economizing on genetic information. Second, there appear to be physical restraints which prevent the tight packing of subunits required by tetrahedral and octahedral symmetry.

Symmetry of an icosahedron

An icosahedron is made up of 20 triangular faces—five at the top, five at the bottom and 10 around the middle—with 12 vertices (Fig. 3.4c). Each triangle is symmetrical and can be inserted in any orientation (Fig. 3.4a). An icosahedron has three axes of symmetry: fivefold, threefold and twofold (Fig. 3.4b).

The construction of more complex icosahedral viruses

The combination of an icosahedral structure linked with evolutionary pressure to use small repeating subunits to form virus particles imposes a restriction on the achievable size of the virion. As the size of the particle defines the maximum size of the

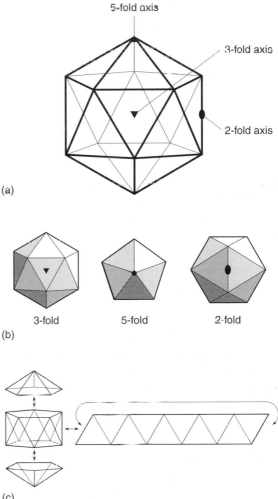

Fig. 3.4 (a) Properties of a regular icosahedron. Each triangular face is equilateral and has the same orientation whichever way it is inserted. Axes of symmetry intersect in the middle of the icosahedron. There are 12 vertices, which have fivefold symmetry, meaning that rotation of the icosahedron by one-fifth of a revolution achieves a position such that it is indistinguishable from its starting orientation; each of the 20 faces has a threefold axis of symmetry and each of the 30 edges has a twofold axis of symmetry—see (b). (c) The icosahedron is built up of five triangles at the top, five at the bottom and a strip of ten around the middle. (Copyright 1991, from *Introduction to Protein Structure* by C. Branden & J. Tooze. Reproduced by permission of Routledge, Inc., part of The Taylor & Francis Group.)

nucleic acid that can be packed within it, it appears that viruses should not be able to package long genomes. However, many viruses have capsids that are large enough to accommodate very long genomes and contain more than 60 subunits. How is this apparent paradox solved? If 60*n* subunits are put on the surface of a sphere, one solution is to arrange them in *n* sets of 60 units, but the members of one set would not be equivalently related to those in another set. For example, consider the arrangement of the subunits in Fig. 3.5. If all the subunits, represented by open and closed circles, are identical, then those represented by closed circles are related equivalently to those represented by open circles. However, open circle units do not have the same spatial arrangement of neighbours as closed circle units and so cannot be equivalently related. Of course, if the structure were built out of *n* different subunits, there would be no conceptual difficulty and, indeed, no problem. However, this would require the

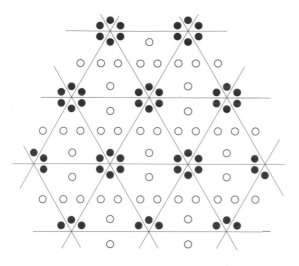

Fig. 3.5 Spatial arrangement of two identical sets of subunits. Note that any member of the set represented by closed circles does not have the same neighbours as a member of the other set represented by open circles.

Fig. 3.6 An example of a geodesic dome—the US Pavilion at Expo '67 in Montreal. (Courtesy of the US Information Service.)

virus to encode many more structural proteins. Accepting the restriction that we must build the structure out of identical subunits, how can we regularly arrange more than 60 asymmetrical subunits? The solution to the problem was inspired by the geodesic domes constructed by Buckminster Fuller (Fig. 3.6). The design of Fuller's domes involves the subdivision of the surface of a sphere into triangular facets, which are arranged with icosahedral symmetry. The device of triangulating the sphere represents the optimum design for a closed shell built of regularly bonded identical subunits. No other subdivision of a closed surface can give a comparable degree of equivalence. Thus, this is a minimum-energy structure and hence gives a further reason for the preponderance of icosahedral viruses.

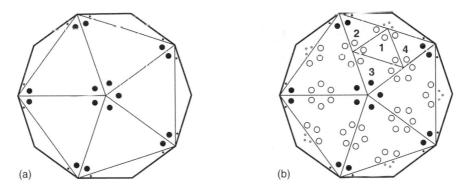

Fig. 3.7 Arrangement of 60n identical subunits on the surface of an icosahedron. (a) n = 1 and the 60 subunits are distributed such that there is one subunit at the vertices of each triangular face. Note that each subunit has the same arrangement of neighbours and so all the subunits are equivalently related. (b) n = 4. Each triangular face is divided into four smaller, but identical, equilateral triangular facets and a subunit is again located at each vertex. In total, there are 240 subunits. Note that, in contrast to the arrangements shown in Fig. 3.5, each subunit, whether represented by an open or closed circle, has the identical arrangement of neighbours—see the face in which triangles 1–4 have been drawn. However, as some subunits are arranged in pentamers and others in hexamers, the members of each set are only 'quasi-equivalently' related.

The triangulation of spheres

It is possible to enumerate all the ways in which this subdivision can be carried out, but, before doing so, let us consider one simple example. If we start with an icosahedron and arrange the subunits around the vertices, there will be 12 groups of five subunits (Fig. 3.7a). Now we can subdivide each triangular face into four smaller and identical equilateral triangular facets, and incorporation of subunits at the vertices of those smaller triangles gives a structure containing a total of 240 subunits (Fig. 3.7b). At the vertices of each of the original icosahedron faces, there will be rings of five subunits, called *pentamers* (solid circles). However, at all the other vertices generated by the triangular facets there will be rings of six subunits, called *hexamers* (open circles). As some of the subunits are arranged as pentamers and some of their colleagues as hexamers, it should be apparent that they cannot be equivalently related—hence the use of the term *quasi-equivalence*, although this still represents the minimum-energy shape.

The original face of an icosahedron can also be divided into three equilateral triangles by distortion of the planar face, and each newly generated triangular face can be further divided by triangulation, and so on. The resultant triangulated structure retains the elements of symmetry found in an icosahedron, but is more accurately referred to as an icosadeltahedron.

The ways in which each triangular face of the icosahedron can be subdivided into smaller, identical equilateral triangles are governed by the laws of solid geometry. These can be calculated from the expression $T = Pf^2$, where T, the *triangulation number* is the number of smaller, identical equilateral triangles; $f = 1, 2, 3, 4$, etc.; and P is

Table 3.1 Values of capsid parameters in a number of icosahedral viruses. The value of T was obtained from examination of electron micrographs, thus allowing the values of P and f to be calculated.

P	f	T $(=Pf^2)$	No. of subunits (60T)	Example
1	1	1	60	Tobacco necrosis virus satellite virus
3	1	3	180	Tomato bushy stunt virus, picornaviruses*
1	2	4	240	Sindbis virus
1	4	16	960	Herpesviruses
1	5	25	1500	Adenoviruses†

*In fact picornaviruses have a pseudo-T = 3 structure (see below).
†See text.

given by the expression $h^2 + hk + k^2$. In this expression, h and k are any pair of integers without common factors, i.e. h and k cannot be multiplied or divided by any number to give the same values.

For viruses examined so far, the values of P are 1 ($h = 1$, $k = 0$), 3 ($h = 1$, $k = 1$) and 7 ($h = 1$, $k = 2$). Representative values of T are shown in Table 3.1. Once the number of triangular subdivisions is known, the total number of subunits can easily be determined because it is equal to 60T.

$T = 1$: the smallest virus—tobacco necrosis virus satellite virus

So far we have not defined what is meant by the *subunit* used in the construction of a virus particle. In its simplest form one subunit is one protein. However, no independently replicating virus is known to consist of only 60 protein subunits, although satellite viruses do (see Table 3.1). These viruses encode one coat protein but depend on co-infection with a helper virus to provide missing replicative functions. The single-stranded RNA genome of tobacco necrosis virus satellite virus is 1239 nucleotides. Presumably the volume of a 60-subunit structure is too small to accommodate the larger genome that is needed by an independent virus. The virion is only 18 nm in diameter, compared with 30 nm in small independent viruses.

Detailed determination of the structure of a virus depends primarily on being able to grow crystals of purified virus, although valuable but lower-resolution information can be obtained from cryoelectron microscopic examination of virus frozen in vitreous ice. We do not completely understand the conditions for crystallization of virus particles or proteins, and many will not form crystals at all. Large stable crystals are required which are then bombarded with X-rays. These are diffracted by atoms within the virion, and the image is captured on film. Knowledge of the amino acid sequence of the proteins that form the particle makes it possible to determine the three-dimensional crystal structure. X-ray analysis gives resolution to around 0.3 nm and cryoelectron microscopy to around 15 nm. Both processes require use of high-powered computers to make the necessary calculations and for image reconstruction.

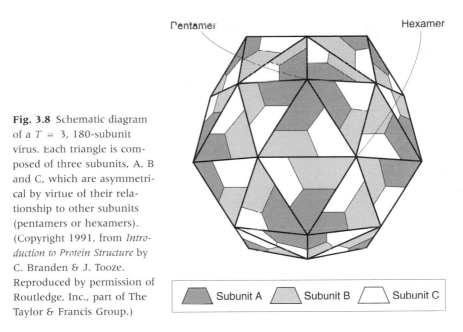

Fig. 3.8 Schematic diagram of a $T = 3$, 180-subunit virus. Each triangle is composed of three subunits, A, B and C, which are asymmetrical by virtue of their relationship to other subunits (pentamers or hexamers). (Copyright 1991, from *Introduction to Protein Structure* by C. Branden & J. Tooze. Reproduced by permission of Routledge, Inc., part of The Taylor & Francis Group.)

The morphological units seen by electron microscopy are called capsomers and *the number of these need not be the same as the number of protein subunits*. The numbers of morphological units seen will depend on the size and physical packing of the subunits and on the resolution of electron micrographs. A repeating subunit may consist of a *complex of several proteins*, such as the four structural proteins of poliovirus (see below and Section 11.4), or a *fraction of a protein*, such as the adenovirus hexon protein, half of which is considered to be a single repeating subunit.

$T = 3$: the molecular basis for quasi-equivalent packing of chemically identical polypeptides—tomato bushy stunt virus

Some plant viruses achieve a structure with $T = 3$ and 180 subunits (Fig. 3.8) while encoding only a single virion polypeptide. They compensate for the physical asymmetry of quasi-equivalence by each polypeptide adopting one of three subtly different conformations. The virion polypeptide of tomato bushy stunt virus has three domains—P, S and R (Fig. 3.9a): this is folded so that P and S are external and hinged to each other, whereas R is inside the virion and has a disordered structure. An arm a connects S to R and h connects S and P (Fig. 3.9b).

Each triangular face is made of three identical polypeptides, but these are in different conformations to accommodate the quasi-equivalent packing, e.g. the C subunit has the S and P domains oriented differently from the A and B subunits (Fig. 3.9c), whereas the arm a is ordered in (c) and disordered in (a) and (b) (not shown). The S domains form the viral shell with tight interactions, whereas the P domains (total = 180) interact across the twofold axes of symmetry to form 90 dimeric protrusions. This virion is 33 nm in diameter and can accommodate a single-stranded RNA genome

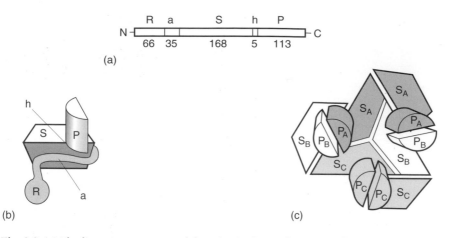

Fig. 3.9 (a) The linear arrangement of domains in the single virion polypeptide of tomato bushy stunt virus. (b) Conformation of the polypeptide. The S domain forms the shell of the virion, whereas P points outwards and R is internal. (c) This shows a triangular face, composed of subunits A, B and C, and the interaction of the P domains to form dimeric projections (see text). (Copyright 1991, from *Introduction to Protein Structure* by C. Branden & J. Tooze. Reproduced by permission of Routledge, Inc., part of The Taylor & Francis Group.)

about four times larger than that of the satellite viruses. Thus, a larger particle can be achieved without any more genetic cost.

$T = 3$: with icosahedra constructed of four different polypeptides—picornaviruses

Picornaviruses are made of 60 copies of each of the four polypeptides VP1, VP2, VP3 and VP4. VP4 is entirely internal. The repeating subunit of picornaviruses is the complex of VP1, VP2 and VP3. This should generate a $T = 1$ particle. However, despite the significant differences in amino acid sequence, the proteins adopt very similar conformations and, in geometric terms, appear as separate repeating subunits. For this reason the assembled picornavirus particles appear to have a $T = 3$ structure; more accurately this is a pseudo-$T = 3$ (compare Figs 3.12 and 3.8).

The pentamers contain 15 polypeptides, with five molecules of VP1 forming a central vertex. These pentamers are found in the cell and are the building blocks from which the virion is assembled. Use of three polypeptides gives a chemically more diverse structure and may be an adaptation to cope with the immune system of animal hosts.

A common structure of plant and animal virion proteins: the antiparallel β barrel

All the virus proteins considered so far—the single proteins of tobacco necrosis virus satellite virus, tomato bushy stunt virus and the VP1, VP2 and VP3 of picornaviruses—have the same antiparallel β-barrel structure, of a type sometimes called a 'jelly roll'. Its

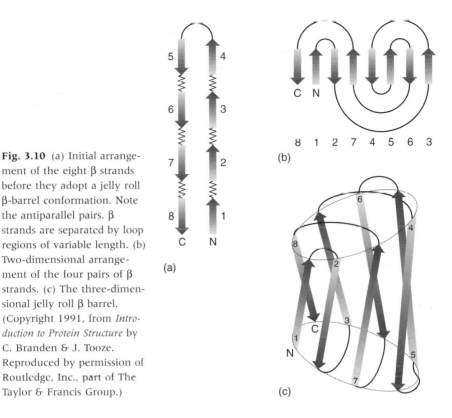

Fig. 3.10 (a) Initial arrangement of the eight β strands before they adopt a jelly roll β-barrel conformation. Note the antiparallel pairs. β strands are separated by loop regions of variable length. (b) Two-dimensional arrangement of the four pairs of β strands. (c) The three-dimensional jelly roll β barrel. (Copyright 1991, from *Introduction to Protein Structure* by C. Branden & J. Tooze. Reproduced by permission of Routledge, Inc., part of The Taylor & Francis Group.)

formation from a linear polypeptide can be visualized in three stages (Fig. 3.10): first, as a hairpin structure, where β strands are hydrogen bonded to each other—1 with 8, 2 with 7, 3 with 6 and 4 with 5 (Fig. 3.10a); second, these pairs are arranged side by side, so that further hydrogen bonds can be formed by newly adjacent strands (e.g. 7 with 4) (Fig. 3.10b); and third, the pairs wrap around an imaginary barrel (Fig. 3.10c). The eight β strands are arranged in two sheets, each composed of four strands: strands 1, 8, 3 and 6 form one sheet and strands 2, 7, 4 and 5 the second sheet. The dimensions are such that each protein forms a wedge, and these are the structures assembled into virus particles (Fig. 3.11).

The attachment site of picornaviruses

Crystallographic, biochemical and immunological data have together identified a depression within the β barrel of VP1, which is thought to be the attachment site of picornaviruses. There are 60 subunits and 60 attachment sites per virion. Apart from its intrinsic interest, the structure of the attachment site is important as the prime target for antiviral drugs which can stop attachment of virus to the host cell. The arrangement of the β strands of VP1 is such that an annulus is formed around each fivefold axis of symmetry (Fig. 3.12). In the rhinoviruses (common cold viruses), this is particularly deep and is called a 'canyon'. The canyon lies within the structure of the β barrel. Amino acid residues within the canyon are invariant, as expected from their requirement to interact with the cell receptor, whereas amino acid residues on

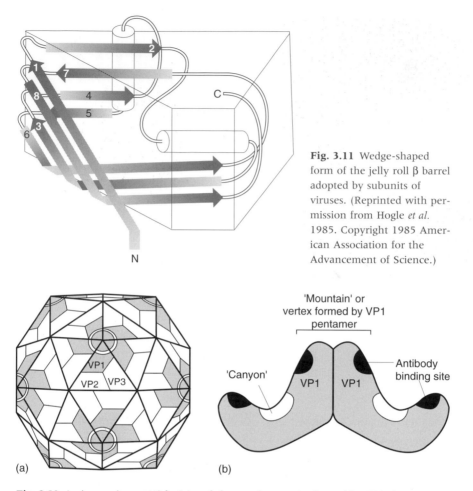

Fig. 3.11 Wedge-shaped form of the jelly roll β barrel adopted by subunits of viruses. (Reprinted with permission from Hogle *et al.* 1985. Copyright 1985 American Association for the Advancement of Science.)

Fig. 3.12 A picornavirus particle (a) and the attachment site formed by VP1 shown as an annulus around the fivefold axis of symmetry. (From Smith *et al.* 1993.) (b) A vertical section through a VP1 pentamer where the cross-section of the annulus is referred to as the 'canyon'. (Reprinted with permission from Luo *et al.* 1987. Copyright 1987 American Association for the Advancement of Science.)

the rim of the canyon are variable. Only these latter amino acid residues interact with antibody. It is thought that the floor of the canyon has evolved so that it physically cannot interact with antibody. This avoids immunological pressure to accumulate mutations in order to escape from reaction with antibody, because these would at the same time render the attachment site non-functional, and hence be lethal to the virus.

An unknown structure: viruses with 180 + 1 subunits and no jelly roll β barrel—RNA bacteriophages

The leviviruses are 24 nm icosahedral RNA bacteriophages, and include MS2, R17 and β. They encode two coat proteins. There are 180 subunits of one of these arranged

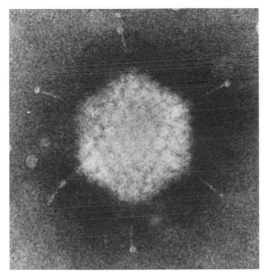

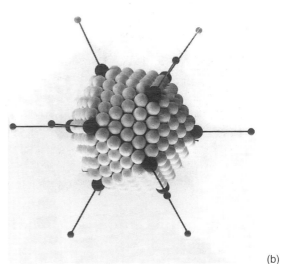

(a)

(b)

Fig. 3.13 The structure of adenoviruses. (a) Electron micrograph of an adenovirus. (b) Model of an adenovirus to show arrangements of the capsomers. (c) Schematic diagram to show the arrangements of the subunits on one face of the icosahedron. Note the subdivision of the face into 25 smaller equilateral triangles. (Photographs courtesy of N G. Wrigley.)

(c)

with $T = 3$, but only a single copy of the second 'A' protein in each particle. This is the attachment protein. It is not known how the single subunit is incorporated into the particle. The main coat protein does not form a jelly roll β barrel as the others described above, but instead has five antiparallel β strands arranged like the vertical elements of battlements. Two subunits interact to form a sheet consisting of 10 antiparallel β strands.

$T = 25$: more complex animal virus particles—adenoviruses

Careful examination of electron micrographs of adenoviruses shows that $T = 25$. In addition there are 240 hexamers, 12 pentamers and a fibre projecting from each vertex of the virus (Fig. 3.13a,b). The fibres, the pentamers and the hexamers are all constructed from different proteins. We are thus faced with the problem of arranging not one, but three, different proteins in a regular fashion, while adhering to the design principles outlined above. This can be achieved by arranging the pentamers and the fibres at the vertices of the icosahedron and the hexamers on the faces of the icosahedron (Fig. 3.13b). However, the formula 60T gives the number of subunits as 1500 (see Table 3.1). How is this difference resolved? The 240 hexons are composed of three identical polypeptides and each functions as *two* repeating subunits. Thus, there are 1440 hexon subunits. The 12 pentons are formed from five identical polypeptides and each functions as a subunit, making 60. The 1440 hexon subunits + 60 penton subunits make up the 1500 predicted subunits. Actually the hexamers are not all spatially equivalent, because those surrounding the vertex pentamers contact five other hexons, whereas the others contact six hexons.

Double-shelled particles: a capsid within a capsid—reoviruses

A different and very complex structural arrangement is found in another class of isometric viruses, the reoviruses, which are composed of a capsid within a capsid. The diameters of the inner and the outer capsid are 51 nm and 73 nm, respectively. Reovirus synthesizes 11 polypeptides, of which eight are located in the virion, three forming the outer capsid and five the inner capsid (Table 3.2). Both capsids have icosahedral symmetry (Fig. 3.14) and the outer has a $T = 13$ symmetry and 780 (13 × 60) subunits. However, it is not certain how the 600 molecules of protein μ1 and the 600 molecules of σ3 are arranged in the outer capsid to form the subunits. The third

Table 3.2 Reovirus polypeptides and their location in the outer or inner capsid of the virion, and some of their properties.

Location in virion or non-structural	Protein	Encoding RNA segment	Number of polypeptides per virion	Function
Outer	μ1	M2	600	Main structural element; role in entry
Outer	σ3	S4	600	Main structural element
Outer, vertex	σ1	S1	36–48	Attachment protein; serotype determinant
Inner	λ1	L3	120	Main structural element
Inner	σ2	S2	150	Stabilizes λ1
Inner, vertex	λ2	L2	60	Forms turret; capping and export of mRNA
Inner	μ2	M1	12	Not known
Inner	λ3	L1	12	RNA-dependent RNA polymerase
Non-structural	μ$_{NS}$	M3	0	Binds ssRNA; role in secondary transcription
Non-structural	σ$_{NS}$	S3	0	Binds ssRNA
Non-structural	σ1$_s$	S1	0	Not known

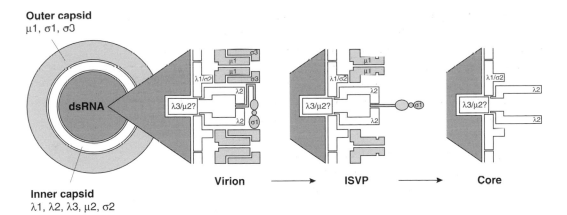

Fig. 3.14 The double capsid structure of reovirus showing the location of polypeptides in the virion. Here a section of the virion has been taken through the fivefold axis of symmetry of the vertex. To demonstrate protein function, two intermediates that occur during the virus uncoating process are also shown: the intermediate subviral particle (ISVP) and the viral core. In the ISVP, σ1 has achieved an extended conformation and σ3 has been lost from the outer capsid. Some molecules of σ1 may also be extended in the virion, but this is seen more frequently in the intermediate. The core is formed by the loss of the entire outer capsid and the change in conformation of the turret protein λ2 to form the channel through which mRNA synthesized in the particle will escape into the cell cytoplasm. (From Nibert *et al.* 1996.)

outer capsid protein σ1 is the attachment protein that is located at each of the 12 vertices (Fig. 3.14). It is supported by the five molecules of protein λ2 vertices, which are part of the inner capsid (although not long ago they were attributed to the outer capsid). The structure of the inner capsid was recently solved to 0.36 nm resolution. Three of the five inner capsid proteins (λ1, σ2 and λ2) are symmetrically arranged: the main skeleton is formed by protein λ1 built into 12 decamers, each of which surrounds a vertex. This is clamped into position by protein σ2. Protein λ2 forms a turret at each vertex, which surrounds the outer core protein σ1. The turret is seen as such only after the outer core has been stripped away. It has an important role in capping viral mRNAs, which are all made within the particle, and acts as a conduit for the export of mRNA from the particle (see Section 10.3). The 10 double-stranded RNAs are tightly coiled like the DNA inside phage heads. Each RNA molecule is thought to be associated with a transcriptase complex and tethered near a vertex.

3.4 Enveloped viruses

Although they appear complex, these viruses have a conventional isometric or helical structure that is surrounded by a membrane—lipid bilayer of 4 nm thickness containing proteins. Examples include many of the larger animal viruses and a few plant and bacterial viruses. Traditionally, these viruses were distinguished from nucleocapsid

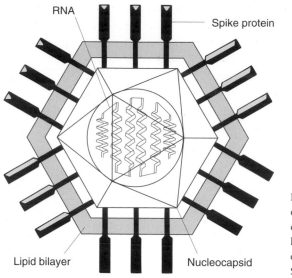

RNA

Spike protein

Lipid bilayer

Nucleocapsid

Fig. 3.15 Sindbis virus: an enveloped icosahedron. The core is $T = 3$, but the envelope is $T = 4$. (See text for the explanation.) (Courtesy of Dr S. D. Fuller.)

viruses by treatment with detergents or organic solvents, which disrupts the membrane and destroys infectivity. Thus, they were sometimes referred to as 'ether-sensitive viruses'. The envelope, which is derived from host cell membranes, is obtained by the virus *budding* from cell membranes, but most viruses contain no cell proteins. How cell proteins are excluded and why retroviruses, the exception, do not exclude cell proteins from their virions is not understood (see Section 11.6).

An isometric core surrounded by an isometric envelope: Sindbis virus

Sindbis virus (a togavirus) has an icosahedral nucleocapsid that comprises a single protein, surrounded by an envelope from which viral spike proteins protrude. The core has $T = 3$ and 180 subunits, exactly like tomato bushy stunt virus described above. Surprisingly, the envelope also has icosahedral symmetry, but to everyone's surprise this is $T = 4$ and has 240 subunits. This apparent paradox was resolved when it was found that the two structures are complementary, so that the internal ends of the spike proteins fit exactly into holes between the subunits of the nucleocapsid (Fig. 3.15). So far, this and its near relations are the only enveloped viruses that have been found to have a geometrically symmetrical envelope.

A helical core surrounded by an approximately spherical envelope: the influenza viruses

One of the best studied groups of enveloped viruses is the influenza viruses. The helical core is composed of ribonucleoprotein, itself composed of a flexible rod of RNA and viral nucleoprotein (NP). This is constructed as described earlier in Section 3.2, and arranged in a twisted hairpin structure. The genome is segmented RNA and there are eight separate core structures. Each is associated with a transcriptase complex. The core is contained within a layer of matrix (M1) protein, which is itself within an lipid envelope that is roughly spherical, and often described as pleomorphic (Fig. 3.16).

In electron micrographs (Fig. 3.16), a large number of protein spikes, projecting about 13.5 nm from the viral envelope, can be observed. These spikes, which have an overall length of 17.5 nm, are transmembrane glycoproteins similar to those of the cell. The spike layer consists solely of virus-specified glycoproteins, and comprises about 800 haemagglutinin and 200 neuraminidase proteins. Neuraminidase spikes are arranged non-randomly in clusters on the virion surface. The haemagglutinin functions in attachment and fusion entry, and the neuraminidase in release of infecting virus that has not been endocytosed or of progeny virus that is attached to the cell surface (see Section 5.1). The spikes are morphologically distinct. Figure 3.17 shows the structure of the influenza virus haemagglutinin. Also in the membrane are a few molecules of an ion channel protein called M2. This allows the passage of protons into the core and is necessary for secondary uncoating.

A helical core surrounded by a non-spherical envelope: the rhabdoviruses

This group of viruses has a helical core consisting of nucleoprotein and an envelope similar to the influenza viruses. What is distinctive about these viruses is that they are not isometric but bullet shaped or bacilliform (rounded at both ends). These are unique morphologies. There are both plant and animal rhabdoviruses, and the latter are all bullet shaped. The envelope contains a dense layer of spikes comprising just one protein, the viral attachment protein, G. The matrix protein underlies the membrane. The unusually rigid structure of the rhabdoviruses, compared with the influenza viruses, may result from the interaction between the spike protein G and the matrix protein. Otherwise the influenza and rhabdoviruses have a fundamentally very similar structure (see Fig. 3.16e,f). The length of the bullet is controlled by the size of the RNA genome, because very small defective–interfering rhabdovirus RNAs form small bullet-shaped particles. Nothing is known of the structural geometry involved in the formation of rhabdovirus particles.

3.5 Viruses with head–tail morphology

The head–tail architectural principle is unique to bacterial viruses (Fig. 3.18), but many have other morphologies (Table 3.3). There is a large variation on this structural theme and bacteriophages can be subdivided into those with short tails, those with long non-contractile tails and those with complex contractile tails (see Chapter 22). A number of other structures, such as base plates, collars, etc., may also be present. Despite their complex structure, the design principles involved in head–tail phages are identical to those outlined earlier for the viruses of simpler architecture. Heads usually possess icosahedral symmetry, whereas tails usually have helical symmetry. All other struc-

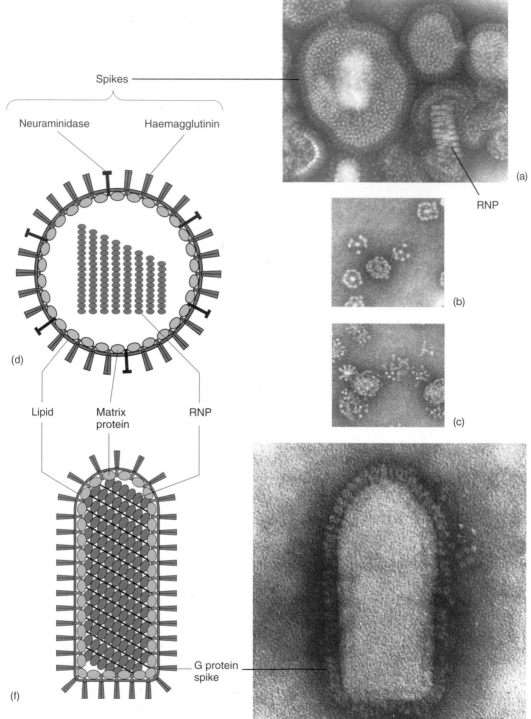

Spikes

Neuraminidase   Haemagglutinin

(a)

RNP

(b)

(c)

(d)

Lipid   Matrix protein   RNP

G protein spike

(f)

(e)

tures, base plates, collars, etc., also possess a defined symmetry. The evolution of this elaborate structure may be connected with the way in which these bacterial viruses infect susceptible cells (see Chapter 5). In brief, the phage attaches to a bacterium via its tail, lyses a hole in the cell wall and inserts its DNA, which is tightly packed into the phage head, into the cell using the tail as a conduit.

3.6 Occurrence of different virus morphologies

The different virus morphologies discussed above do not occur with equal frequency among animal, plant and bacterial viruses (see Table 3.3). There are relatively few icosahedral viruses in bacteria, helical viruses are found almost exclusively in plants, enveloped icosahedral viruses are unique to animals, whereas enveloped helical viruses are widely distributed. Finally, head–tail virus morphology is found only in bacteria. Unfortunately, there is almost no information as to why this should be so.

3.7 Principles of disassembly: particles are metastable

In ending this discussion about virus particles, it is important to remember that all particles have to be constructed so that they can be disassembled, to permit the

Table 3.3 Distribution of the various types of structure among virus families of animals, plants and bacteria.

Type of structure	Animals	Plants	Bacteria
Non-enveloped icosahedral	Common	Common	Uncommon: micro-, levivi- and corticoviruses
Non-enveloped helical	Not known	Common	Rare: rudiviruses
Enveloped: icosahedral	Uncommon: asfar-, toga-, flavi- and arteriviruses	Not known	Rare: cystoviruses
Enveloped: helical	Common	Uncommon: bunya- and rhabdoviruses	Rare: lipothrixviruses
Head–tail	Not known	Not known	Common

Fig. 3.16 (*Facing page*) Influenza A virus (an orthomyxovirus) and vesicular stomatitis virus (a rhabdovirus): viruses with enveloped helical structures. Although their morphology is different, these viruses are constructed in the same way. (a) Electron micrograph of influenza A virus showing the internal helical ribonucleoprotein (RNP) core and the surface spikes. (b) Aggregates of purified neuraminidase. (c) Aggregates of purified haemagglutinin. Note the triangular shape of the spikes when viewed 'end-on'. (d) Schematic representation of the structure of influenza virus. (e) Electron micrograph of vesicular stomatitis virus. (f) Schematic representation of the structure of vesicular stomatitis virus. (Electron micrographs courtesy of N. G. Wrigley and C. J. Smale.)

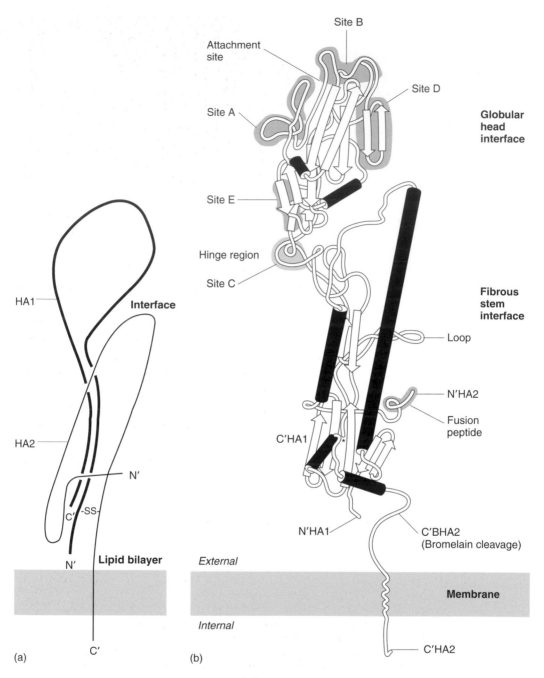

Fig. 3.17 The influenza virus haemagglutinin (HA). This is a homotrimer but only a monomer is shown here. The HA is synthesized as a single polypeptide which is proteolytically cleaved into the membrane-bound HA2 and the distal HA1. (a) An outline structure showing that HA1 and HA2 are both hairpin structures. (b) The crystal structure. The globular head of HA1 bears all the neutralization sites (A–E; shaded) and is made of a distorted jelly roll β barrel similar to most of the icosahedral viruses. (Reprinted with permission from *Nature* (Wiley *et al.* 1981). Copyright 1981 Macmillan Magazines Limited.)

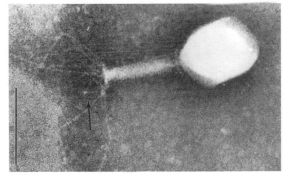

Fig. 3.18 Electron micrograph of bacteriophage T2. (Courtesy of L. Simon.) Six long tail fibres are evident. Tail pins cannot be seen but a short fibre (indicated by the arrow) can be seen. The bar is 100 nm.

genome to enter a new target cell. This is supremely important to the virus particle because it has only the one chance to do this successfully and hence propagate its genome. The idea is developing that the particle is in fact metastable, i.e. it can spontaneously descend to a lower energy level and, in doing so, releases its genome (see Wien *et al.* 1996). Not surprisingly there are a number of fail-safe devices that tell the virus when it is safe to let go the genome. One of the simplest systems is used by enveloped animal viruses such as human immunodeficiency virus 1. This undergoes a succession of interactions between cell receptors and virus envelope protein binding sites, similar to people using a series of passwords to gain entry to a high-security establishment. If everything is in order, the metastable envelope protein then undergoes profound rearrangements which allow a hidden hydrophobic segment to insert into the cell membrane. This initiates fusion of the lipid bilayer of the virus with that of the cell plasma membrane, and the virus genome automatically enters the cell cytoplasm. However, if the sequence of passwords proves incorrect, the virus detaches from the cell and the process is repeated. Mechanisms of entry are discussed in Chapter 5.

3.8 Further reading and references

Branden, C. & Tooze, J. (1998) *Introduction to Protein Structure* (2nd edn). New York: Garland Publishing.

Butler, P. J. G. (1984) The current picture of the structure and assembly of tobacco mosaic virus. *Journal of General Virology* **65**, 253–279.

Caspar, D. L. D. & Klug, A. (1962) Physical principles in the construction of regular viruses. *Cold Spring Harbor Symposium of Quantitative Biology* **27**, 1–24.

Eiserling, F. A. (1979) Bacteriophage structure. In: *Comprehensive Virology*, Vol. 13, pp. 543–580. Fraenkel-Conrat, H. & Wagner, R. R. (eds). New York: Plenum.

Finch, J. T. & Holmes, K. C. (1967) Structural studies of viruses. In: *Methods in Virology*, Vol. 3, pp. 352–474. Maramarosch, K. & Koprowski, H. (eds). London: Academic Press.

Harrison, S. C. (1984) Structure of viruses. In: *The Microbe*, Vol. 36, *Part 1*, pp. 29–73. Cambridge: Cambridge University Press.

Hogle, J. M. (ed.) (1990) Virus structure. *Seminars in Virology* **1**(6), 385–487.

Hogle, J. M., Chow, M. & Filman, D. J. (1985) Three-dimensional structure of poliovirus at 2.9 Å. *Science* **229**, 1358–1365.

Klug, A. (1983) Architectural design of spherical viruses. *Nature (London)* **303**, 378–379.

Luo, M., Vriend, G., Kamer, G. *et al.* (1987) The atomic structure of mengo virus at 3 Å resolution. *Science* **235**, 182–191.

McKenna, R., Xia, D., Willingmann, P., Ilag, L. L., Krishnaswamy, S., Rossmann, M. G., Olson, N. H., Baker, T. S. & Incardona, N. L. (1992) Atomic structure of single-stranded DNA bacteriophage φX174 and its functional implications. *Nature (London)* **355**, 137–143.

Nibert, M. L., Schiff, L. A. & Fields, B. N. (1996) Reoviruses and their replication. In: *Virology*, pp. 1557–1623. Fields, B. N., Knipe, D. M. & Howley, P. M. (eds). Philadelphia: Lippincott-Raven.

Smith, T. J., Olson, N. H., Cheng, R. H., Liu, H., Chase, E. S., Lee, W. M., Leippe, D. M., Mosser, A. G., Rueckert, R. R. & Baker, T. S. (1993) Structure of human rhinovirus complexed with Fab fragments from a neutralizing antibody. *Journal of Virology* **67**, 1148–1158.

Stuart, D. (1993) Virus structures. *Current Opinion in Structural Biology* **3**, 167–174.

Wien, M. W., Chow, M. & Hogle, J. M. (1996) Poliovirus: new insights from an old paradigm. *Structure* **4**, 763–767.

Wiley, D. C., Wilson, I. A. & Skehel, J. J. (1981) Structural identification of the antibody-binding sites of Hong Kong influenza haemagglutinin and their involvement in antigenic variation. *Nature (London)* **289**, 373–378.

Also check Chapter 22 for references specific to each family of viruses.

Chapter 4
Viral nucleic acids

The nucleic acid of a virus contains all the information needed to produce new viral particles. Some of this information is used directly to make virion components and some to make proteins or to provide signals that allow the virus to subvert the biosynthetic machinery of a cell and redirect it towards the production of virus. Whereas the standard form of genetic material in living systems is double-stranded DNA, viruses contain a diverse array of nucleic acid forms and compositions. This chapter reviews the use of nucleic acid content as part of a basis for grouping viruses, and describes some of the peculiarities of viral nucleic acids, to set the scene for later chapters which consider how various types of viral genome are replicated.

4.1 Types of nucleic acids found in viruses

The nature of a particular nucleic acid sample is assessed by determining its base composition, sensitivity to DNase or RNase, buoyant density, etc. Single-stranded nucleic acids are distinguished from double-stranded ones by the absence of a sharp increase in absorbance of ultraviolet light upon heating and the non-equivalence of the molar proportions of adenine (A) and thymine (T) (or uracil [U]) or guanine (G) and cytosine (C). From these types of analysis, it appears that there are four possible kinds of viral nucleic acid: single-stranded DNA, single-stranded RNA, double-stranded DNA and double-stranded RNA. Each kind of genome is found in many virus families, which between them contain members that infect a diverse array of organisms, including animals, plants and bacteria.

4.2 Classifying viruses—the Baltimore scheme

Viruses exhibit great diversity in terms of morphology, genome structure, mode of infection, host range, tissue tropism, disease (pathogenesis), etc. Although, in principle, each of these properties can be used to place viruses into groups, grouping viruses on the basis of these parameters does not give a good basis for unifying discussions of virus replication processes. For example, many viruses with a similar morphology have quite different mechanisms of replication and may cause a wide range of different diseases in their respective hosts. Such schemes offer little opportunity to infer anything about the fundamental nature of the virus in question. To provide this, Nobel

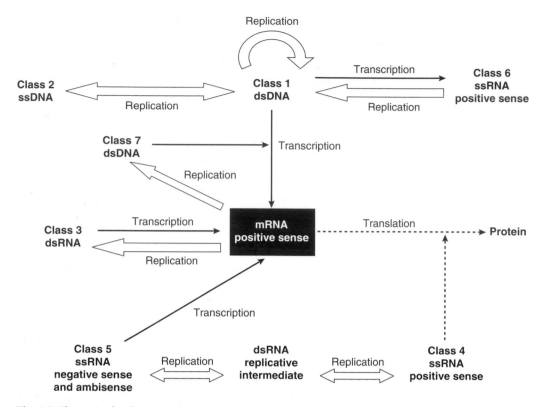

Fig. 4.1 The revised Baltimore scheme (see text for details): solid arrows, transcription; open arrows, replication; dashed arrows, translation.

laureate David Baltimore proposed a scheme that encompassed all viruses, based on the nature of their genomes and their modes of replication and gene expression. The International Committee on Taxonomy of Viruses (ICTV) uses these, together with other parameters, to place viruses into families and genera.

The revised Baltimore scheme is based on the fundamental importance of messenger RNA (mRNA) in the replication cycle of viruses (Fig. 4.1). Viruses do not contain the molecules necessary to translate mRNA and rely on the host cell to provide these. They must therefore synthesize mRNAs that are recognized by the host cell ribosomes. Accordingly, viruses are grouped according to their mechanism of mRNA synthesis and their replication strategy. By convention all mRNA is designated as positive (or 'plus') sense RNA. Strands of viral DNA and RNA that are complementary to the mRNA are designated as negative (or 'minus') sense and those that have the same sequence are termed positive sense. In this way, seven classes of viruses can be distinguished, and each is commonly referred to by the nature of the virus genomes in that class.

Class 1 consists of viruses that have double-stranded (ds) DNA genomes. In this class, the designation of positive and negative sense is not meaningful because mRNAs may come from either strand. Transcription can occur using a process similar to that found in the host cells.

Class 2 consists of viruses that have single-stranded (ss) DNA genomes. The DNA can be of positive or negative sense, depending on the virus being studied. For viruses in class 2, the DNA must be converted to a double-stranded form before the synthesis of mRNA can proceed.

Class 3 consists of viruses that have dsRNA genomes. Most viruses of this type have segmented genomes but mRNA is synthesized only from one template strand of each segment. The process of transcription from a dsRNA genome can be envisioned as occurring using a mechanism similar to that for transcription from a dsDNA genome. However, the enzymes necessary to carry out such a process do not exist in uninfected cells. Consequently, these enzymes must be encoded by the virus genome, packaged in the virion, and carried into the cell by the virus to initiate the infectious process.

Class 4 consists of viruses with ssRNA genomes of the same (positive) sense as mRNA and which can be translated. Synthesis of a complementary strand, generating at least a partially dsRNA intermediate, precedes synthesis of mRNA. As with the class 3 viruses, the enzymes required for RNA synthesis are not present in non-infected cells, and are virus encoded. However, they are not carried in the particle but synthesized after infection is initiated.

Class 5 consists of viruses that have ssRNA genomes which are complementary in sequence to the mRNA and are known as negative-strand RNA viruses. Synthesis of mRNA occurs by transcription from the genome strand and requires novel virus-encoded enzymes. Some class 5 viruses also use the newly synthesized 'antigenome' RNA strand as a template for production of an mRNA. These are referred to as 'ambisense' viruses. Generation of new virus genomes requires the synthesis of a partially dsRNA intermediate, the positive-sense strand of which is used as a template for replication. Again, viral RNA-synthesizing enzymes are not present in non-infected cells, and are virus encoded. These are packaged, and carried into the cell by the virion where they initiate the infectious process.

Class 6 consists of viruses that have positive-sense ssRNA genomes and that generate a dsDNA intermediate as a prelude to replication. This is carried out by a virus-coded enzyme (reverse transcriptase) not found in non-infected cells and carried in the virion.

Class 7 comprises some dsDNA viruses, termed 'reversiviruses', that have been transferred from class 1 into a new class 7. This is based on their novel replication strategy via a positive-sense ssRNA intermediate and a reverse transcriptase. This is the inverse of, but otherwise very similar to, class 6 viruses.

The Baltimore scheme has both strengths and weaknesses as a tool for understanding virus properties. A particular strength is that assignment to a class is based on fundamental, unchanging characteristics of a virus. Once assigned to a class, certain predictions about the molecular processes of nucleic acid synthesis can be made, such as the requirement for novel virus-encoded enzymes. A weakness is that, although it brings together viruses with similarities of replication mechanism, the scheme takes no account of their biological properties, e.g. bacteriophage T2 and variola virus (the cause of smallpox) are classified together in class 1, although they are totally dissimilar in structure and biology. Similarly, the identification of a positive-sense RNA genome is not sufficient to classify the virus unambiguously because viruses of classes 4 and 6 have similar forms of genome nucleic acids. For this reason, a number of classification schemes must be used together for full understanding of the relationships between viruses. (See Chapter 22 for a summary of the properties of the virus families referred to in this text, with selected references.)

4.3 Viral genomes—DNA

Viral DNA genomes may exist either as linear molecules or as circles. For some viruses, the genome exists in one or other of these forms throughout the life cycle. For others, the process of replication requires interconversion between linear and circular forms. Circular and linear molecules can be distinguished by their sensitivity to exonucleases, which require a free 5' or 3' terminus on which to act, and by their sedimentation and electrophoretic behaviour. As well as varying in form, viral DNA genomes differ considerably in size, ranging from the hepadnaviruses at around 3000 base-pairs (3 kbp) to the poxviruses, typically around 200 kbp. The largest genome sequenced so far (fowlpox) is 290 kbp.

Virus families that have genomes comprising either ssDNA or dsDNA in either circular or linear form have been identified. An example of a virus with a circular single-stranded genome is bacteriophage φX174. The geminiviruses, which infect plants, also have this type of genome. Recently, a completely novel human virus, TTV, that has this genome type has been described and tentatively assigned to the circovirus family. Viruses with linear ssDNA genomes are exemplified by the parvovirus family, which has members that infect various animal species. There are a far greater number of viruses that have dsDNA genomes. The papovaviruses (infecting mammals) and the baculoviruses (infecting insects) have this genome type in circular form. The hepadnaviruses (infecting vertebrates) and the caulimoviruses (infecting plants) represent a special case of the circular dsDNA genome type, in which one of the two strands contains a break, making an incomplete circle. Viruses with linear dsDNA genomes include the adenoviruses and herpesviruses (infecting vertebrates), the poxviruses (infecting both vertebrates and invertebrates) and various bacteriophages, such as T4, T7 and λ.

4.4 Unusual features of viral DNA genomes

Terminal redundancy (direct repeats)

Experiments to characterize the genetic material of herpes simplex virus (HSV) showed that its DNA carries direct repeats at its ends. This is referred to as terminal

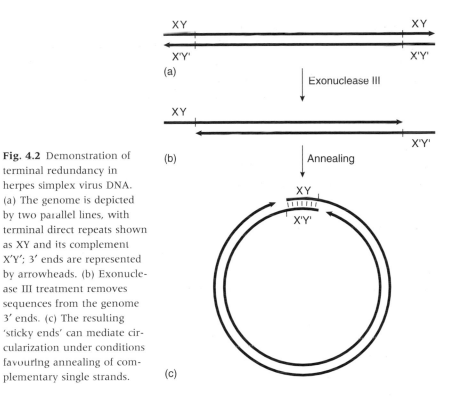

Fig. 4.2 Demonstration of terminal redundancy in herpes simplex virus DNA. (a) The genome is depicted by two parallel lines, with terminal direct repeats shown as XY and its complement X'Y'; 3' ends are represented by arrowheads. (b) Exonuclease III treatment removes sequences from the genome 3' ends. (c) The resulting 'sticky ends' can mediate circularization under conditions favouring annealing of complementary single strands.

redundancy. When the DNA was treated with a strand-specific exonuclease to produce single-stranded segments at both ends of the DNA, followed by incubation under conditions that allowed any complementary single-stranded sequences to anneal together, electron microscopy showed that the DNA formed circles. The explanation of this result is that the genome, being terminally redundant, gives single-stranded 'sticky' ends (Fig. 4.2 and see Section 4.5) when digested with the exonuclease, enabling the molecules to circularize during the incubation period. This terminal redundancy is quite small (about 500 bp) and does not include any genes.

Terminal redundancy (inverted repeats)

Treatment of adenovirus DNA with exonuclease does not allow the formation of dsDNA circles. This means that the single-stranded termini exposed by nuclease digestion cannot base pair, and therefore that the ends of adenovirus DNA are not terminally repetitious in the manner observed for herpesviruses. However, when adenovirus DNA is denatured with alkali and then neutralized, both of the individual DNA strands are capable of forming single-stranded circles. As these circles are always of unit length, they must be formed by interaction between the 3' and 5' termini of the same strand, i.e. there must be an inverted terminal repetition. Base pairing between these termini produces a duplex projection or 'panhandle' on a single-stranded circle (Fig. 4.3).

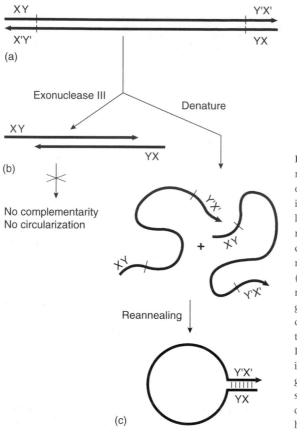

Fig. 4.3 Inverted terminal repeat sequences in adenovirus DNA. (a) The genome is depicted by two parallel lines, with inverted terminal repeats shown as XY and its complement X'Y'; 3' ends are represented by arrowheads. (b) Exonuclease III treatment removes sequences from the genome 3' ends, but this does not reveal complementary single-strand regions. (c) Denaturation and reannealing of the undigested genome allows formation of single-stranded circles with double-stranded 'panhandle' projections.

Another family of viruses whose genomes show inverted terminal repeats are the poxviruses. These repeats extend for some 10 kbp and contain several genes; however, the repeats cannot be revealed by the experiment described for adenovirus DNA because the ends of the poxvirus DNA are covalently closed. This means that the two pairs of nucleotides at the ends of the molecule are each linked by a standard 5'–3' phosphodiester bond so that there are no free 5' or 3' ends on the molecule. When the DNA is completely denatured, a circular single-stranded molecule is generated.

Other repeated sequences

Herpes simplex virus type 1 (HSV-1) DNA is, as already noted, terminally redundant (see Fig. 4.2). However, the DNA is considerably more complex than this. When the isolated single strands are self-annealed, two kinds of molecules of particular interest can be seen by electron microscopy (Fig. 4.4). The first of these, designated type A, has one terminus annealed to an internal region to form a structure consisting of a small single-stranded loop, a short double-stranded region and a long single-stranded tail. Molecules designated type B have both termini annealed to internal sequences

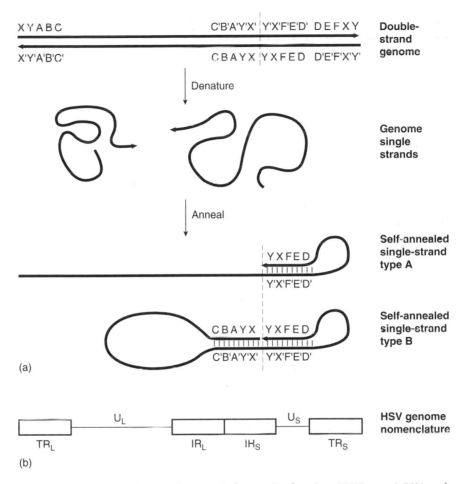

Fig. 4.4 (a) Detection of inverted repeats in herpes simplex virus (HSV) type 1 DNA and the proposed structure of the DNA. Sequences and their complements are represented by A, B, etc. and A', B', etc. Note that the sequence XY appears at both ends of the molecule (terminal redundancy: direct repeat—see Fig. 4.2). The two forms of self-annealed strand are shown for the upper strand of the double-stranded genome. Exactly the same forms may be adopted by the lower strand. (b) Conventional nomenclature to describe the HSV genome: L, long; S, short; U, unique sequence; TR, terminally repeated sequence; IR, internally repeated sequence. The genome is not drawn to scale; U_S should represent about 8.5% and U_L about 71% of the total length.

to yield a structure consisting of long and short single-stranded loops bridged by a double-stranded region. These results indicate that the two terminal sequences are each also found internally in the form of adjacent inverted repeats, which separate the body of the molecule into two unequal parts—the so-called long and short unique DNA segments, U_L and U_S (Fig. 4.4). Thus, the molecule has terminal redundancy (direct repeats at each end as already discussed) embedded within longer sequences, which are unique to each end but which are also repeated internally in the molecule.

The herpesvirus family is large and diverse, and its various members show a corresponding diversity of genome organizations. Nevertheless, each has direct sequence repeats at its termini, often together with either the same or other repeated sequences at internal locations.

4.5 Interconverting linear and circular molecules

Single-stranded ('sticky') ends

DNA purified from bacteriophage λ particles forms a single band when analysed on a sucrose gradient and can therefore be judged to be homogeneous. However, when the DNA is heated, but only to temperatures below those that are necessary to separate the DNA strands, and then slowly cooled, two new components appear, one sedimenting 1.13 times faster and the other 1.41 times faster than native λ DNA. These new forms disappear after reheating and then quick cooling, suggesting that hydrogen bonding is involved in their formation. In fact, there are stretches of 12 bases of ssDNA at either end of the linear molecule that are complementary. When the ends of the same molecule base pair, they form a circle; when the ends of two molecules interact in the same way, a dimer is formed (Fig. 4.5). These molecular forms can be observed directly by examining samples in the electron microscope. The capacity to circularize is essential in the replication of bacteriophage λ DNA.

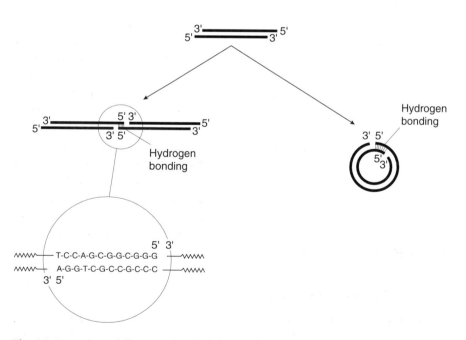

Fig. 4.5 Formation of dimers and circles of phage λ DNA after incubation under conditions favouring annealing. The base composition of the 'sticky ends' is also shown. In λ these are extensions to the 5′ ends of the genome. Exactly the same phenomenon would be observed with 3′ extensions.

The linear herpes simplex virus DNA also circularizes as an essential step in its replication. This is mediated via complementary single nucleotide 3' overhangs on the genome ends, which are probably brought together through the action of a protein.

Circular permutation

The phage T4 genome is a linear DNA molecule and yet its genetic map is circular. The answer to this apparent paradox is that the genes in bacteriophage T4 are circularly permuted. This situation arises because, during replication, DNA is produced as a long concatemer of genomes arranged in a head-to-tail array. Genome length molecules are then cleaved from this concatemer, although this process is not site specific. Moreover, the genome length is greater than the length of one unique set of genes so, wherever cleavage occurs, some genes will be present twice. Which genes these are will depend on where in the concatemer cleavage takes place and will be different for each individual genome molecule (Fig. 4.6). The result of this genome organization is that, if a preparation of T4 linear DNA is denatured and then annealed, circles are formed which can be observed using the electron microscope. These circles result from pairing of complementary strands from different molecules to give linear molecules with sticky ends, which then enable the molecules to circularize, in the manner already discussed for phage λ.

Circularization through protein linkers

Examination of partially disrupted adenoviruses by electron microscopy reveals circular DNA, whereas DNA extracted by treatment with proteolytic enzymes, detergent and phenol (i.e. fully deproteinized) consists of linear duplex molecules. Adenovirus DNA does contain inverted terminal repetitions (see Fig. 4.3), but these do not permit circularization of duplexes. Instead, the explanation for the circular DNA molecules observed, and for their disruption when protein is degraded during DNA purification, is the presence of a protein linked covalently to the 5' end of each DNA strand. This protein mediates circularization of the genome by binding non-covalently to the 3' terminus at the opposite end. When this covalently attached protein is preserved during DNA isolation, the purified DNA sediments in sucrose gradients faster than DNA obtained by conventional extraction methods (Fig. 4.7a) and electron microscopy reveals that up to 90% of the DNA is in the form of circles and oligomers. Treatment of these circles and oligomers with detergent or proteolytic enzymes converts them to linear duplex monomers (Fig. 4.7b).

4.6 Viral genomes—RNA

Viral RNA genomes exist only as linear molecules; infectious circular RNA molecules have been characterized but form the genomes of a specialized type of agent, the viroid, described in Section 4.7. As described in Section 4.2, RNA genomes can be single stranded or double stranded and the former may be of either positive (mRNA like) or negative sense. An unusual feature of many RNA viruses is that their genomes

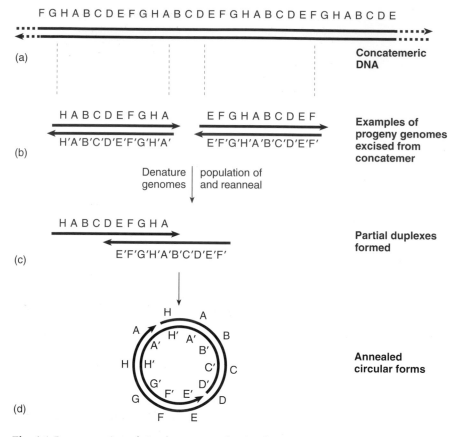

FGHABCDEFGHABCDEFGHABCDEFGHABCDE

(a) Concatemeric DNA

HABCDEFGHA   EFGHABCDEF

H'A'B'C'D'E'F'G'H'A'   E'F'G'H'A'B'C'D'E'F'

(b) Examples of progeny genomes excised from concatemer

Denature | population of
genomes | and reanneal

HABCDEFGHA

E'F'G'H'A'B'C'D'E'F'

(c) Partial duplexes formed

(d) Annealed circular forms

Fig. 4.6 Demonstration of circular permutation in phage T4. (a) Replication produces concatemeric DNA. (b) Genome length molecules are cleaved sequence non-specifically from the concatemer. The genome length is greater than the length of one set of genes. (c, d) A population of genome molecules, if denatured and allowed to reanneal, will form a variety of partial duplexes which can circularize. Different genes are indicated by A, B, etc. and their sequence complements as A', B', etc.

consist of multiple segments, analogous to chromosomes of host cells; these viruses must ensure that at least one copy of each segment is present in the mature particle to generate a full complement of genes. Viruses with segmented genomes include reoviruses and influenza viruses. During replication, the RNA molecules remain linear, and covalently closed circular molecules are never observed. The RNA genomes of different viruses vary greatly in size, although they do not display the range seen with DNA virus genomes. The largest RNA virus genomes known are those of the coronaviruses, which are approximately 30 000 nucleotides (nt.) in length, with the smallest entire animal virus RNA genomes being those of the picornaviruses at approximately 7500 nt. Several bacteriophages have RNA genomes smaller than this, one of the smallest being MS2 with a single-stranded positive-sense RNA genome of 3569 nt. All enzymes involved in RNA synthesis, whether of virus or host-cell origin, are unable

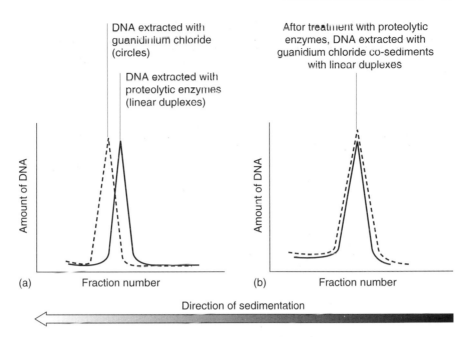

Fig. 4.7 Sucrose gradient sedimentation of DNA extracted from adenovirus by two different methods. (a) Separation of circles and oligomers from linear duplexes. (b) Conversion of circles and oligomers to linear duplexes by treatment with proteolytic enzymes.

to 'proofread' (i.e. to correct incorrectly inserted bases). This contrasts with DNA synthesis in which correction may occur. This is likely to place a limit on the maximum size of an RNA-based genome because, as the size of the genome increases, the probability of it containing a mutation in an important region will also increase.

RNA genomes frequently have unusual structures at one or other of their termini. These may take the form of covalently attached proteins at the 5′ end in the case of picornaviruses such as poliovirus, long homopolymeric polyadenylate (polyA) tracts at the 3′ ends of many positive-sense RNA genomes such as those of picornaviruses, alphaviruses, flaviviruses and coronaviruses, and near-perfect inverted repeats at the termini of most negative-sense RNA genomes such as those of paramyxoviruses and influenza viruses. These structures play important roles in the replication of the RNA molecules (see Chapter 7).

4.7 Unusual features of viral RNA genomes

Segmented genomes

Electrophoretic analysis of RNA extracted from reovirus particles showed that the dsRNA genome consists of 10 different segments of RNA falling into three size classes: L (large—L1–L3 of approximately 4500 bp), M (medium—M1–M3 of approximately 2300 bp) and S (small—S1–S4 of approximately 1200 bp). These 10 segments showed no base sequence homology in hybridization tests and so could not have arisen by random fragmentation of the genome. Nucleotide sequence analysis of all 10 segments has confirmed that they are unique. The sum total of the lengths of the 10 segments corresponds to the size of the viral genome as estimated by chemical means, suggesting that the intact virus contains one copy of each segment. The original observation raised many questions about the accuracy of the experimental procedure, with the possibility that the genome existed in the intact particle as a single molecule, which was susceptible to breakage at fixed points during extraction, rather than as a collection of 10 different segments. However, this and other alternative models were discounted by evidence from experiments in which the free 3′ ends of the RNA in intact virus particles were radioactively labelled before purification and electrophoretic separation. All 10 segments were labelled equally, indicating that they exist as individual molecules in the virion.

Other RNA viruses that have segmented genomes include phage φ6, which also has dsRNA, and the negative-sense ssRNA animal orthomyxoviruses, bunyaviruses, arenaviruses and members of several plant virus families. However, there is an important difference; in animal viruses all segments are present within one particle but in plant viruses the segments are distributed separately in different particles, e.g. the two RNAs of the comovirus cowpea mosaic virus are contained in separate particles and the four RNAs of brome mosaic virus are contained in three particles, all of which are essential for infectivity. Frequently, the smallest RNA of multicomponent plant viruses duplicates information contained in one of the other segments and is co-encapsidated (Fig. 4.8). The significance of this is not known. In brome mosaic virus, segment 4 is an exact copy of part of segment 3. The requirement for two or more virus particles

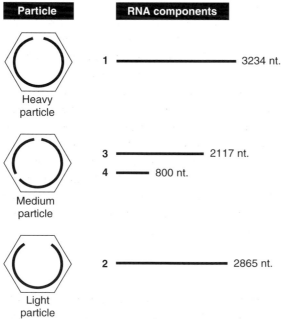

Particle   **RNA components**

Heavy particle — 1 ———————— 3234 nt.

Medium particle — 3 ———————— 2117 nt.
 4 ——— 800 nt.

Light particle — 2 ———————— 2865 nt.

Fig. 4.8 Relationships between the encapsidated RNAs and virus particles of brome mosaic virus. The RNAs are numbered **1–4** on the basis of size. The smallest RNA (**4**) is an mRNA transcribed from RNA **3**. These RNAs are packaged together in the medium-density particle. Segments **1** and **2** are packaged separately into heavy and light particles, respectively. The three particles can be separated by velocity gradient centrifugation in sucrose gradients.

to infect a single cell simultaneously for the infection to begin means that the inter-action between the virus and host does not follow single-hit kinetics. Consequently, such infections are possible only when the virus is spread by a route in which the number of particles transmitted is high. For plant viruses, spread of infection is usually mediated by insects that carry the virus from plant to plant during feeding.

RNAs of satellite viruses, satellite nucleic acids and viroids

Satellite viruses and satellite nucleic acids depend on a co-infecting helper virus to provide essential functions. The satellite genome does not have any significant homol-ogy with that of the helper virus. Satellite viruses and satellite nucleic acids differ in that satellite viruses encode their own coat protein, whereas satellite nucleic acids rely on the helper virus to provide coat protein to encapsidate their genome. The presence of a satellite virus or satellite nucleic acid may affect the replication of the helper virus, and may also increase or decrease the severity of the disease(s) caused by the helper virus. Satellite viruses and satellite nucleic acids with DNA and RNA genomes have been identified. The RNA genomes can be very small indeed, ranging from approxi-mately 220 nt. to 1800 nt. Satellite viruses and satellite nucleic acids with linear dsDNA, linear ssDNA, linear dsRNA, linear ssRNA and circular ssRNA genomes have been characterized (see Section 22.5).

A satellite nucleic acid that is important because of its association with human disease is the RNA-containing hepatitis delta virus (HDV). Its helper is the DNA-containing hepatitis B virus (class 7—HBV), and HDV is associated with enhanced pathogenicity of HBV. The HDV genome is a circular ssRNA which is extensively base paired and appears as a rod-like or dumb-bell-shaped structure by electron microscopy. HBV provides the structural proteins that encapsidate the HDV genome and allow it to spread. HDV has a negative-sense genome and encodes delta antigen from a tran-scribed mRNA, analogous to gene expression by class 5 viruses (see Chapter 10).

Viroids are common agents of disease in plants, and their infectious material is a single circular ssRNA molecule with no protein component. Unlike satellite nucleic acids, viroids have no helper and, with genomes ranging in size from 246 to 370 nt., they are the smallest self-replicating pathogens known. Up to 70% of the nucleotide bases in the genome RNAs are base paired, and resemble the genome of HDV seen by electron microscopy (Fig. 4.9). Sequence analysis of viroid RNA has shown that no proteins are encoded by either the viroid genome or the antigenome. It is thought that the diseases associated with viroids result from the RNA interfering with essen-tial host-cell mechanisms. Viroids and HDV appear to be replicated by the cellular DNA-dependent RNA polymerase II, which normally recognizes a DNA template. Replication of viroid RNA is described in Section 7.3 (Chapter 7).

Terminal caps

Messenger RNAs are synthesized, translated and possibly also degraded in a 5'→3' direction. Consequently, their 5' termini are of some interest, particularly with regard

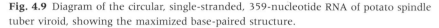

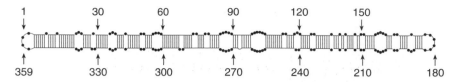

Fig. 4.9 Diagram of the circular, single-stranded, 359-nucleotide RNA of potato spindle tuber viroid, showing the maximized base-paired structure.

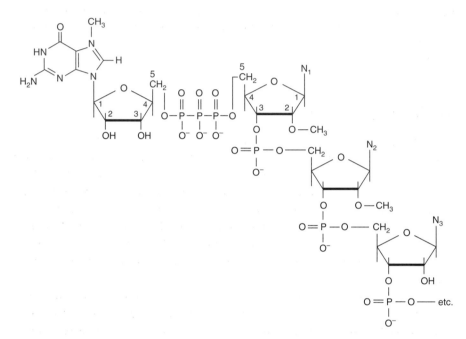

Fig. 4.10 Structure of a capped RNA molecule. Note especially the 5′–5′ phosphodiester linkage. N_1, N_2 and N_3 can be any of the four nucleic acid bases. The 2′-O-methyl group on the ribose of N_1 nucleotide is always present (cap 1 and 2 structures) and the methyl group on N_2 is present in cap 2 structures only.

to their ability to regulate gene expression (see Chapter 10). In prokaryotes, and viruses of prokaryotes, the 5′ end of many mRNAs is a triphosphorylated purine corresponding to the residue that initiated transcription. In contrast, most eukaryotic cellular and viral mRNAs, as well as native nucleic acid from some RNA viruses of eukaryotes, are modified at the 5′ end. The modification consists of a 'cap' that protects the RNA at its 5′ terminus from attack by phosphatases and nucleases, and promotes mRNA function at the level of initiation of translation.

The general structural features of the 5′ cap are shown in Fig. 4.10. The terminal 7-methylguanine and the penultimate nucleotide (the first nucleotide copied from the genome template) are joined by their 5′-hydroxyl groups through a triphosphate bridge. This 5′–5′ linkage is inverted relative to the normal 3′–5′ phosphodiester bonds in the remainder of the polynucleotide chain and is formed post-transcriptionally.

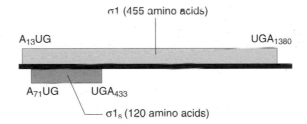

Fig. 4.11 Arrangement of the overlapping open reading frames (ORFs) of the reovirus S1 gene mRNA. The relative locations and protein products of the two ORFs are indicated. The numbers refer to the nucleotide positions of the first nucleotide of the relevant codon.

4.8 Overlapping genes

Many virus genomes have been completely sequenced and the list continues to grow rapidly. A complete genome sequence can be scrutinized systematically for information on its overall genetic organization, e.g. it is possible to locate potential protein-coding regions by identifying all possible initiation codons (AUG) and then scanning the sequence downstream for the next occurrence of a termination codon (UGA, UAA or UAG). This is especially useful when splicing is not used extensively or at all during viral gene expression (see Chapters 9 and 10). Confirmation of the potential utility of open reading frames (ORFs) requires knowledge of the transcribed regions of the virus genome.

From the earliest sequence analyses of virus genomes, it has been known that ORFs in different reading frames may overlap each other and examples of this are described below. As well as enhancing the coding capacity of these viruses by using the same sequence to encode two or more different proteins, the utilization of alternative ORFs also allows an extra method of control of gene expression, as described in Chapter 10.

Several animal virus genomes encode mRNAs that contain more than one ORF and each ORF may be used, usually independently of the other. One example is the mRNA transcribed from the S1 segment of the reovirus genome which contains a large ORF with an AUG translation initiation codon located near the 5′ end of the mRNA. This ORF directs the synthesis of a structural protein of 455 amino acids called σ1. Further from the 5′ end is a second ORF, which begins with another AUG codon and which is in a different reading frame (Fig. 4.11). Occasionally, ribosomes that have failed to initiate translation at the first AUG do so at the second AUG, producing a protein of 120 amino acids called σ_s. As σ1 and σ_s are encoded from different reading frames, they are unique and do not share any amino acid sequences.

A more complex situation is seen with certain paramyxoviruses, such as Sendai virus, in which two proteins are generated by translation of overlapping ORFs. In addition, further proteins are generated by utilizing alternative AUG initiation codons within the second ORF. The additional proteins from the second ORF share the same carboxyl-terminal amino acid sequences. Yet another protein is produced by initiation at an ACG codon, instead of the more conventional AUG, generating a protein from the second ORF that has additional residues at the amino terminus. This is described in detail in Chapter 10.

RNA splicing adds further to the possibilities of protein synthesis from overlapping or adjacent ORFs. This can bring together segments of reading frames to generate novel proteins and is seen with mRNAs that direct synthesis of the T antigens of SV40 and polyomavirus, and in the expression of the bovine papillomavirus genome (see Chapter 9).

4.9 Viral genome diversity and replication strategy

Understanding the structural peculiarities of viral nucleic acids is important for several reasons. First, these features may help the virus to subvert the cell's biosynthetic machinery and redirect it to the production of new virus. Second, models for the replication of these nucleic acids must take into account their structural peculiarities. In many cases, these unusual features are crucial to successful replication. The various strategies that are adopted by viruses to achieve replication of their genetic material are discussed in Chapters 6, 7 and 8.

4.10 Further reading and references

Baltimore, D. (1971) Expression of animal virus genomes. *Bacteriological Reviews* **35**, 235–241.

Flores, R., Di Serio, F. & Hernández, C. (1997) Viroids: the noncoding genomes. *Seminars in Virology* **8**, 65–73.

Karayiannis, P. (1998) Hepatitis D virus. *Reviews in Medical Virology* **8**, 13–24.

van Regenmortel, M. H. V., Fauquet, C. M., Bishop, D. H. L., Carstens, E., Estes, M. K., Lemon, S., Maniloff, J., Mayo, M. A., McGeoch, D. J., Pringle, C. R. & Wickner, R. (1999) Virus taxonomy. Classification and nomenclature of viruses. *Seventh Report of the International Committee on Taxonomy of Viruses.* New York: Academic Press.

Symons, R. H. (ed.) (1990) Viroids and related pathogenic RNAs. *Seminars in Virology* **1**(2), 75–162.

The website of the International Committee on Taxonomy of Viruses (ICTV) www.ncbi.nlm.nih.gov/ICTV/

The Genbank sequence database (includes virus genome sequences) www.ncbi.nlm.nih.gov/Entrez/index.html

Chapter 5

The process of infection: I. Attachment of viruses and the entry of their genomes into the target cell

The process of infection begins with the coming together of a virus particle and a susceptible target cell, but this union occurs by different means with each of the three types of virus, namely, bacteriophages, plant viruses and animal viruses. The initial interaction of animal viruses with animal cells occurs by simple diffusion, because particles the size of a virus are in constant Brownian motion when suspended in liquid. Diffusion of bacteriophage is probably also the force influencing their union with bacterial cells. In most cases plants become infected with viruses after mechanical damage to the plant, very often as a result of the wind, or by the activities of virus-carrying pathogens. Consequently, the way in which the union of virus and cell occurs is not so important in plant systems. However, when viruses are transmitted in plants as a result of grafting, diffusion through the vascular system is most likely to be responsible. However, plants are a special case because all cells directly communicate with their neighbours, and functionally a plant behaves as a single cell or a syncytium.

Virtually all data for the attachment and entry of animal viruses come from experiments with cells cultured *in vitro*. It is worth remembering that the *in vivo* situation, where the virus is infecting differentiated cells that may be part of a multicellular tissue structure, may not take place by exactly the same process. The efficiency of infection, for example, may be less.

Terminology relating to attachment can be confusing. Consequently, we use the terms 'virus attachment protein' (of the virus) and 'cell receptor' (of the cell) to describe the interacting components. In some situations, it is useful to refer to cell receptor sites, which may be multivalent and consist of several cell receptors.

5.1 Infection of animal cells

The attachment of animal viruses to target cells

Animal cells are bounded by a lipid bilayer (the plasma membrane) into which is inserted a variety of proteins by which the cell communicates with its environment. A cell cannot be infected unless it expresses the molecule that serves as a receptor for that particular virus on its outer surface. These are usually proteins, but carbohydrates and, very occasionally, lipids are also used (Table 5.1). Receptors are molecules that have a role in the normal functioning of a cell, and viruses have evolved to take

Table 5.1 Some cell surface molecules used by animal viruses as receptors.

Molecule	Normal function	Virus	Type of receptor
Protein			
ICAM-1	Adhesion to other cells via CD54	Rhinoviruses (most but not all)	Primary
CAR (Coxsackie–adenovirus receptor)	A unique protein of unknown function	Adenoviruses, Coxsackie B viruses	Primary
$\alpha_v\beta_x$-Integrin	Adhesion to other cells via vironectin	Adenoviruses	Coreceptor
$\alpha_v\beta_6$-Integrin	Adhesion to other cells via vironectin	Foot-and-mouth disease virus	Primary
CD4	Ligand for MHC II on T helper cells	HIV-1, HIV-2, SIV	Primary
CCR5	Binds chemokines	HIV-1, HIV-2, SIV	Coreceptor
CXCR4	Binds chemokines	HIV-1, HIV-2	Coreceptor
CD155	Not known	Poliovirus	Primary
MHC I	Ligand for CD8; presents peptides to T cells	Human cytomegalovirus	Primary
MHC II	Ligand for CD4; presents peptides to T cells	Lactate dehydrogenase-elevating virus	Primary
CR2	Receptor for complement component C3d	Epstein–Barr virus	Primary
IgA receptor	Binds IgA for transport across the cell	Hepatitis B virus	Primary
Phosphate transporter	Transports phosphate	Some retroviruses	Primary
Virus-specific IgG bound to cells by Fc receptors (see text)	Binds to virus	Dengue virus *in vivo* and many others *in vitro*	Primary
β-Adrenergic receptor	Binds the hormone adrenaline (epinephrin)	Reoviruses	Primary
Acetylcholine receptor	Binds the acetylcholine neurotransmitter	Rabies virus	Primary
Carbohydrate			
N-Acetylneuraminic acid (only when terminal in the carbohydrate moiety)	Part of the carbohydrate moiety of glycoproteins and glycolipids. Gives cells much of their negative charge.	Influenza virus A, B, C, paramyxoviruses, polyomavirus, reoviruses, encephalomyocarditis virus	Primary
Heparan sulphate	Extracellular matrix glycosaminoglycan	HIV-1, herpes simplex virus, dengue virus, Sindbis virus, cytomegalovirus, adeno-associated virus, respiratory syncytial virus, foot-and-mouth disease virus	Coreceptor
Lipid			
Phosphatidylserine or phosphotidylserine	Constituent of lipid bilayer	Vesicular stomatitis virus	Primary

CR, complement receptor; HIV, human immunodeficiency virus; ICAM, intercellular adhesion molecule; IgA, immunoglobulin A; MHC, major histocompatibility complex; SIV, simian immunodeficiency virus.

advantage of these. Unequivocal identification of protein receptors has come from expressing the gene for the receptor in a cell to which a virus does not normally bind or, alternatively, by blocking virus attachment with a monoclonal antibody to the receptor. Usually receptors are highly specific for a particular virus. One notable exception is the sugar, N-acetylneuraminic acid, which often forms the terminal moiety of a carbohydrate group of a glycoprotein or glycolipid, and is used as a receptor by members of several different families of viruses. Some viruses can use different types of molecule as primary receptors (e.g. reoviruses bind to the β-adrenergic receptor or N-acetylneuraminic acid—Table 5.1).

Viruses bind up to three different types of receptor molecule on the cell surface in succession. These are low-affinity receptors, primary receptors and co- or secondary receptors. In principle, receptors serve to overcome any repulsive forces that may exist between the virus and the cell, which is negatively charged, and to trigger the release of the viral genome into the cell. The first receptor is a high-abundance molecule that has a low-affinity interaction with the virus. This serves to get the virus out of fluid bathing the cell and in intimate contact with its surface. Several viruses, including possibly human immunodeficiency virus type 1 (HIV-1), use heparans whereas others use the sugar, N-acetylneuraminic acid, as their first receptor (see Table 5.1). This serves as the sole receptor for the influenza viruses. HIV-1 has three receptors; it binds the low-affinity receptor, heparan, via a cellular protein, cyclophilin A—an essential component of the HIV-1 particle. This interaction allows the virus to make the initial contact with the cell, and the opportunity of contacting the primary receptor, which may be far less abundant. If the virus does not find a primary receptor molecule, it will dissociate from the cell completely and the process begins again. The search process consists of the virus rolling along the cell surface, dissociating from and reassociating with the low-affinity receptor, until it comes in contact with the primary receptor. For HIV-1 this is the CD4 protein, and virus is only able to bind CD4 if it has first bound to heparan. Presumably the latter causes a conformational rearrangement that reveals the CD4-binding site on the virus-encoded envelope protein, gp120. The CD4-gp120 binding is a high-affinity interaction. In turn, binding to CD4 causes further conformational changes to gp120, which exposes another site that binds to the co-receptor, CCR5 or CXCR4. This series of dependent binding steps is akin to a fail-safe system, telling the virus that it has bound to a cell that will be able to replicate its genome.

Some viruses use non-neutralizing, virus-specific antibody as a surrogate receptor for cells that carry Fc receptors in their plasma membranes (see Table 5.1), a process known as *antibody-dependent enhancement* of infectivity. This is widespread in cell culture but rare *in vivo*, the classic example being Dengue fever virus (a flavivirus). This virus normally causes a mild subclinical or febrile illness in humans but in the presence of antibody is able to infect macrophages and to cause life-threatening disease.

Entry of animal virus genomes into their target cells

The plasma membrane that surrounds the cell is a very mobile and active structure. It comprises a lipid bilayer into which are inserted a variety of proteins. The

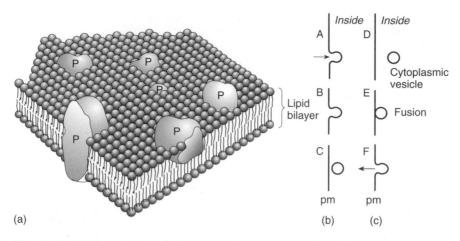

Fig. 5.1 (a) 'Lipid sea' model of plasma membrane structure devised by Singer and Nicholson. (Reprinted from Singer & Nicholson (1972) *Science* **175**, 720–731. Copyright 1972 American Association for the Advancement of Science.) Proteins (P) may span the lipid bilayer or may not, and are free to move laterally like icebergs. (b) Endocytosis by a plasma membrane (pm) inwards (A, B, C) and (c) exocytosis outwards (D, E, F).

membrane has been compared to a sea and the proteins to icebergs that can move laterally (Fig. 5.1a). Cells are constantly taking samples of their immediate environment by endocytosis, during which the membrane invaginates and a vesicle is pinched off into the cytoplasm, or carrying out the reverse process to export substances from the cell such as enzymes, hormones or neurotransmitters (Fig. 5.1b,c).

A virus, attached to cells as described above, may now need to recruit further receptors before it is able to enter the cell and/or uncoat. As the receptors in the lipid membrane are mobile, both the attached virus and free receptors can move laterally and find each other by random collision. This recruitment of a finite number of receptors is the final signal for the entry/uncoating stage to begin, because, in addition to allowing the virus to find the correct target cell, interaction with receptors prepares the virus particle for uncoating and the entry of the viral genome into the cell cytoplasm. Uncoating happens in two ways: some enveloped viruses, such as HIV-1, uncoat by *fusion* of the lipid bilayer of the virus with that of the plasma membrane, whereas other enveloped viruses and all non-enveloped viruses are taken up into the cell by *receptor-mediated endocytosis.*

Entry of virus genomes into the cell by fusion of viral and cell membranes at neutral pH

Membrane fusion is a universal biological phenomenon that occurs in a myriad of processes from fertilization to the trafficking of membrane vesicles within the cell. However, lipid bilayers are immensely stable structures, composed of two monolayers arranged with their long hydrophobic chains inside and their polar head groups exposed on their two surfaces to an aqueous environment. Fusion is a far from

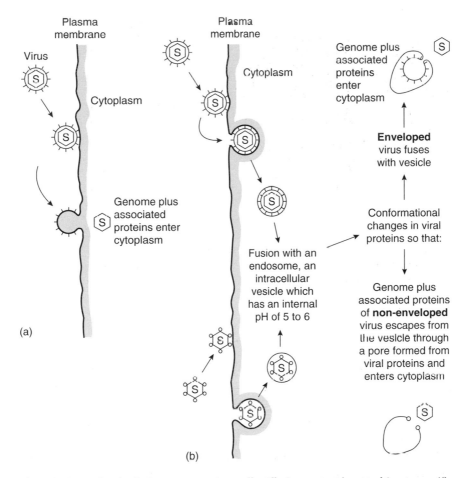

Fig. 5.2 Entry of animal virus genomes into cells. All viruses start by attaching to specific receptors on cells. (a) Entry by fusion of the lipid bilayers of an enveloped virus and the plasma membrane at neutral pH. (b) Entry by endocytosis, followed by a fusion of the vesicle with an endosome and a decrease in the pH of the endosome. This promotes conformational changes in viral proteins. For enveloped viruses (upper panel), this leads to fusion of the lipid bilayers of the virus and the endosome. For nucleocapsid virus particles (lower panel), the low pH causes conformational changes in viral proteins. This results in the insertion of newly exposed hydrophobic regions into the lipid bilayer of the vesicle, and the escape through these of the viral genome and associated proteins into the cytoplasm.

spontaneous process. For fusion to occur, the viral and cell bilayers first disrupt at a common point and so expose their hydrophobic interiors in an incompatible environment, and then combine and re-form as one bilayer. It is still not understood how this occurs. Direct fusion of a viral lipid bilayer with a plasma membrane to which the virus has become attached (Fig. 5.2a) is initiated by the triggering of conformational changes in the metastable envelope proteins as a result of the series of virus–receptor interactions as described above. This activates the hydrophobic segment of the envelope protein from its normally concealed position close to the virion membrane. Some of these bend towards the plasma membrane and some to the viral membrane itself. Fusion has two essential steps. In the first a small number of envelope proteins combine together to form a hydrophobic channel between the two membranes and, in the second, insertion of the viral hydrophobic peptides into the cell membrane and into its own membrane disrupts both bilayers. The disrupted lipid flows through the hydrophobic channel and allows the two bilayers to fuse together, so that the viral and cell membranes become one. Once this has happened, the genome plus associated proteins are automatically in the cell cytoplasm. HIV-1 and members of the paramyxovirus family infect cells in this way.

Entry of virus genomes into the cell by receptor-mediated endocytosis followed by disruption of the virion at low pH

All other animal viruses are first taken up by the cell in an endocytic vesicle by receptor-mediated endocytosis, but formally the virus is still on the *outside* of the cell because it is bounded by what was plasma membrane (Fig. 5.2b). Release of the genome into the cytoplasm is dependent on a decrease in pH of the internal environment of the endocytic vesicle to pH 5–6. This is achieved by fusion of the endocytic vesicle with an intracellular vesicle called an endosome. Protons are concentrated in the endosome by the action of a membrane-bound protein that pumps protons into the lumen of the vesicle. With enveloped viruses, the low pH initiates conformational changes in the viral envelope proteins, which promote fusion of the viral lipid bilayer with that of the endosome, exactly as described above. The only difference is the way in which the envelope proteins are activated. The low pH also brings about conformational changes in the capsids of non-enveloped viruses. This exposes hydrophobic regions of viral proteins, which then form a channel in the wall of the endosome, through which the genome and associated proteins enter the cytoplasm (Fig. 5.2b).

The complex structure of uncoated genomes and secondary uncoating

It should be pointed out that, regardless of the nature of the virus particles, what enters the cell is a nucleoprotein structure and not naked nucleic acid. In addition, either the genome that enters the cell is complexed as viral nucleoprotein, or it rapidly associates with cellular structures such as polymerases or ribosomes for the next phase of multiplication. The viral proteins associated with the genome include enzymes for nucleic acid synthesis, and internal structural virus proteins that form the core of some of the more complex viruses. Often there is also *secondary uncoating* of the viral nucleocapsid inside the cell, during which some viral nucleocapsid proteins are removed or rearranged. Equally it is a common misconception that viral nucleic acids are stretched out (as illustrators frequently draw them). In fact they form double-stranded secondary structures wherever regions of base complementarity can be brought together. Thus, together with their associated proteins, their structure will be complex.

The inefficiency of the infectious process

One of the striking aspects of infection of animal cells by viruses is its inefficiency. In poliovirus infections, the majority of the RNA of the infecting virus is degraded as a result of interaction with cells. This has several causes, such as failure of virus–cell interaction to lead on to successful infection. Here, instead of changes in the virion capsid structure leading to release of the RNA genome into the cytoplasm and its complexing with ribosomes and later with replicase proteins, it is hydrolysed by marauding ribonucleases. Such failure to initiate infection accounts, at least in part, for the very high physical particle:infectious particle ratio of 1000:1 or so. It follows that most of these 'non-infectious' particles are in fact potentially infectious, but fail to complete the infectious process. Neither electron microscopy nor biochemical studies can distinguish between these non-infectious particles and infectious particles, which both contain the complete virus genome, and this is a serious difficulty in attempting to define the infectious entry pathway for a virus.

5.2 Infection of plants by viruses

The infection of plant cells by viruses differs considerably from the infection of animal cells. Most plants have rigid cell walls of cellulose and consequently viruses must be introduced into the host cytoplasm by some traumatic process. Thus, plants are infected either with the help of *vectors*, namely animals that feed on plants, or invading fungi, or by mechanical damage caused by the wind or passing animals, all of which allow viruses to enter directly into cells. In the laboratory, the application of carborundum to leaves increases the number of lesions produced during a local lesion assay of virus-containing material and mimics natural mechanical damage (see Fig. 1.1b). Many plants that become infected naturally do so because virus-carrying animals feed upon them. However, this transmission is not a casual process that occurs whenever any animal chances to feed on an uninfected host plant after feeding on an infected one. Rather, the transmission of most plant viruses is a highly specific process, requiring the participation of particular animals as vectors. Although some viruses, such as tobacco mosaic virus (TMV), require no vector and can be transmitted mechanically, most have a specific association with their animal vectors, which may include leafhoppers, aphids, thrips, whiteflies, mealy bugs, mites and nematodes. It should be pointed out that most of these animals feed by piercing plant tissues with their mouthparts and not by biting them. Biting insects have limitations as vectors, because biting not only damages leaves excessively, but is a method most likely to succeed only with viruses that can be spread by mechanical inoculation.

Once introduced into the plant, uncoating takes place in the cytoplasm through the capsid binding to cytoplasmic proteins and the influence of divalent cations such as calcium (see Wilson 1985, 1992). Once released the genome can move around the plant as a result of the connections (plasmodesmata) between plant cells that make a plant functionally unicellular. This can be shown by infecting plants with infectious genomes from a virus that has a mutation in its coat protein which makes this non-functional. The signs of disease on the leaves show that the virus spreads as normal. What plant viruses have is a unique *viral movement protein* that forms channels along the plasmodesmata and facilitates the transmission of viral genomes to adjacent cells.

5.3 Infection of bacteria

Bacteria, like plants, have a cell wall that a virus must breach in order to infect the cell. Some carry apparatus that introduces their genome into the cell while the coat protein stays outside. Thus the classic experiment of Hershey and Chase was successful (see Section 1.3). Had they studied an animal virus their results would have been different.

Attachment of bacteriophages to the bacterial cell

Most bacteriophages attach to the cell wall, but there are other cell receptors on the pili, flagella or capsule of the host. Most of the tailed bacteriophages attach to the cell wall and do so by the tip of their tail. In those phages with a head much larger than the tail, diffusion will tend to cause the tail to oscillate more than the head, making it more likely that the tail will collide first with the cell. The chemical nature of the cell receptor on the cell wall has been elucidated for some salmonella phages. Figure 5.3 shows the structure of the O antigen of wild-type *Salmonella typhimurium* and certain cell wall mutants derived from it. When the mutants are tested for sensitivity to different phages, it becomes apparent that each has a characteristic sensitivity pattern, e.g. the presence of the O-specific side-chain makes a cell resistant to phages 6SR, C21, Br60 and Br2, but sensitive to P22 and Felix O. The absence of the O-specific side-chain, as in *rfb* T mutants, now makes the cell sensitive to Felix O, 6SR, Br60 and Br2, and resistant to P22 and C21. Thus, if the lipopolysaccharide is the cell receptor for a particular phage, it is possible to determine precisely the composition and structure of the receptor by testing the different mutants for sensitivity to that phage. Experiments of this type are of interest not only to virologists but also to cell wall chemists. For example, suppose we wished to isolate an *rfc* mutant of *Salmonella* sp. In the absence of a selective procedure, this would be a tedious task. However, if a bacterial lawn showing plaques of P22 is incubated for several days, small colonies arise within the plaques, which represent the growth of phage-resistant mutants. It is clear that these resistant mutants could belong to any one of the mutant classes shown in Fig. 5.3. However, by simultaneously selecting for resistance to 6SR, C21, Br60 and Br2, we can eliminate all but the *rfc* class, which should be sensitive to Felix O.

The best-documented example of phage attachment is that of phage T2 and T4. These viruses have a complex structure, including a tail, base plate, pins and tail fibres (see Chapter 3). The initial attachment of these phages to the receptors on the bacterial surface is made by the distal ends of the long tail fibres (Fig. 5.4). The long tail fibres, which make the first attachment, bend at their centre and their distal tips contact the cell wall only some distance from the midpoint of the phage particle. After attachment, the phage particle is brought closer to the cell surface. When the base

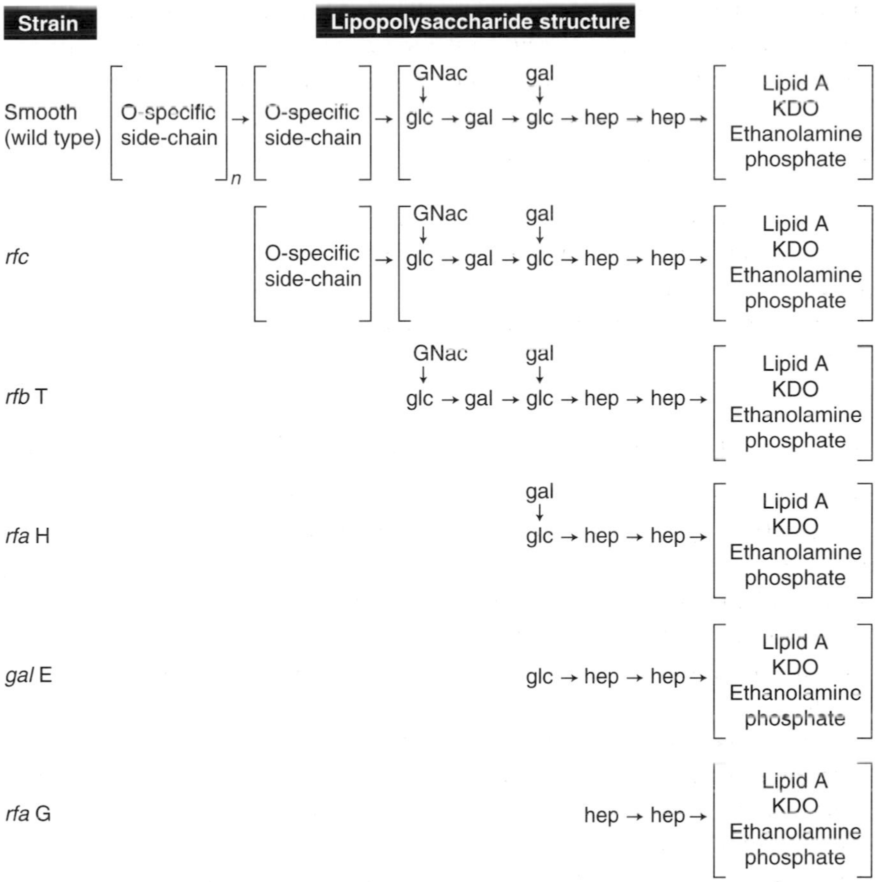

Strain **Lipopolysaccharide structure**

Smooth (wild type): [O-specific side-chain]ₙ → [O-specific side-chain] → [GNac↓, gal↓; glc → gal → glc → hep → hep →] [Lipid A, KDO, Ethanolamine phosphate]

rfc: [O-specific side-chain] → [GNac↓, gal↓; glc → gal → glc → hep → hep →] [Lipid A, KDO, Ethanolamine phosphate]

rfb T: GNac↓, gal↓; glc → gal → glc → hep → hep → [Lipid A, KDO, Ethanolamine phosphate]

rfa H: gal↓; glc → hep → hep → [Lipid A, KDO, Ethanolamine phosphate]

gal E: glc → hep → hep → [Lipid A, KDO, Ethanolamine phosphate]

rfa G: hep → hep → [Lipid A, KDO, Ethanolamine phosphate]

Phage

	P22	Felix O	6SR	C21	Br60	Br2
Smooth (wild type)	S	S	R	R	R	R
rfc	R	S	R	R	R	R
rfb T	R	S	S	R	S	S
rfa H	R	R	R	S	S	S
gal E	R	R	R	S	S	R
rfa G	R	R	R	S	S	S

Fig. 5.3 Correlation of phage sensitivity with lipopolysaccharide structure of the cell wall. The upper part of the diagram shows the structure of the lipopolysaccharide in different mutants of *Salmonella typhimurium*. Abbreviations: glc, glucose; gal, galactose; GNac, *N*-acetylglucosamine; hep, heptose; KDO, 2-keto-3-deoxyoctonic acid. The table in the lower part shows the sensitivity of the different mutants to several bacteriophages. Abbreviations: S, sensitive; R, resistant.

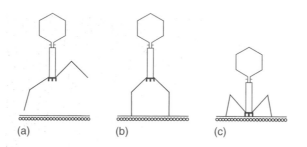

Fig. 5.4 Major steps in the attachment of bacteriophage T4 to the cell wall of *Escherichia coli*. (a) Unattached phage showing tail fibres and tail pins (compare Fig. 3.18). (b) Attachment of the long tail fibres. (c) The phage particle has moved closer to the cell wall and the tail pins are in contact with the wall.

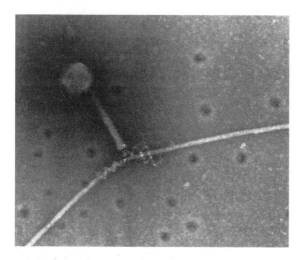

Fig. 5.5 Attachment of a single bacteriophage χ to the filament of a bacterial flagellum. (Courtesy of J. Adler.)

plate of the phage is about 10 nm from the cell wall, contact is made between the short pins extending from the base plate and the cell wall, but there is no evidence that the base plate itself attaches to the cell wall.

Other attachment sites for bacteriophages

Not all tailed phages attach to the cell wall. Some, such as phage χ (chi) and PBSI, attach to flagella. The tip of the tail of these phages has a kinky fibre on it, which wraps around the filament of the flagellum and the phage then slides down the filament till it reaches the base of the flagellum (Fig. 5.5). Other tailed phages attach to the capsule of the cell.

The other important cell receptors are on the sex pili. Bacteria that harbour the sex factor (F) or certain colicins or drug resistance factors produce pili, and two classes of phage have been shown to attach to these pili. The filamentous ssDNA phages attach to the tips of the pili, whereas many types of spherical RNA phages attach along the

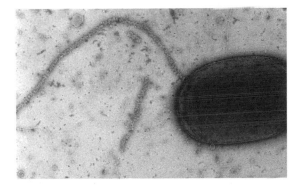

Fig. 5.6 Attachment of many spherical RNA phages (one is arrowed) to the sex pilus of *Escherichia coli*. (Courtesy of C. C. Brinton.)

pili (Fig. 5.6). These phages are particularly useful to microbial geneticists because they offer a ready means of establishing whether cells harbour pili of these types.

Entry of the genomes of bacteriophages into bacterial cells

The head–tail bacteriophages

The experiments of Hershey and Chase (see Section 1.3) indicated that mainly nucleic acid entered the cell during infection of the head–tail bacteriophage, T2. The way in which this occurs has now been elucidated and is a complex but fascinating story, which can only briefly be sketched here. The tail of the bacteriophage is contractile and in the extended form consists of 24 rings of subunits surrounding a core. Each ring consists of a number of small and large subunits. After attachment, the tail contracts, resulting in a merging of the small and large subunits to give 12 rings of 12 subunits. The tail core, which is not contractile, is pushed through the outer layers of the bacterium with a twisting motion, and contraction of the head results in the injection of the DNA into the cell. This process is probably aided by the action of the lysozyme that is built into the phage tail. There are 144 molecules of adenosine 5′-triphosphate (ATP) built into the sheath and the energy for contraction most probably comes from their conversion to adenosine diphosphate (ADP). The phage has been likened to a hypodermic syringe. The various steps in penetration are shown in Fig. 5.7.

The RNA bacteriophages

Most bacteriophages do not possess contractile sheaths (see Chapter 22), and the way in which their nucleic acid enters the cell is not known. Hershey–Chase-type experiments with the filamentous DNA phages, which attach to the sex pili, suggest that both the DNA and the coat protein enter the cell and it has been postulated that, after attachment, the phage–pilus complex is retracted by the host cell. However, when similar experiments are performed on the RNA phages, which also attach to the sex pili, the results obtained are similar to those for T2, i.e. the phage coat protein does not enter the cell. Indeed, after attachment, the RNA phage particles are rapidly eluted. Examination of the eluted phage by centrifugation through sucrose gradients shows

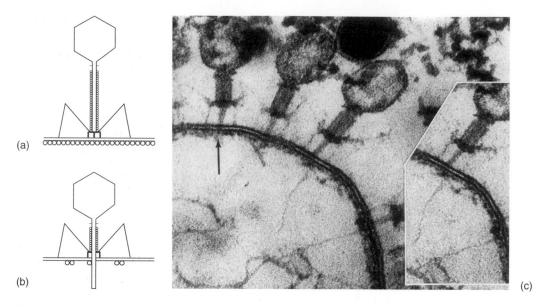

Fig. 5.7 Schematic representation of the mechanism of entry of the phage T4 genome into the bacterial cell wall. (a) The phage tail pins are in contact with the cell wall and the sheath is extended. (b) The tail sheath has contracted and the phage core has penetrated the cell wall; phage lysozyme has digested away the cell beneath the phage. (c) Electron micrograph of T4 attached to an *E. coli* cell wall, as seen in thin section. The core of one of the phages can be seen to penetrate just through the cell wall (arrow). Thin fibrils extending on the inner side of the cell wall from the distal tips of the needles are probably DNA.

that 70% sediment as intact 78S particles, whereas the remaining 30% lack RNA and sediment as 42S particles. ('S' stands for Svedberg, the unit of sedimentation rate, and a particle that sediments rapidly has a high S value. The rate of sedimentation of a particle is complex but depends mainly on its size and density.) Treatment of the eluted 78S particles with ribonuclease (RNase) converts them to 42S particles, whereas normal phages which have not been allowed to interact with pili remain RNase resistant. Thus interaction with the pilus causes conformational changes to the virus particle, which permit the penetration of RNase. An analogous sequence of events occurs when poliovirus interacts with cells carrying the appropriate receptors, except that the RNA of the particles is still RNase resistant.

As the A protein of the ssRNA phages is involved in attachment, it is likely that the RNase sensitivity of the eluted 78S particles stems from loss of this protein. It is, in fact, possible to deduce the course taken by the A protein by locating the radioactivity of [³H]histidine-labelled phage after its interaction with host cells. Such experiments are based on the fact that the coat protein of the RNA phages lacks histidine, whereas the A protein contains approximately 4.5 residues of this amino acid per molecule. When [³H]histidine-labelled phages are allowed to attach and elute from pili, the histidine label is completely absent from the 42S particles and only a small amount is present in the 78S particles. It can thus be concluded that the A protein is released from the phage particles after attachment. Further experiments show that the viral RNA and the A protein are taken up by the cell in approximately equimolar amounts, and that their kinetics of penetration are similar. It is likely that, in a successful infection, the A protein enters the cell, taking the RNA with it.

5.4 Prevention of the early stages of infection

One of the goals of studying the cell–virus relationship is to develop methods of aborting viral infections, particularly those of human beings and domestic animals; how better to achieve this than by preventing virus from attaching and releasing its genome. The main approach has been and still is to raise an immune response by means of immunization (see Chapter 14 and Sections 16.1–16.7). Here antibodies are most important, especially if they are already present to meet the invaders. Cellular receptors are mainly proteins or glycoproteins, which are components of the cell and only incidentally serve the needs of the virus. Thus, attacking these may endanger the cell itself. As to viruses themselves, there are an increasing number of atomic resolution, three-dimensional structures of virions/coat proteins being revealed, so the search for antiviral drugs that inhibit the attachment and uncoating processes is moving from random screening to rational development. However, an insuperable problem for antiviral drugs in the *prevention* of disease is that no one knows that they are infected until clinical signs and symptoms appear, and by then the infection is well advanced. Antiviral drugs are discussed in Section 16.8.

For the protection of plants, it should be apparent that the best preventive measure is a reduction in the number of appropriate vectors, but this is not necessarily an easy task. The selection of genetic resistance to disease is another. Bacteriophages can also cause problems, because they are capable of infecting certain organisms used in industrial fermentations, resulting in cell lysis and loss of product. The best measure to adopt here is the use of resistant strains, or simply to reduce the concentration of divalent cations, because the latter are frequently important for phage attachment.

5.5 Further reading and references

Bishop, N. E. (1997) An update on non-clathrin-coated endocytosis. *Reviews in Medical Virology* 7, 199–209.

Carrasco, L. (1995) Modification of membrane permeability by animal viruses. *Advances in Virus Research* 45, 61–112.

Dimmock, N. J. (1982) Initial stages of infection with animal viruses. *Journal of General Virology* 59, 1–22.

Hernandez, L. D., Hoffman, L. R., Wolfsberg, T. G. & White, J. M. (1996) Virus–cell and cell–cell fusion. *Annual Review of Cell and Developmental Biology* 12, 627–661.

Lindberg, A. A. (1977) Bacterial surface carbohydrate and bacteriophage adsorption. In: *Surface Carbohydrates of Prokaryotic Cells*, pp. 289–356. Sutherland, I. W. (ed.). London: Academic Press.

Marsh, M. & Pelchen-Matthews, A. (1993) Entry of animal viruses into cells. *Reviews in Medical Virology* 3, 173–185.

Marsh, M. & Pelchen-Matthews, A. (1994) The endocytic pathway and virus entry. In: *Cellular Receptors for Animal Viruses*, pp. 215–240. Wimmer, E. (ed.). Cold Spring Harbor, NY: Cold Spring Harbor Press.

Melikyan, G. B. & Chernomordik, L. V. (1997) Membrane rearrangements in fusion mediated by viral proteins. *Trends in Microbiology* **5**, 349–355.

Nomoto, A. (ed.) (1992) Viral receptors and cell entry. *Seminars in Virology* **3**(2), 77–133.

Schneider-Schaulies, J. (2000) Cellular receptors for viruses: links to tropism and pathogenesis. *Journal of General Virology* **81**, 1413–1429.

Singer S. J. & Nicholson G. L. (1972) The fluid mosaic model of the structure of cell membranes. *Science* **175**, 720–731.

Smythe, E. & Warren, G. (1991) The mechanism of receptor-mediated endocytosis. *EMBO Journal* **202**, 689–699.

White, J. M. (1990) Viral and cellular membrane fusion reactions. *Annual Review of Physiology* **52**, 675–697.

Wilson, T. M. A. (ed.) (1992) Early events in RNA virus infection. *Seminars in Virology* **3**(6), 419–527.

Wilson, T. M. A. (1985) Nucleocapsid disassembly and early expression of positive-strand RNA viruses. *Journal of General Virology* **66**, 1201–1207.

Also check Chapter 22 for references specific to each family of viruses.

Chapter 6
The process of infection: IIA.
The replication of viral DNA

The basic mechanism whereby a new strand of DNA is synthesized in a cell is the same, regardless of whether the DNA is of cellular or viral origin. A DNA polymerase copies a template strand, beginning from a pre-existing nucleic acid 3' end that is already base paired to the template (the primer) and moving in a 5' → 3' direction. These features of DNA synthesis led to the model, proved by the classic Meselson–Stahl experiment, of semiconservative replication of double-stranded (ds) DNA, i.e. each daughter molecule contains one parental strand and one newly synthesized strand. In the context of viral DNA replication, these features also mean that all single-stranded (ss) DNA genomes are replicated via a double-stranded intermediate.

The fact that all DNA synthesis shares these fundamental features has meant that viral DNA molecules, which can be manipulated with ease, have often been studied by biochemists attempting to unravel the mysteries of DNA replication. However, although these studies have added greatly to our understanding of this process, they have also revealed that viruses employ this basic mechanism in many different ways, in each case to suit the peculiarities of their genome structures (see Chapter 4). This chapter reviews the fundamental mechanism of DNA synthesis in the context of viral replication and describes the genome replication strategies of key virus families.

6.1 The universal mechanism of DNA synthesis

DNA strand polarity and the nature of DNA polymerases

The successful replication of a dsDNA molecule requires that two daughter strands be synthesized. As these must base pair with the two template strands, which are antiparallel, the daughter strands must also be antiparallel. Thus, if an enzyme complex were simply to move along a dsDNA template synthesizing two daughter strands, one of these strands would need to be made with 5' → 3' polarity and the other with 3' → 5' polarity. However, no DNA polymerase yet characterized has the capacity to synthesize DNA in the 3' → 5' direction. All are incapable of adding deoxyribonucleotides to 5' termini.

One solution to the problem is for synthesis to proceed in the 5' → 3' direction along one parental strand, the *leading* strand, and for discontinuous 5' → 3' synthesis to occur in the opposite direction along the other, or *lagging*, strand (Fig. 6.1). In essence,

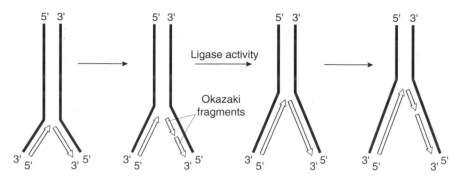

Fig. 6.1 Model for replication of double-stranded (ds) DNA through discontinuous synthesis of the lagging strand. Both strands are synthesized in a 5′ → 3′ direction but only one strand (the leading strand) is synthesized continuously, the other being formed by ligation of a series of short DNA molecules—'Okazaki fragments' (see text). Solid lines, parental DNA; open lines, new DNA.

the enzyme complex moves forwards, copying the leading strand template and revealing an increasing length of uncopied lagging strand template. Once a sufficient length of this template is available, a new daughter strand is initiated and copied from it, synthesis continuing until the polymerase comes up against the 5′ end of the previously synthesized lagging strand segment. The lagging strand fragments produced by this discontinuous synthesis are then joined together by a DNA ligase. This is the solution that is adopted in all organisms other than viruses, and by some viruses too, and is discussed later in the context of SV40 genome replication (see Section 6.2).

Evidence for this discontinuous synthesis model was obtained in a classic experiment by Okazaki. A culture of *Escherichia coli* that had been infected with phage T4 was pulsed with radioactive thymidine, and the resulting labelled DNA examined by velocity sedimentation in a sucrose gradient. Immediately after the pulse, most of the label was found in DNA fragments (since termed 'Okazaki fragments') approximately 1000–2000 nucleotides (nt.) long. One minute later, the radioactivity had been chased into material that sedimented much faster, as would be expected if the fragments had been covalently linked together by ligase action.

The alternative solution to the direction of synthesis problem, which is adopted by some viruses, is to separate in time the production of the two daughter strands. Once the requirement to have both strands made by the same enzyme complex is removed, then it is clearly straightforward for both the daughter strands to be made by continuous synthesis in the 5′ → 3′ direction. An example of a virus employing this replication mechanism is adenovirus, which is discussed in Section 6.4. The problem with this mechanism, which is probably the reason why evolution has not used it more generally, is that synthesis of one strand is necessarily delayed relative to the other and therefore, in the interim period after the first replicating enzyme has passed through the template duplex and copied the first strand, there are large amounts of ssDNA (the second template strand) exposed. As ssDNA promotes recombination by

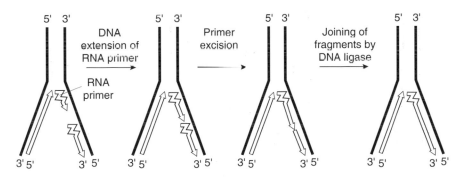

Fig. 6.2 Involvement of an RNA primer in the synthesis of Okazaki fragments. The primer is subsequently excised and replaced by DNA before the fragments are joined by DNA ligase. Solid lines, parental DNA; open lines, new DNA.

strand invasion, such a mechanism would carry a cost of genetic instability which would be unsustainable for large genomes.

DNA strand initiation and the need for a primer

For any DNA replication mechanism, it is clear that new strands must be initiated at least to begin the process and, if the discontinuous mechanism applies, repeatedly throughout the process as well. This raises a second problem because DNA polymerases are unable to start DNA chains *de novo*; all require a hydroxyl group to act as a primer from which to extend synthesis. The general solution to this problem is to invoke the action of an RNA polymerase to provide the primer, because these do not themselves require a primer to start synthesis. The enzyme synthesizes a short RNA primer, copying the DNA template, and this is then extended by DNA polymerase (Fig. 6.2). In support of this model is the observation that the antibiotic rifampicin, which inhibits the enzyme RNA polymerase in *E. coli*, inhibits the conversion of single-stranded M13 DNA to the double-stranded form. No such inhibition is observed in an *E. coli* mutant with a rifampicin-resistant RNA polymerase. A similar mechanism operates in general eukaryotic replication but, instead of employing the same RNA polymerase as is used for transcription, a dedicated enzyme known as a primase is used. Primase and DNA polymerase α enzymes form a complex that initiates replicative DNA synthesis. Once they have served their purpose, RNA primers are excised by specific enzymes which recognize and degrade RNA that is duplexed with DNA. The gaps so created are then filled by continuing 5′ → 3′ DNA synthesis from the adjacent fragment and the fragments finally ligated together (Fig. 6.2).

Why the requirement for an RNA primer?

Why has evolution generated this complex process of RNA-primed DNA synthesis, which could have been avoided if DNA polymerases were able to initiate strands *de novo?* The answer probably lies in a key evolutionary benefit of using DNA rather than RNA as genetic material, which is that the fidelity of DNA replication is very much greater (one error in 10^9–10^{10} base-pair replications) than that of RNA replication (one error in 10^3–10^4). The improved fidelity of DNA replication arises from the ability of DNA polymerases to 'proofread' the DNA that they have just synthesized, e.g. phage T4 and *E. coli* DNA polymerases have a very strong requirement for a Watson–Crick base-paired residue at the 3′-OH primer terminus to which they are about to add a nucleotide. When confronted with a template–primer complex with a terminal mismatch because of a previous error in synthesis, these polymerases make use of their built-in 3′ → 5′ exonuclease activities to clip off unpaired primer residues by hydrolysis, until a base-paired terminus is created. This self-correcting feature allows DNA polymerase to select for the proper template base pairing of each added nucleotide in a separate backward reaction, in addition to its strong selection for base pairing of nucleoside triphosphates during the initial polymerization. In contrast, RNA polymerases are not self-correcting, because such activity is incompatible with an ability to initiate synthesis of new molecules without a properly base-paired primer, an essential feature of RNA polymerases. However, as RNA transcripts are continually turned over and have no long-term role in the organism, errors in their synthesis can be tolerated. This, however, has consequences for those viruses that have RNA genomes (see Chapter 7).

RNA primers and the 'end-replication' problem

The requirement for a primer for DNA synthesis creates a difficulty in achieving complete replication of linear molecules. If synthesis in the 5′ → 3′ direction is initiated with an RNA primer that is later digested away, there is no mechanism for filling the gap left by the primer at the 5′ end of the leading strand (Fig. 6.3); to fill this gap would require 3′ → 5′ synthesis and we know that this cannot occur. When the repli-

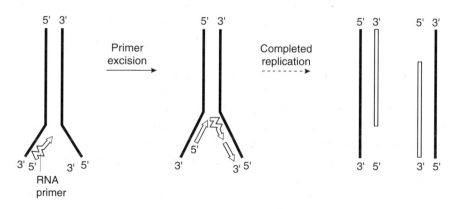

Fig. 6.3 The problem of replicating the ends of linear DNA molecules through the use of RNA primers. Once the first primer has been excised, there is no mechanism for filling the gap. Solid lines, parental DNA; open lines, new DNA.

cating fork reaches the other end of the template, a similar problem arises with the lagging strand. Without a solution, the net result would be synthesis of two daughter molecules, each with a 3′ single-stranded tail. If such molecules were to undergo further rounds of replication, smaller and smaller 3′-tailed duplexes would result. This of course is a universal problem in biology, which is solved in eukaryotes through the use of telomeres at the ends of each linear chromosome. These are specialized sequences that are replicated reiteratively by the enzyme telomerase, so as to maintain the chromosome length from one generation to the next. In prokaryotes, by contrast, there is a simpler solution, which many viruses have also adopted—the genome is circular. On such a molecule, the initiating primer can be excised and replaced by extension from the 3′ end of the new fragment eventually synthesized adjacent to it.

Viruses have found various ways to initiate DNA replication so as to solve the 'end-replication' problem in the context of their particular genome structure. As already noted in Chapter 4, many viral genomes are circular whereas others circularize during the replication cycle. These viruses can use RNA primers and excise them without facing a problem in repairing the gaps. Other viruses have evolved unique strategies to solve the 'end-replication' problem without the use of RNA primers. Examples of all these strategies are considered in the following sections.

6.2 Replication of circular dsDNA—papovaviruses

Of the viruses infecting eukaryotes, SV40 has perhaps the best understood replication mechanism. Moreover, this mechanism is widely held to be a very close parallel, in its detailed biochemistry, to that employed generally by eukaryotic cells. The overall characteristics of SV40 replication are summarized in Fig. 6.4. The circular genome has a single origin of replication, at which new synthesis initiates coordinately on both strands using RNA primers. These two newly initiated strands are then extended away from the origin in opposite directions, forming the leading strands of two diverging replication forks. As synthesis proceeds, the lagging strand templates are revealed at each fork (note that different strands of the template constitute the lagging strand template at the two forks) and these are then copied back by discontinuous synthesis (see Section 6.1). Progress of the two forks around the template gives an intermediate structure (Fig. 6.4c) with the appearance of the Greek letter θ ('theta'), which is known as a theta-form intermediate. Ultimately, the two forks converge again on the opposite side of the template circle from the origin and, when they meet, both strands will have been completely copied. The two daughter molecules are, like the parent, closed circular duplexes. Initially, they are topologically linked; this follows necessarily from the helical, closed circular nature of the template, one strand of which ends up in one daughter and one in the other. The final stage in SV40 replication is the separation of these circles.

Through work in the laboratories of Kelly, Stillman and other workers, the complete replication of circular DNA molecules, directed by the SV40 origin and using only purified, characterized proteins, has been achieved in the test tube. The 10 proteins needed are listed in Table 6.1, with their activities in the replication process. It

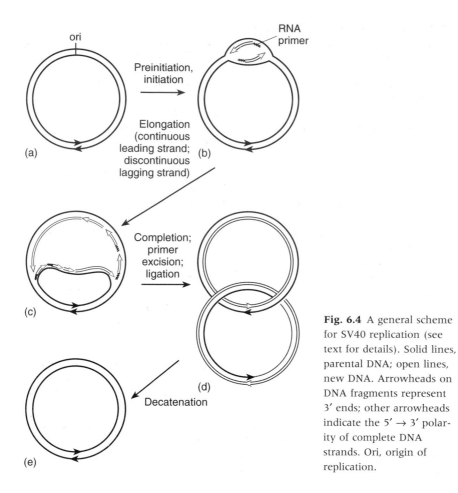

Fig. 6.4 A general scheme for SV40 replication (see text for details). Solid lines, parental DNA; open lines, new DNA. Arrowheads on DNA fragments represent 3′ ends; other arrowheads indicate the 5′ → 3′ polarity of complete DNA strands. Ori, origin of replication.

is noteworthy that only one of these proteins is encoded by the virus; all the others are provided by the cell and perform the same functions in SV40 replication as they are believed to do in host-cell replication. The one viral protein, large T antigen (see Chapter 9), provides the key initial recognition of the origin DNA sequence, begins the separation of the DNA strands with its helicase activity and then recruits cellular ssDNA-binding protein and DNA polymerase α/primase complex through protein–protein interactions; this sets the scene for the initiation reaction. After initiation, T antigen continues to play a role as a helicase, unwinding the template ahead of each replication fork.

The mechanism of papillomavirus replication closely follows that of SV40. In bovine papillomavirus and other papillomavirus infections E1 protein functions analogously to the SV40 large T antigen, except that its sequence-specific DNA-binding activity is very weak, and is revealed only in the presence of the E2 protein, which has much stronger specific DNA-binding activity (see Section 9.4). E1 and E2 bind as a complex to the origin of replication of BPV. Thereafter, E1 provides DNA helicase activity whereas all other replication functions are taken from the host.

Table 6.1 Proteins involved in SV40 replication.

Protein	Function(s)	Role in SV40 replication
Large T antigen	Sequence-specific DNA binding; DNA helicase	Initial recognition of the viral replication origin Unwinding of the DNA template duplex
RP-A	Single-stranded DNA binding	Stabilization of unwound DNA in the replication bubble at initiation, and within the replication forks
DNA polα/primase	Complex of activities synthesizing primers for initiation of new DNA strands and production of short DNA stretches	Initiation and short distance extension of the leading strands and each lagging strand fragment
DNA polδ	Extension of DNA strands; highly processive* in association with accessory factors	Processive* extension of leading strands and each lagging strand fragment
PCNA and RF-C	Accessory factors for DNA polδ	Increased processivity* of DNA polδ
Topoisomerase I	Relieving torsional stress in duplex DNA molecules	Removing the excess supercoiling which builds up in the DNA template ahead of the replication fork due to helicase action
RNaseH	Degradation of RNA within RNA : DNA hybrids	Removal of primers
DNA ligase	Joining DNA ends	Linking up lagging strand fragments after primer removal and fill-in synthesis
Topoisomerase II	Separating topologically linked DNA molecules	Separating the daughter duplexes at the end of the replication process

*Polymerase processivity describes how far the polymerase is likely to travel on a given template molecule before dropping off and having to rebind the template (or another template) molecule to recommence synthesis.

6.3 Replication of linear dsDNA that can form circles—herpes simplex virus and bacteriophage λ

Two well-characterized examples of viruses that circularize linear genomes in order to replicate are discussed here—bacteriophage λ and herpes simplex virus (HSV). In both viruses, the linear genome from the virion forms a closed circle immediately upon infection, through the interaction of cohesive ends (see Section 4.5). Subsequently, once replication is in progress, much of the viral DNA in the infected cells is found to be in the form of concatemers. These are DNA oligomers of high relative molecular mass, which, when analysed by restriction digestion, produce relatively little of the fragments deriving from the ends of the linear genome as compared with fragments coming from the middle; this result means that the concatemers comprise covalently joined genomes. These concatemers are believed to be the initial product of DNA synthesis because, in radioactive labelling experiments with λ, pulse-labelled DNA from infected cells is found initially as concatemers and can be chased into linear monomers

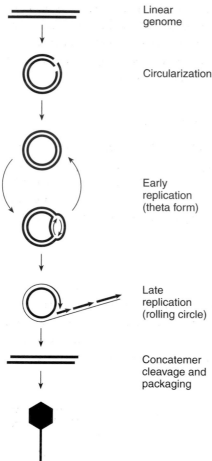

Linear
genome

Circularization

Early
replication
(theta form)

Late
replication
(rolling circle)

Concatemer
cleavage and
packaging

Fig. 6.5 A model for the replication of bacteriophage λ DNA (see text for details).

(progeny genomes). Formation of concatemeric product from a circular DNA template is a hallmark of a mechanism termed 'rolling circle' replication.

Figure 6.5 summarizes the replication cycle of λ. During the period of early replication, λ DNA replicates in a circular form, usually bidirectionally, to generate circular progeny. This amplification of replication templates resembles the scheme used by SV40. Late replication is initiated by the conversion of replication to the rolling circle mechanism. This synthesis generates a concatemeric tail, which is cleaved during packaging into molecules of the correct length and possessing the same cohesive ends as the linear genome that initiated infection. Both phases of replication employ RNA primers to initiate new strands whenever initiation is required.

The replication of HSV is believed to follow a similar path to that of λ. The rolling circle phase of its replication is well characterized, but the existence of a preceding 'circle amplification' phase can only be inferred at present. However, another herpesvirus, Epstein–Barr virus (EBV), does have a clearly defined mechanism for replication as a circle, supporting the notion that HSV also uses this strategy. During its circle replication phase, EBV uses a single viral protein to recognize its origin of replication and bring the cell replication machinery to it, in much the same way as SV40 does. During the rolling circle phase of replication as defined in HSV, there is a greater contribution from viral proteins: these provide origin recognition, DNA helicase, ssDNA-binding, primase and DNA polymerase activities. There is also a requirement for some cellular proteins, e.g. topoisomerase, in the replication process.

6.4 Replication of linear dsDNA genomes

As discussed in Section 6.1, the replication of a linear DNA molecule has to solve the 'end-replication' problem. Although conversion of the linear genome into a circle is one way to achieve this, there are other strategies, two of which are exemplified by the adenoviruses and the poxviruses. In each case, replication is achieved without any discontinuous lagging strand synthesis, because the synthesis of the two strands is separated in time.

Adenoviruses

The replication strategy of adenoviruses is summarized in Fig. 6.6. Understanding of the details of this process was made possible through the development of cell-free replication systems for the virus by Kelly and co-workers. We recall that the genome has proteins covalently linked to its 5' ends (the terminal protein) and inverted repeat

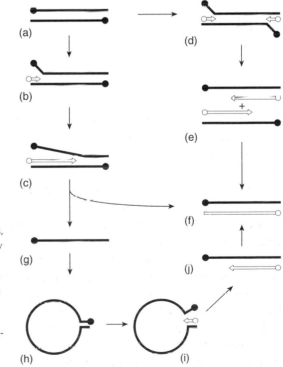

Fig. 6.6 A general scheme for adenovirus DNA replication (see text for details). Solid lines, parental DNA; open lines, new DNA. Arrowheads on new DNA strands represent 3' ends being extended. Filled circles represent the terminal proteins attached to parental DNA 5' ends and the open circles the terminal protein precursor molecules that prime new DNA synthesis.

sequences at each end which contain the origins of replication at their two termini (see Sections 4.4 & 4.5). To initiate replication, the origins are recognized by a complex of two viral proteins: a DNA polymerase (pol) and the terminal protein precursor (pTP). This binding is assisted by two cellular proteins, which bind specifically to sequences adjacent to the origin. These are actually transcription factors that the virus 'borrows' to assist its replication. However, it is not their transcription regulatory activity that is needed. Rather, they are used to alter the conformation of the DNA so as to promote binding of the pTP–pol replication complex.

The bound polymerase initiates synthesis by copying from the 3' end of the template strand, using the pTP that bound with it as a primer (Fig. 6.6b). The first deoxynucleotide is linked by a phospho-ester bond to a serine side chain within the protein. This priming mechanism is the key to adenovirus solving the 'end-replication' problem because there is no RNA primer complementary to viral sequence that later has to be excised and somehow replaced. However, the mechanism could be said to break the rule of universal proofreading of DNA synthesis, because initiation does not require a primer that is base paired to the template; in some way the pTP–DNA interaction must substitute for this requirement. Synthesis then progresses 5' → 3' across the length of the template, displacing the non-template strand that is bound and stabilized by a virus-coded DNA-binding protein (Fig. 6.6c). This process may occur at similar times from the origins at each end of the molecule (Fig. 6.6d), in which case the template duplex falls apart when the replication complexes meet (Fig. 6.6e); replication is then completed by the polymerases moving on to the ends of their respective templates (Fig. 6.6f). When only one origin is used, the non-template strand is completely displaced without being replicated (Fig. 6.6g). Its replication is achieved via formation of a 'pan-handle' intermediate (Fig. 6.6h; see Section 4.4), where the short double-stranded region exactly resembles a genome end and can therefore serve as an origin in the way already described (Fig. 6.6i,j).

Poxviruses

Poxvirus replication has been characterized through studying vaccinia virus (the smallpox vaccine virus). Vaccinia virus uses a different strategy to avoid problems in replicating its genome ends. The current model for its replication is summarized in Fig. 6.7. The process occurs in the cytoplasm of infected cells using exclusively virus-coded proteins. As mentioned in Section 4.4, the linear DNA has lengthy inverted terminal repeats, which are covalently closed at the ends (Fig. 6.7a). Replication is thought to initiate through site-specific recognition and nicking of one template strand within one of the two inverted repeats; the nicking enzyme has not been defined so far. Either or both of these origins of replication may be used simultaneously (Fig. 6.7b). DNA polymerase can then extend from the 3' end that is produced, displacing the non-template strand, until the end of the template is reached (Fig. 6.7c). However, the polymerase does not have to stop at this point, because the terminal repetition means that the molecule can base pair in an alternative way, which once again presents a template for extension (Fig. 6.7d). The newly synthesized strand folds back

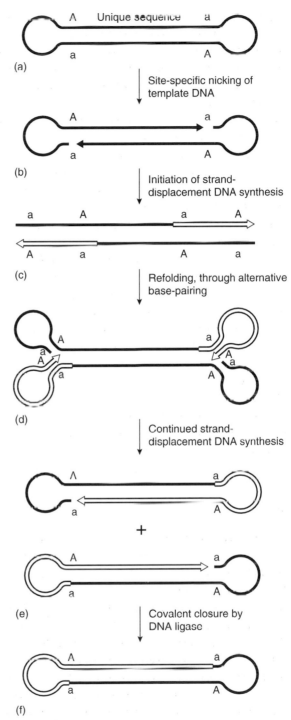

Fig. 6.7 A general scheme for replication of vaccinia virus DNA (see text for details). Solid lines, parental DNA; open lines, new DNA. Arrowheads on new DNA strands represent 3′ ends being extended. Sequence 'A' is the complement of sequence 'a'.

and base pairs to itself, so that the paired 3' end is now directed back towards the centre of the molecule and synthesis continues. As with adenovirus, exactly the same events can proceed from both ends of the duplex template. When the replication forks meet in the centre of the molecule, the two halves fall apart (Fig. 6.7e) and replication is completed when the polymerases run up against the base-paired 5' end and the ends are ligated (Fig. 6.7f). Notice how nicking the parental DNA to provide a primed template complex avoids the need to use RNA primers that would have to be excised later.

What happens if, on a template molecule, only one of the two origins is activated? The replication fork, as it proceeds back through the template, will not meet one coming the other way. Nor will it come up against a base-paired 5' end. Instead, it will continue to synthesize a complementary strand from its template to and then around the covalently closed end of the genome and back along the other side. What results is a concatemer—two unit-length molecules covalently joined. In fact, concatemers can be produced even when both origins are used together. When the polymerase reaches the base paired 5' end ahead of it, it does not have to stop. Its situation exactly resembles that at the initiation event and, just as it did then, it can continue on, displacing the non-template strand and beginning the whole cycle of events again. How then are these concatemers resolved? They can be cleaved by the same site-specific nicking enzyme that created the 3' end on which synthesis initiated. These nicked molecules can then refold into unit-length molecules which can be closed by DNA ligase.

6.5 Replication of single-stranded circular DNA

The genomes of some bacterial viruses and the recently described human TT virus consist of single-stranded circular DNA (see Chapter 4). The first step in replicating such a genome must be the creation of a complementary strand. Once this is achieved, replication mechanisms will resemble those for double-stranded genomes. The best-studied example of single-stranded circular DNA replication is bacteriophage φX174, the replication of which is outlined in Fig. 6.8. The infecting single strand is first converted to a double-stranded replicative form (RF) via RNA-primed DNA synthesis. An RNA primer is synthesized by host RNA polymerase, double-stranded template for it to transcribe coming from a short hairpin duplex formed in the single-stranded genome. The parental RF then replicates to generate progeny RF, but at present it is not clear whether this occurs via a theta-form intermediate (as in SV40) or via a rolling circle (as in HSV and phage λ replication). Once the RF has amplified sufficiently, a single-stranded concatemer is produced from which progeny genomes are excised and circularized.

6.6 Replication of linear ssDNA—parvoviruses

The parvovirus family comprises both autonomous and defective viruses. The autonomous parvoviruses, such as the minute virus of mice (MVM), package a

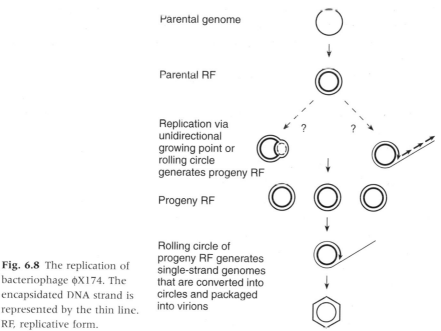

Parental genome

Parental RF

Replication via unidirectional growing point or rolling circle generates progeny RF

Progeny RF

Rolling circle of progeny RF generates single-strand genomes that are converted into circles and packaged into virions

Fig. 6.8 The replication of bacteriophage φX174. The encapsidated DNA strand is represented by the thin line. RF, replicative form.

negative-sense DNA strand whereas members of the dependovirus genus, such as the adeno-associated viruses (AAV), package both positive- and negative-sense DNA strands in separate virions, either one being infectious. These latter viruses are almost completely dependent on co-infection with helper virus for their replication (either adenovirus or various herpesviruses can perform this function).

The essential first step in parvovirus replication (as with φX174) is conversion of the genome to a double-stranded form. Once this is achieved, the replication mechanism is quite similar to that of the poxviruses. Parvovirus genomes contain terminal hairpins (inverted repeats) which provide a base-paired 3'-OH terminus at which DNA elongation can be initiated. Thus RNA primers are not needed at any stage during replication. Although these terminal hairpins have distinct sequences in the autonomous viruses, they are complementary in the defective viruses (Fig. 6.9). As discussed below, this difference explains why the two types of virus differ in the polarities of DNA strand that they package.

Cells infected by either type of parvovirus accumulate double-stranded forms of DNA, a large fraction of which cannot be irreversibly denatured, suggesting that the two strands are covalently linked. These molecules appear to involve positive- and negative-sense DNA strands linked end to end. These observations suggested a model for autonomous parvovirus DNA replication (Fig. 6.10). The base-paired 3' end of the genome (Fig. 6.10a) primes extension synthesis, copying the body of the genome (Fig. 6.10b). Displacement of the base-paired 5' end allows synthesis to continue to the template strand 5' end (Fig. 6.10c,d). The structure now undergoes rearrangement to form a 'rabbit-eared' structure (Fig. 6.10e). This recreates the hairpin originally

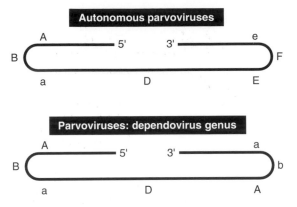

Fig. 6.9 Schematic representation of the genome structures of autonomous and defective parvoviruses. The letters A and a, etc., represent complementary sequences.

present at the 5′ end of the parental genome and also forms a copy of this hairpin at the 3′ end of the complementary strand, which can serve as a primer for continuing synthesis (Fig. 6.10f,g). Further rearrangement and strand displacement yields a dimer-length duplex (Fig. 6.10h). The resulting molecule comprises a single polynucleotide chain from which two viral genomes could be generated by specific endonuclease action. However, in the absence of endonuclease action, continued replication could occur to generate larger multimers.

The dimers and multimers created by the process just described could serve as replicative intermediates from which progeny viral DNA would be excised by displacement synthesis. A nick would be introduced at the 5′ end of a genome within the concatemer by a sequence-specific nicking enzyme (Fig. 6.10i). This activity is now known to be provided by viral proteins; all other activities directly involved in replication appear to come from the host. The 3′-OH terminus then acts as a DNA primer for displacement synthesis of progeny DNA strands, possibly driven by the packaging process (Fig. 6.10i,j). After a complete genome has been displaced, excision of the progeny genome could be completed by another site-specific endonuclease, resulting in the release of an intact virus particle and termination of the displacement synthesis. This model neatly accounts for two experimental observations. First, duplex molecules with single-stranded tails have been observed in cells infected with MVM. Second, no free ssDNA has been detected in such infected cells.

The defective parvovirus AAV differs from MVM in having terminal sequences that not only fold back on themselves to form hairpins, but are also related to each other. The entire hairpin sequence at one end is the complement of the sequence at the other end. The result is that the 5′ ends of both negative- and positive-sense strands are identical and so, within the model just described for linked strand displacement and packaging of single strands, both strands will be produced and packaged equally. It is still unclear why this subset of the parvoviruses shows dependence on helper functions for growth, because these required functions do not necessarily include any that are directly involved in DNA synthesis. The helper functions can be provided by various unrelated viruses and the dependence on them can also be overcome in cell culture by treating cells with genotoxic agents. It is also significant that AAV is now

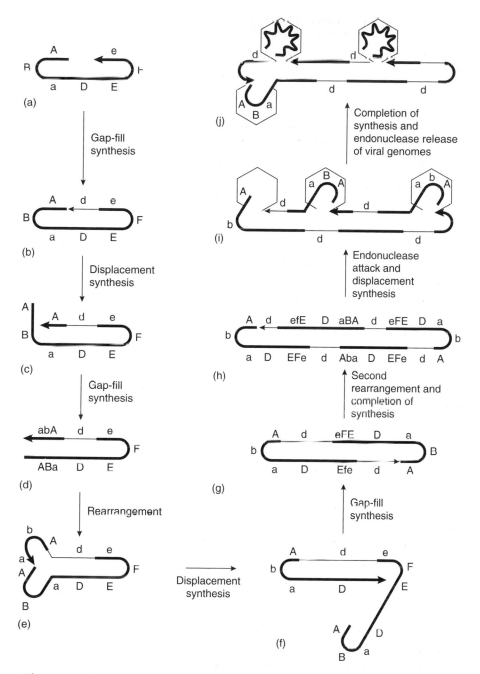

Fig. 6.10 A scheme for the replication of an autonomous parvovirus (see text for details). The letters A and a, etc., represent complementary sequences. Thick lines, parental and genomic DNA; thin lines, replicated genome-copy DNA. Arrowheads represent 3′ ends.

known to be able to establish a latent infection, integrating into a specific site in the host chromosomes; this might be the preferred strategy of the virus for ensuring perpetuation of its genetic material. Thus, 'helper' functions may alternatively be viewed as viral functions that alter the cell environment so as to block establishment or maintenance of latency and hence favour AAV productive replication.

6.7 Dependence versus autonomy among DNA viruses

The autonomy of viruses from their hosts as regards the replication of their DNA varies between wide limits and is a function of the size of the viral genome. At one end of the scale are viruses, such as the poxviruses, with genomes of around 200 kilobase-pairs (kbp) corresponding to 100–200 genes. Such viruses require little more from their host cells than an enclosed environment, protein-synthesizing machinery, a supply of amino acids and deoxyribonucleotide triphosphates, and an energy source. Some may not even require this much; herpes simplex and vaccinia viruses both specify a new thymidine kinase and several other enzymes. At the other end of the scale are viruses, such as MVM and SV40, the genomes of which can specify only a few proteins. As some of these are needed to form the virus coat, not many genes are left to code for functions essential to replication. These viruses rely on the host not only for nucleic acid precursors but also for polymerases, ligases, nucleases, etc.

The extreme in terms of lack of autonomy is represented by viruses that cannot replicate in a host without assistance from another virus. AAV may represent an example, although the basis for its helper dependence is still unclear (see Section 6.6). Another example is phage P4, which is a satellite of phage P2. Unlike most satellite viruses, this phage can replicate its nucleic acid in the absence of helpers. Instead, it lacks all known genes for its morphogenesis and its DNA is packaged in a head composed of helper phage proteins.

6.8 Conclusion

Within the known viruses of Baltimore classes 1 and 2, there is a diverse range of genome structures, which necessitates a diversity of replication strategies. This diversity is further increased through the use of alternative means to overcome the general problem of replicating completely the ends of a piece of DNA, a problem that follows from the fundamental nature of all DNA polymerases. All of the replication mechanisms employ at least one virus-encoded protein, but often the majority of proteins needed come from the host. How much of the replication machinery a virus can provide for itself and how much it has to take from its host essentially depend on its genome size.

6.9 Further reading and references

Boehmer, P. E. & Lehman, I. R. (1997) Herpes simplex virus DNA replication. *Annual Review of Biochemistry* **66**, 347–384.

Challberg, M. (ed.) (1991) Viral DNA replication. *Seminars in Virology* **2**(4).

Kornberg, A. & Baker, T. (1991). *DNA Replication* (2nd edn). San Francisco: W. H. Freeman.

Ogawa, T. & Okazaki, T. (1980) Discontinuous DNA replication. *Annual Review of Biochemistry* **49**, 421–457.

Waga, S. & Stillman, B. (1994) Anatomy of a DNA replication fork revealed by reconstitution of SV40 DNA replication in vitro. *Nature* **369**, 207–212.

Waga, S. & Stillman, B. (1998) The DNA replication fork in eukaryotic cells. *Annual Review of Biochemistry* **67**, 721–751.

Also check Chapter 22 for references specific to each family of viruses.

Chapter 7

The process of infection: IIB.
Genome RNA synthesis by RNA viruses

The synthesis of RNA by RNA viruses involves: (1) replication, which is defined as the production of progeny virus genomes; and (2) transcription to produce messenger RNA (mRNA). The process of transcription for RNA viruses is described in Chapter 10 where it is discussed in terms of gene expression. This chapter focuses on the replication of RNA virus genomes. During the process of replication, as for all other processes that involve synthesis of nucleic acid, the template strand is 'read' by the polymerase travelling in a 3′ → 5′ direction, with the newly synthesized material being produced beginning at the 5′ nucleotide and progressing to the 3′ end. Most RNA viruses can replicate in the presence of DNA synthesis inhibitors, indicating that no DNA intermediate is involved. However, this is not true for the retroviruses (Baltimore class 6) and these will be considered separately (see Chapter 8).

The polymerases that carry out RNA replication are encoded by the virus and are either transported into the cell at the time of infection or synthesized very soon after the infection has begun. Frequently, the polymerases involved in RNA replication are referred to as 'replicases' to differentiate them from the polymerases involved in transcription. However, both processes are carried out by the same enzyme, exhibiting different synthetic activities at different times in the infectious cycle.

7.1 Regulatory elements for RNA virus genome synthesis

Certain features are common in the process of replication for all RNA viruses. To make a faithful copy of the genome, the RNA-dependent RNA polymerase must begin synthesis at the 3′-terminal nucleotide of the template strand. The 3′ terminus must therefore contain a signal to direct initiation of synthesis. As indicated by the principles that underpin the Baltimore scheme (see Chapter 4), all RNA viruses must replicate via a double-stranded (ds) RNA intermediate molecule, e.g. a virus with a single-stranded (ss) RNA genome must produce a dsRNA intermediate by synthesizing a full-length 'antigenome' strand. The antigenome strand will then, in turn, be used as a template to synthesize more genomes for packaging into progeny virions. As before, synthesis using an antigenome as template must begin at the 3′-terminal nucleotide if a faithful, full-length copy is to be made, and the 3′ end of the antigenome must also contain a signal to direct the polymerase to start synthesis. For RNA viruses that contain inverted repeat sequences at the ends of the genome (see Section 4.4), both the genome and antigenome strands will contain the same sequence at the 3′ terminus. For viruses with genomes that do not contain inverted repeat sequences, the initiation of synthesis of antigenome- and genome-sense molecules must each be controlled by different processes. For most viruses, the mechanism of initiation of RNA synthesis during replication is only poorly understood, if at all, although a combination of old and new analyses has identified the termini of RNA viruses as containing the regulatory elements that direct RNA synthesis.

The generation and amplification of defective–interfering virus RNA

All RNA and DNA viruses produce defective–interfering (DI) particles as the result of errors in their nucleic acid synthesis. Here, we consider only DI RNA viruses, about which more is known. DI viruses are mutants in which the genomes have large deletions, leaving RNA that may comprise as little as 10% of the infectious genome from which they were derived. DI viruses are unable to reproduce themselves without the assistance of the infectious parental virus (i.e. they are defective). For this reason, propagation of DI virus is optimal at a high multiplicity of infection when all cells contain an infectious virus genome. DI genomes depress (or interfere with) the yield of infectious progeny, by competing for a limited amount of some product synthesized only by the infectious parent. Interference takes place only when the ratio of DI to infectious genomes reaches a critical level. Up to this point, both infectious and DI genomes are replicated to the fullest extent. Many DI viruses synthesize no proteins and some have no open reading frame. As they depend on parental virus to provide those missing proteins, DI and parental viruses are composed of identical constituents, apart from their RNAs. Thus it is usually difficult to separate one from the other. A notable exception is the DI particle of the rhabdovirus, vesicular stomatitis virus (VSV), whose particle length is proportional to that of the genome. When centrifuged, these short particles remain at the top of sucrose velocity gradients and are thus called T particles to distinguish them from infectious B particles, which sediment to the bottom. Some biological implications of DI particles are discussed in Chapter 13.

A clue leading to one hypothesis explaining how DI RNAs are generated came from electron microscopic examination of genomic and DI RNAs from ssRNA viruses. Both were found to be circularized by hydrogen bonding between short complementary sequences at the termini, forming structures called 'pan-handles' or 'stems' (Fig. 7.1a). The deletion that results in DI RNA may arise when a polymerase molecule detaches from the template RNA strand, and reattaches either at a different point or to the newly synthesized, incomplete strand (Fig. 7.1). Thus the polymerase begins faithfully but fails to copy the entire genome. Most VSV viruses are of the latter type and lack the 3′ end of the standard virus genome. There are no DI viruses known that lack the 5′ terminus. In other DI viruses, parts of the genome can be duplicated, often several times over, during subsequent replication events, making complex structures that bear little resemblance to the standard genome from which they were derived. The three classes of DI genome

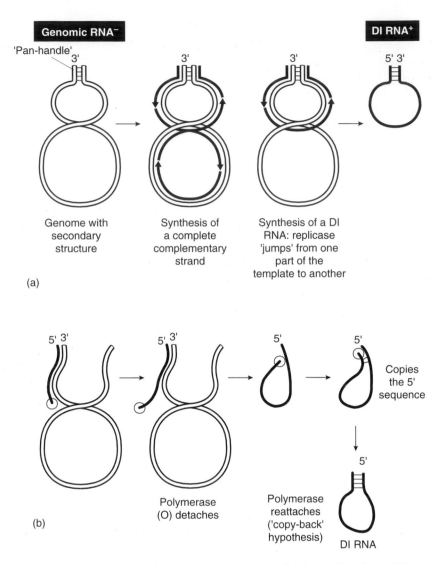

Fig. 7.1 Hypothetical schemes to explain the generation of defective–interfering (DI) RNAs having sequences identical with both the 5′ and 3′ regions of the genome (a) and with the 5′ region only (b).

are summarized in Fig. 7.2. A key point for the DI RNA viruses is that their genomes always contain the same termini as those of the parent virus, or consist of inverted repeats of sense and antisense copies of the 5′ end of the normal genome, whereas the remainder of the genome can be substantially deleted without impairing the ability of the DI genome to replicate. This was taken to indicate that the genome termini are essential and contained the regulatory elements to direct genome synthesis.

Usually, interference occurs only between the DI virus and its parent. This is because DI virus lacks replicative enzymes and requires those synthesized by infectious virus.

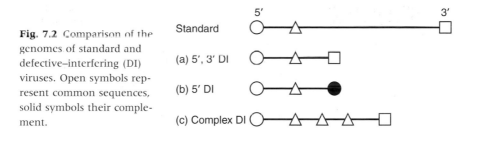

Fig. 7.2 Comparison of the genomes of standard and defective–interfering (DI) viruses. Open symbols represent common sequences, solid symbols their complement.

Specificity resides in the enzymes, which only replicate molecules carrying certain unique nucleotide recognition sequences. Intuitively, it can be seen that in a given amount of time an enzyme will be able to make more copies of the smaller DI RNA. Thus, as time progresses, the concentration of DI RNAs increases relative to the parental RNA in an amplification step. However, this is not the whole story, as some large DI RNAs interfere more efficiently than smaller ones. Such DI RNAs seem to have evolved a polymerase recognition sequence that has a higher affinity for the enzyme than that of the infectious parents and hence confers a replicative advantage. This may indicate that, although the essential minimal sequences for directing genome synthesis are located at the termini, sequences located elsewhere may also play an enhancing role. The only other sequence that all DI genomes retain is a packaging or encapsidation sequence because without this they cannot be recognized by virion proteins and form viral particles.

Reverse genetics of RNA viruses

Over recent years one of the most exciting developments in the field of RNA viruses has been the generation of reverse genetic systems for a wide range of viruses, particularly those with negative-sense ssRNA genomes (see Chapter 21). The absence of a DNA intermediate in the replication cycle of RNA viruses has, until recently, limited research potential because there are no tools to modify RNA molecules in the same way that there are for DNA. However, for many RNA virus systems, it is now possible to generate DNA copies of virus genomes and convert these back into RNA, and ultimately into infectious virus particles. The DNA can be manipulated in a variety of ways, including the generation of deletions or specific mutations that are then mirrored in the synthetic replicas of the virus genomes. Analysis of these mutated genomes has shown, for representatives of all RNA virus families, that the immediate termini of the genome, or each genome segment, contains the elements that are essential to direct RNA synthesis for replication. This is true whether the virus genome contains inverted repeat sequences or different, unique, sequences at the termini of the genome. Mutation of the sequences at the termini have serious deleterious consequences for the ability of the RNA molecules to be replicated by virus proteins.

7.2 Synthesis of the RNA genome of class 3 viruses

Reoviruses and rotaviruses of animals and wound tumour virus of plants and other families, listed in Chapter 22, contain multiple segments of dsRNA. In reoviruses, the genome consists of 10 segments (see Section 4.7). By analogy with DNA replication, this dsRNA could replicate by a semiconservative mechanism such that the complementary strands of the parental RNA duplex are displaced into separate progeny genomes, or the parental genome could be conserved or degraded. In fact, dsRNA genomes are replicated conservatively.

After initiation of infection, several proteins in the reovirus particles are removed by protease digestion during the uncoating process, to form a subviral particle that is found in the cytoplasm where replication takes place. The dsRNA genome is retained within the subviral particle and does not leave it during the infectious cycle. The observation that the genome is not completely uncoated in the infected cell indicated that replication of the dsRNA could not occur in the normal semiconservative way as seen for DNA. The only virus nucleic acid found in the cytoplasm outside the subviral particles is mRNA, which is generated by transcription using the particle-associated RNA-dependent RNA polymerase as described in Chapter 10. As both strands of the genome RNA are retained in the subviral particle, it was clear that the single-stranded mRNA transcripts must be the sole carriers of genetic information from parent to progeny. This means that only one strand of each of the 10 genome segments is used as template and the newly synthesized RNA is then replicated to form a new dsRNA genome segment. The proof of this conservative method of reovirus genome replication was elegantly shown by Schonberg in 1971.

Infected cells were labelled with [^{3}H]uridine for 30 min at various times during the replication phase of the reovirus infectious cycle. Total dsRNA was then isolated and hybridized to excess unlabelled positive-sense RNA which had been prepared *in vitro*. It was proposed that three results could be obtained, depending on the mode of replication (Fig. 7.3).

1 If replication was semiconservative, then the dsRNA should be equally labelled in both strands. Hybridization to an excess of unlabelled positive-sense strands would occur with only 50% of the label.

2 If the negative-sense strand was used as template, then labelled positive-sense RNA would be produced and hybridization would generate dsRNA that did not contain any radioactivity.

3 If the positive-sense strand was used as template, hybridization of the labelled negative strand would result in 100% of the label forming a hybrid.

The result showed that all of the label was associated with the negative strand and thus replication utilizes the positive-sense mRNA as template.

In cells exposed continuously to [^{3}H]uridine, the label was found to be divided between both strands. Varying the time of the short pulse of radioactivity showed that there is a lag between the synthesis of the positive template strand and the onset of replication of the negative-sense strand. Presumably this allows the mRNA to be translated to produce the polymerase before being used as a template in replication.

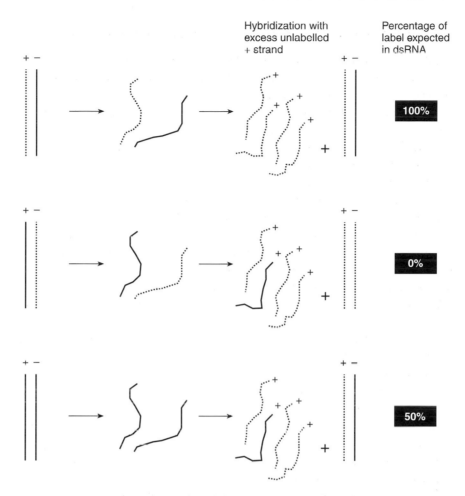

Fig. 7.3 Diagrammatic representation of the experiment by Schonberg to illustrate the conservative mechanism of reovirus genome replication. Radioactively labelled RNA strands are represented by dashed lines and unlabelled strands by solid lines. (See text for details.)

Newly synthesized negative-sense RNA is found only as part of a dsRNA molecule. The dsRNA segments are never found free in infected cells and are always associated with an immature virus particle. Each particle must contain a single copy of each of the 10 reovirus genome segments. The mechanism by which a virion specifically packages one of each of the 10 RNA segments is not yet understood.

7.3 Synthesis of the RNA genome of class 4 viruses

Although there are many differences in the details of the replication cycles of class 4 viruses in terms of gene expression (see Sections 10.4–10.6) and assembly, the process by which their positive-sense ssRNA genomes are replicated is very similar. A great deal of information is available about the mechanism of picornavirus genome replication and this is described here in detail. In principle, the process used by picornaviruses is applicable to all class 4 viruses, with the generation of the same type of intermediate molecules *in vivo* and *in vitro*. A similar process is also applicable to the replication of the coronavirus subgenomic mRNAs described in Section 10.6.

As the picornavirus genome RNA is the only nucleic acid that enters the infected cell, it must act as a template for both translation and replication, just as for the reovirus mRNAs. Translation is necessary as a first step in the infectious cycle to produce the polymerase. Aspects of translation of picornavirus genome RNA are discussed in Section 10.4. Replication takes place on smooth cytoplasmic membranes in a replication complex and, as indicated earlier, must involve the generation of a dsRNA intermediate. Initiation of replication is at the 3′ end of the positive-sense genome/mRNA polyA tail, the presence of which is essential for replication to occur. The initiation process is not yet understood, nor how the VPg protein becomes attached to the 5′ end of the newly synthesized RNA. One possibility is that the attachment of VPg and initiation of RNA synthesis are coupled. If so, this presumably happens on initiation of both positive and negative strand synthesis, because both strands have VPg at their 5′ ends. Once RNA synthesis begins, the polymerase proceeds along the entire length of the template RNA. The replication process requires concurrent protein synthesis because addition of protein synthesis inhibitors, even when replication has started, inhibits any further rounds of replication.

Analysis of virus-specific RNA isolated from infected cells has shown that the replication complex contains an RNA molecule that is partially ssRNA and partially dsRNA. This is called the replication intermediate (RI). Negative-sense RNA is only ever found in association with positive-sense RNA in the RI. Deproteination and ribonuclease (RNase) treatment of the RI generates an entirely dsRNA the same length as the virus genome. This is called the replicative form (RF). The RF is produced by removing the ssRNA tails found in the RI (Fig. 7.4). A dsRNA–RF complex is also found in poliovirus-infected cells treated with inhibitors of host-cell RNA polymerase, but it is not known whether this is involved in the replication of the virus RNA or is an artefact of the drug treatment.

The suggested mode of replication is that, initially, the positive-sense genome ssRNA is used as a template to generate negative-sense, antigenome RNA with VPg at the 5′ end. In the RI, the nature of the association of the positive and negative strands is not known, but it is likely that they are only loosely linked and the two strands may associate only at the region where synthesis is occurring. Once the polymerase complex has moved along the template, the 3′ end will become available for a further round of replication, even though the preceding complex has not yet completed copying the template. In fact, multiple initiations may occur on the template before the first replication complex has completed its work, and it is estimated that up to five functional replication complexes may be associated with the picornavirus template at any one time (Fig. 7.4). The newly synthesized RNA is in turn used as template for production of new copies of the genome RNA, with VPg attached to the 5′ end, in a similar way. As much more positive-sense genome ssRNA than negative-sense ssRNA is

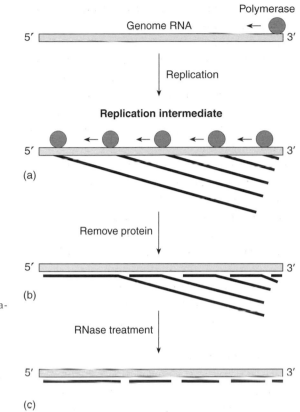

Fig. 7.4 (a) Proposed structure for a molecule of replicative intermediate before deproteinization, (b) after deproteinization and (c) the effect of treating deproteinized replication intermediate (RI) with ribonuclease (RNase).

produced, the synthesis process must be biased, or asymmetrical. Unlike the negative-sense RNA, the completed positive-sense RNA is released from the replication complex for packaging into virions, used as template for replication or for translation (with concomitant loss of VPg).

7.4 Synthesis of the RNA genome of class 5 viruses

Class 5 viruses can be divided into two main categories: those with genomes consisting of a single molecule and those with segmented genomes. The former are grouped together in a taxonomic order, the Mononegavirales, indicating that they have a single, negative-sense ssRNA molecule as genome. The Mononegavirales include the paramyxo-, rhabdo-, filo- and bornavirus families. The viruses with segmented genomes include the orthomyxo-, arena- and bunyaviruses. The details of many aspects of RNA replication of class 5 viruses are not yet known, but the general principles appear to be common to all irrespective of the number of molecules that make up the genome. As the genome RNA is complementary in base sequence to the mRNA, class 5 virus RNA synthesis can occur only using a pre-existing RNA-dependent RNA polymerase present in the virus particle. Polymerase activity can be detected for some class 5 viruses after partial disruption of the virus with detergent in the presence of the four ribonucleoside triphosphates and appropriate ions. Viruses 'activated' in this way in the absence of whole cells synthesize RNA *in vitro* at a linear rate for at least 2 h. In general, the *in vitro* systems yield only mRNA, not positive-sense antigenome RNA. However, the virus-associated polymerase is responsible for both replication and transcription in infected cells.

Replication of the RNA genome of non-segmented class 5 viruses: rhabdoviruses

Most information is available about the process of replication of rhabdoviruses, especially vesicular stomatitis virus (VSV), and they have been used to generate a model of replication applicable to many other class 5 viruses, with either a single or multiple genome segments including all Mononegavirales, and the bunya- and arenaviruses.

The genome of VSV consists of the ssRNA closely associated with three virus proteins in a helical complex. The most abundant is the nucleoprotein (NP) with smaller amounts of a phosphoprotein (P), and only a few molecules of a large (L) protein. The L protein is the catalytic component of the replication complex responsible for carrying out the RNA synthesis, but the NP and P proteins are essential for its activity. The complex of NP, P and L proteins together with genomic RNA, referred to as the nucleocapsid, also carries out transcription to produce mRNA (see Section 10.10). It is not known what causes the complex to transcribe mRNA at some times and to replicate the genome at others. The relative amounts of the three proteins appear to be critical for the function of the nucleocapsid.

The replication of class 5 viruses requires continuous protein synthesis and addition of inhibitors of protein synthesis results in an immediate cessation of replication. Class 4 viruses are the same in this respect. Consequently, virus mRNA and protein synthesis occur before the onset of replication. As for all nucleic acid synthesis, replication begins at the 3′ end of the negative-sense RNA template and, continuing to the 5′ end, generates a positive-sense, antigenome RNA. During replication the polymerase ignores the signals for termination of mRNA synthesis, which are recognized during transcription.

Replication of rhabdoviruses is similar to that of picornaviruses, with production of an RI molecule. The RI is closely associated with the three replication proteins and antigenome RNA is found only in the RI. Purification of the RI and treatment with RNase generates a double-stranded RF RNA analogous to that generated from the picornavirus RI (Fig. 7.4).

The antigenome RNA is used as template to produce negative-sense ssRNA for progeny virus genomes. The negative-sense RNA is found as a nucleocapsid structure which can be used in further rounds of replication before incorporation into virus particles. As considerably more negative-sense RNA is produced than positive-sense RNA, the replication process must be asymmetrical to favour production of one strand over the other.

Replication of the RNA genome of segmented class 5 viruses: orthomyxoviruses

As for the Mononegavirales, the genome RNA of class 5 viruses which are made up of multiple segments is also present as helical nucleocapsid structures. The major protein of the influenza virus nucleocapsid is the NP. The NP protein interacts directly with the genome RNA, binding to the sugar phosphate backbone and leaving the nucleotide bases exposed on the surface of the structure. The location of the other nucleocapsid structure proteins PA, PB1 and PB2 (which are also involved in transcription—see Chapter 10) is less clear, but they are thought to be associated with the nucleotide bases on the outside of the helix. Each segment replicates independently.

The positive-sense RNA which is generated from the genome template is probably initiated *de novo*. This contrasts with the use of a primer derived from the host-cell mRNA during transcription of mRNA (see Section 10.8). The reason for the different activities of the nucleocapsid complex during replication and transcription are not known. The replication complex is dependent on the continued synthesis of at least one viral protein. This is similar to the situation for the other RNA viruses.

During replication, the synthesis of RNA does not stop in response to the polyadenylation signal in the genome segments, but continues to the end of the template molecule. The production of the positive-sense RNA is thought to occur by way of an anti-termination event, i.e. in the absence of a cap structure on the RNA being synthesized, NP protein acts in an unknown way to prevent termination at the polyadenylation signal, before the end of the template is reached. This may be achieved by direct interaction of the NP with the RNA and the PB proteins in the nucleocapsid structure. This complex may be different depending on whether or not a cap has been used to initiate RNA synthesis. However, the details of this process are not yet fully understood.

The newly synthesized positive-sense RNA, which is present in nucleocapsid complexes, is used as templates for the production of negative-sense ssRNA, which can be used for further rounds of replication before formation of new progeny virions.

The mechanism used to control the acquisition of influenza virus genome segments by influenza viruses is unknown. It is possible that each particle can package more than one copy of each segment into new virus particles and recombinant viruses generated *in vitro* can be forced to accept additional segments. Such potential flexibility would mean that there is less need for specificity in packaging to ensure that a complete complement of genome segments is present, but the limit of the number of segments that can be packaged, and the relevance to the situation *in vivo*, are not clear.

7.5 Synthesis of the RNA genome of viroids and hepatitis delta virus

The covalently closed ssRNA genomes of plant viroids and hepatitis delta virus (HDV) do not encode any proteins. When the RNA enters the cell, either as naked RNA in the case of the viroids, or through the action of hepatitis B virus structural proteins in the case of HDV, it must be replicated by the proteins already present. For the viroids this means that the host plant cell enzymes must be used, whereas for HDV, which can replicate only if the cell is already infected with hepatitis B virus, host enzymes, possibly with the assistance of hepatitis B virus proteins, must be used. In both cases the enzyme most likely to be responsible for replicating the genome RNA is the host-cell DNA-dependent RNA polymerase. It is not known how this enzyme functions on an RNA template, but this may be, at least in part, the result of the extensive base-pairing structure of the genome RNA (see Section 4.7).

The replication of viroid and HDV genome RNA begins by adopting a rolling circle mechanism as described for circular DNA (see Section 6.3). The RNA polymerase II begins replication at a precise point on the genome, but the nature of the initiation event is not known. The replication process generates linear concatemeric RNA, of opposite sense to the genome, from which genome-length RNA molecules must be excised. The excision relies on the action of an unusual RNA sequence within the newly synthesized molecule, called a ribozyme. These ribozyme sequences adopt a complex three-dimensional structure and autocatalytically cleave the RNA at a specific site, generating genome-length, linear, ssRNA molecules. These molecules are then converted into circles. For HDV, the ribozyme also appears to have RNA ligase activity and can form the antigenome circles. It is not known how circular plant viroid antigenome RNAs are formed. In the covalently closed circular conformation, the RNA adopts an extensively base-paired rod-like structure, analogous to that of the infecting genome, and the ribozymes cannot adopt their active conformation. Consequently, the circular RNA is not cleaved. The circular antigenome RNA is used as a template to produce more genomes by the rolling circle model. The genome-sense RNA also contains a ribozyme sequence that cleaves the concatemer to produce linear genome-length molecules that are circularized as before.

7.6 Further reading and references

Braam, J., Ulmanen, L. & Krug, R. M. (1983) Molecular model of a eukaryotic transcription complex: functions and movements of influenza P proteins during capped RNA-primed transcription. *Cell* **34**, 609–618.

Conzelmann, K.-K. (1998) Nonsegmented negative-stranded RNA viruses: genetics and manipulation of viral genomes. *Annual Review of Genetics* **32**, 123–162.

Curran, J. & Kolakofsky, D. (1999) Replication of paramyxoviruses. *Advances in Virus Research* **54**, 403–422.

Dimmock, N. J. (1991) The biological significance of defective interfering viruses. *Reviews in Medical Virology* **1**, 165–176.

Karayannis, P. (1998) Hepatitis D virus. *Reviews in Medical Virology* **8**, 13–24.

Lai, M. M. C. (1995) The molecular biology of hepatitis delta virus. *Annual Review of Biochemistry* **64**, 259–286.

Lai, M. M. C. & Cavanagh, D. (1997) The molecular biology of coronaviruses. *Advances in Virus Research* **48**, 1–100.

Marriott, A. C. & Easton, A. J. (2000) Paramyxoviruses. In: *Reverse Genetics of RNA Viruses*. *Advances in Virus Research* **53**, 312–340.

Portela, A., Zurcher, T., Nieto, A. & Ortin, J. (1999) Replication of orthomyxoviruses. *Advances in Virus Research* **54**, 319–348.

Roux, L., Simon, A. E. & Holland, J. J. (1991) Effects of defective interfering viruses on viral replication and pathogenesis *in vitro* and *in vivo*. *Advances in Virus Research* **40**, 181–211.

Taylor, J. M. (1992) The structure and replication of hepatitis delta virus. *Annual Review of Microbiology* **42**, 253–276.

Also check Chapter 22 for references specific to each family of viruses.

Chapter 8

The process of infection: IIC.
The replication of RNA viruses with a
DNA intermediate and vice versa

The idea of DNA synthesis by viruses with RNA genomes, reverse transcription, once regarded as heresy to the doctrine of information flow from DNA to RNA to protein, now has an established place in molecular biology. Indeed, it is now known that many mammalian DNA sequences (mammalian pseudogenes, some highly repetitive sequences and certain types of transposable elements) have been created by reverse transcription. The first viruses in which such activity was described belonged to the retrovirus family, known for many years as the RNA tumour viruses (see Section 17.6). The family now includes the major pathogen human immunodeficiency virus (HIV) type 1 (see Chapter 18). Reverse transcription has also been found to be an essential part of the replication of the hepadnaviruses of animals and the plant caulimoviruses.

8.1 Retroviruses

The unexpected involvement of DNA in retroviral replication

Early studies showed that DNA played a critical role in the multiplication of RNA tumour viruses. Infection by these viruses could be prevented by inhibitors of DNA synthesis added during the first 8–12 h after exposure of the cells to the virus. Also, the formation of virions was sensitive to actinomycin D, suggesting a requirement for DNA-dependent RNA synthesis. Finally, Rous sarcoma virus infection of cells (being a non-lytic virus, the cells do not die) conferred on them stably inheritable changes to their appearance and growth properties (known as transformation—see Section 17.1), and the details of these features were virus-strain specific. Viral sequences in a heritable (i.e. DNA) form offered the only explanation for this observation. These data led Howard Temin to propose his 'provirus' theory, which postulated the transfer of the information of the infecting RNA to a DNA copy (the provirus), which then served as a template for the synthesis of progeny viral RNA. It is now clear that this theory is correct, and that the proviral DNA becomes integrated into the genome of the infected cell (an integrated provirus).

Temin's theory required the presence in infected cells of an RNA-dependent DNA polymerase or 'reverse transcriptase'. At that time, no enzyme had been found in any type of cell that could synthesize DNA from an RNA template. It was clear that, if such an enzyme existed, the RNA tumour virus must induce its synthesis soon after infection or else carry the enzyme into the cell as part of the virion. A search was begun for a reverse transcriptase in retrovirus particles, and David Baltimore and Temin each independently reported the presence of such an enzyme in 1970, work for which they were subsequently awarded the Nobel prize. The enzyme that Baltimore and Temin discovered has since become a cornerstone of all molecular biology investigations because it provides the means to produce copy or cDNA from mRNA in the laboratory.

Properties of a retrovirus genome

Early studies of Rous sarcoma virus (RSV) showed that its genomic RNA sedimented at 70S. However, when denatured, the RNA sedimented at 35S, indicating that there were two molecules present of equal size. Electron microscopy showed two RNA molecules that were linked together as in Fig. 8.1. Initially by an oligonucleotide mapping technique, and subsequently by complete sequencing, these molecules were shown to be identical. Thus, each virion contains two copies of the genome and hence is diploid; the reasons for this are considered below. Structurally, the viral genome resembles a typical eukaryotic messenger RNA, being positive sense, having a polyadenylate (polyA) tract at the 3' terminus and a cap structure at the 5' terminus.

The gene order in RSV is shown in Fig. 8.1. The *gag* gene encodes three group-specific antigens, now known to be the major internal virion proteins. The *pol* gene encodes three enzymes present in the virion—a reverse transcriptase, an integrase and a protease—whereas the *env* gene encodes the virion envelope proteins. These three genes are present in this order in all retroviruses. The fourth gene, *src*, was originally acquired from the host and is not needed for virus multiplication; it is the key gene in transformation and carcinogenesis caused by RSV (see Section 17.6). Other retroviruses since characterized generally lack an *src* gene or anything equivalent, but may carry several other genuine viral genes which contribute in various ways to the life cycle. None of these additional genes encodes a protein that plays any direct role in viral genome replication, so this process may be understood by referring to the simpler viruses of the family.

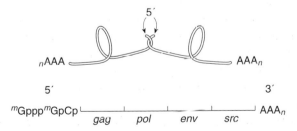

Fig. 8.1 Top: a retrovirus RNA genome, showing identical molecules held together at their 5' ends. Each has an intramolecular loop. Bottom: genetic map of Rous sarcoma virus (not to scale); the *gag, pol* and *env* genes are common to all retroviruses.

Comparing the structures of the genome RNA and proviral DNA

Aligning the sequence of a proviral DNA molecule with the genomic RNA from which it derives gives a first insight into what has to be achieved during reverse transcription (Fig. 8.2). The sequences immediately adjacent to the cap and polyA tail on the RNA are the same, i.e. there are direct repeats, R, at each end of the genome. However, in the provirus, these R sequences are now internal to the molecule, i.e. additional sequence has been added outside each R sequence in the double-strand DNA provirus, compared with the genomic RNA. At the outer ends of these added sequences are short *inverted* repeats. A scan of the genomic RNA sequence reveals that these are also present here, but at internal locations in the RNA, not at the ends. In the provirus, each of these has been duplicated to create the proviral termini.

Where does the remainder of the sequence added to the ends of the provirus come from? Again, it is found in the genome internal to the R repeats. The sequence U_5, which is copied to the 3' end of the provirus, lies just inside the 5' end of the genome and conversely U_3, which is copied to the 5' end, lies originally just in from the 3' end of the genome. Therefore the process of reverse transcription has to duplicate sequences from one end of the genome and place the copy at the other end. In doing so, it creates long, directly repeated sequences at each end of the provirus,

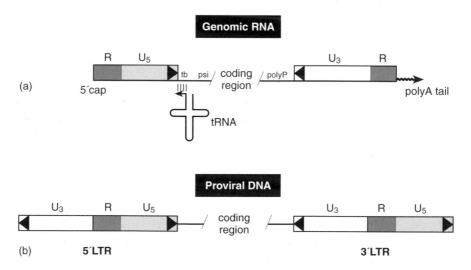

Fig. 8.2 Comparison of the structures of retrovirus genome RNA (a) and the proviral DNA created from it by reverse transcription (b). U_5 and U_3 are unique sequences at the 5' and 3' ends of virion RNA; R is a directly repeated sequence at the RNA termini. Short inverted repeat sequences are represented as ◄►; tb is the binding site for a transfer RNA and polyP is a polypurine region, both significant in reverse transcription (see text and Fig. 8.3). Long-terminal repeats (LTRs) comprise reiterations of the sequences U_3, R, U_5; psi is the specific packaging signal for RNA genomes. Not to scale.

comprising the elements U_3, R and U_5. These sequences are known as the long-terminal repeats or LTRs.

Properties of reverse transcriptase

The active forms of reverse transcriptase (RT) protein vary between retroviruses. In avian retroviruses, RT is composed of two subunits: α (relative molecular mass, M_r of 60 000) and β (M_r 90 000), which comprises the α polypeptide with integrase still linked to its C terminus. In HIV, RT is also a heterodimer, but integrase is fully cleaved from RT and instead the subunits differ by the presence or absence of the RNaseH domain of RT. In each case, the full-length RT subunit provides three enzymatic activities: (1) synthesis of DNA from an RNA template; (2) synthesis of DNA from a DNA template; and (3) digestion of the RNA strand from RNA–DNA hybrids (RNaseH activity).

The RT activity of the protein, like all other DNA polymerases, requires a primer from which to initiate DNA synthesis (see Section 6.1). The primer used to reverse transcribe the viral RNA is a cellular 4S transfer RNA (tRNA), of which there is one molecule per genome RNA that enters the cell in the viral particle. The 3' end of the tRNA is base paired to the genomic RNA near to the 5' terminus, exactly adjacent to (and in effect defining) the sequences that are destined to be duplicated at the 3' end of the provirus after reverse transcription is complete (see Fig. 8.2). Each retrovirus contains a specific type of tRNA, e.g. RSV has tryptophan-tRNA and Moloney murine leukaemia virus has proline-tRNA. It is not known how these tRNAs are specifically selected, although it is probable that the specific sequence of the genomic RNA at the tRNA-binding site is the important determinant.

The model for reverse transcription to form a provirus

Figure 8.3 presents a model for proviral DNA synthesis, given its structure relative to that of the genome. Evidence for the model comes from detection of the short negative- and positive-sense DNAs predicted in steps (b) and (f). Initial synthesis by RT from the tRNA primer 3' end copies to the 5' end of the template (Fig. 8.3b). As it does so, the RNaseH activity of the protein degrades the template, leaving the newly synthesized complement to the 5'-end R sequence free to base pair with another R sequence (Fig. 8.3c). This could be at the 3' end either of the same RNA molecule (as shown) or of the second genome copy. Synthesis of the negative strand can then proceed along the body of the genomic RNA and, as this occurs, RNaseH continues to degrade the template RNA (Fig. 8.3d).

A key question is how synthesis of the positive strand (or second strand) of the provirus is started. Given the comparison of genome and provirus, it can be predicted that the initiation event must occur immediately 5' to (and therefore defining) the sequences from inside the 3' end of the genome, which are ultimately copied to the 5' end of the provirus. The primer for this initiation is provided by a fragment of the largely degraded genome RNA. All retroviruses have at this location a conserved

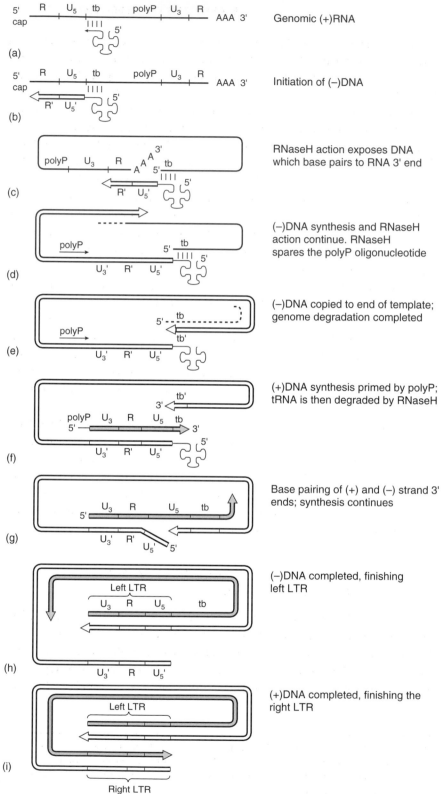

(a) Genomic (+)RNA

(b) Initiation of (–)DNA

(c) RNaseH action exposes DNA which base pairs to RNA 3' end

(d) (–)DNA synthesis and RNaseH action continue. RNaseH spares the polyP oligonucleotide

(e) (–)DNA copied to end of template; genome degradation completed

(f) (+)DNA synthesis primed by polyP; tRNA is then degraded by RNaseH

(g) Base pairing of (+) and (–) strand 3' ends; synthesis continues

(h) (–)DNA completed, finishing left LTR

(i) (+)DNA completed, finishing the right LTR

purine-rich region (polyP in Fig. 8.2), which is resistant to RNaseH degradation and so can provide the required primer (Fig. 8.3d,e). Synthesis then proceeds rightwards to the end of the newly synthesized minus strand, which is being used as a template (Fig. 8.3f).

A second template transfer is now required for RT to continue positive strand synthesis. At the very 5′ end of the negative strand is the tRNA that originally primed its synthesis. RT uses the 3′ segment of this (which had been paired with the genome originally and therefore has the exact complementary sequence) as template (Fig. 8.3f). In the meantime, synthesis of the negative strand has continued to the end of its template (Fig. 8.3e), which will be the sequence that originally bound the tRNA primer (as an RNA–RNA hybrid, neither will have been degraded by RNaseH). Thus, the 3′ ends of the negative- and positive-strand DNAs are complementary; their base pairing gives both polymerases the template strands that they need to complete provirus synthesis (Fig. 8.3g,h,i).

It would, of course, have been easier to explain reverse transcription if nature had put the tRNA primer at the 3′ end of the genome! However, this would have made the provirus an exact copy of the genome, i.e. it would not have had LTRs. As discussed below, these LTRs are essential to the generation of progeny viral RNA genomes from the provirus. To function in vivo, RT must enter a cell with the genomic RNA. This indicates that reverse transcription occurs without full uncoating of the genome and explains why the genome, although having all the features of mRNA, is never translated.

The existence of template jumps during reverse transcription suggested a possible explanation for the presence of two genome RNA copies in the particle. Perhaps the RT had to jump from the 5′ end of one genome copy to the 3′ end of the other. Although this is an attractive theory, the experimental evidence is that proviral DNAs can be formed solely by intramolecular jumps. It is probable, however, that the possibility of intermolecular jumps offers a way of minimizing the effects of genome damage on virus viability and so confers an evolutionary advantage.

Integration of the provirus into cellular DNA and production of progeny genomes

Reverse transcription converts retroviral RNA into linear double-stranded DNA in the cytoplasm. This then migrates, as a complex with proteins including the virus-coded integrase enzyme (termed the 'preintegration complex'), to the nucleus where it is integrated into cellular DNA. Successful integration requires the five-base-pair (bp) inverted repeat sequence at the proviral termini (see Fig. 8.2) and the integrase

Fig. 8.3 (*Facing page*) Scheme for the synthesis of retroviral linear DNA by reverse transcriptase (see text for details). Thin lines, RNA; open lines, negative-strand DNA; shaded lines, positive-strand DNA. Arrowheads represent 3′ ends. Abbreviations and other symbols as for Fig. 8.2.

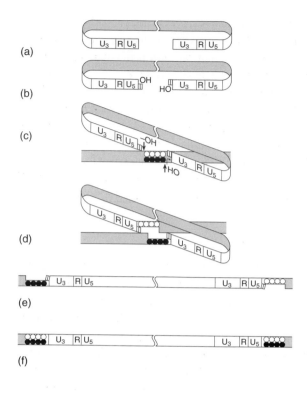

(a)

(b)

(c)

(d)

(e)

(f)

Fig. 8.4 Integration of retroviral DNA into the host genome. (a) Linear proviral DNA, synthesized as in Fig. 8.3, is attacked by the integrase, which cleaves two bases from each 3′ end (b). (c,d) This molecule migrates to the nucleus and the integrase cleaves the cellular DNA by catalysing an attack on it by the viral 3′-OH groups. (e,f) DNA repair reconstitutes the double strand. (From Whitcomb and Hughes 1992.)

enzyme. As with RT, this enzyme must enter the cell with the infecting particle to be able to act. Evidence for these requirements comes from deletion mutagenesis, because removal of these repeat sequences results in no integration.

There are three steps in the integration process (Fig. 8.4). First, two bases are removed from both 3′ ends of the linear proviral DNA molecule by the action of the viral integrase (Fig. 8.4b). Second, the 3′ ends are annealed to sites a few (four to six) bases apart in the host genome (Fig. 8.4c), which is then cleaved with the ligation of the proviral 3′ ends to the genomic DNA 5′ ends (Fig. 8.4d). This part of the integration reaction requires no input of energy from ATP, because the energy of the cleaved bonds is used to create the new ones (i.e. an exchange reaction), and thus is reversible. Third, gaps and any mismatched bases are repaired by host DNA-repair functions (Fig. 8.4e); this renders the integration irreversible. Infected cells typically contain 1–20 copies of integrated proviral DNA. There are no specific sites for integration, although there is some preference for relatively open regions of chromatin. Integration occurs only in cells that are actively moving through the cell cycle.

New viral RNA genomes are transcribed from integrated DNA. In all respects, this process resembles the production of cellular mRNA; cellular DNA-dependent RNA polymerase II (RNA pol II) transcribes the provirus, and the primary transcript is capped and polyadenylated by host-cell enzymes. It is important to realize the crucial significance to genome production of the LTRs created during reverse transcription. RNA pol II promoters lie upstream of the transcription start point which must, if the progeny genomes are to resemble exactly the parental ones, be placed at the 5' end of the R element within the left-hand LTR. If the provirus were simply a copy of the genome, there would be no viral sequences upstream of this start point to provide the promoter, and so genome synthesis would depend on fortuitous integration adjacent to a host promoter. However, with the creation of the LTRs, the U_3 element, which is placed in front of the required transcription start, can provide the promoter elements. Equally, at the other end of the genome, the polyA addition site must be fixed exactly at the 3' end of the R element within the right-hand LTR. As sequences both upstream and downstream of a polyadenylation site are important in determining its position, it is again essential that proviral sequences extend beyond the intended genome 3' end; these sequences are provided by the U_5 element within the right-hand LTR.

The spumaviruses—a special case

The spumavirus genus of the retrovirus family (also known as the foamy viruses) has only recently been studied in detail. Most work has been done on human foamy virus, although this virus is actually a chimpanzee virus that crossed into humans as a dead-end zoonotic infection; there is no evidence for the existence of a genuine human foamy virus. Unlike the standard retrovirus replication cycle, reverse transcription in spumaviruses at least begins (and may even be completed) within assembling progeny particles before they are released from a cell. Thus the genetic material within spumavirus particles may be DNA rather than RNA. In this sense, the spumaviruses somewhat resemble the hepadnaviruses, which are also reverse-transcribing viruses (below). Other details relating to gene expression (see Section 9.9) also suggest that the spumaviruses have similarity to both standard retroviruses and hepadnaviruses.

8.2 Hepadnaviruses

Human hepatitis B virus (HBV) is the pre-eminent member of this family. There are currently estimated to be 300 million infectious carriers of the virus worldwide. The virus causes acute liver disease, chronic disease and liver cancer and is, with smoking, one of the leading causes of cancer in humans (see Section 17.7). Chapter 16 records the excellent progress that has been made in anti-HBV vaccines.

Hepadnavirus replication

HBV has a DNA genome composed of a linear negative strand, which is covalently linked to a protein at its 5' end, and a complementary positive strand, which is incomplete (Fig. 8.5a). It is possible to give each strand a polarity because all the genes are arranged in the same direction. The positive-sense strand always overlaps the 5'-3' junction of the negative-sense strand and acts as a cohesive end to circularize the genome. Molecular analysis of HBV replication has been slow in coming because the virus still cannot be grown in culture. However, a scheme for replication has been

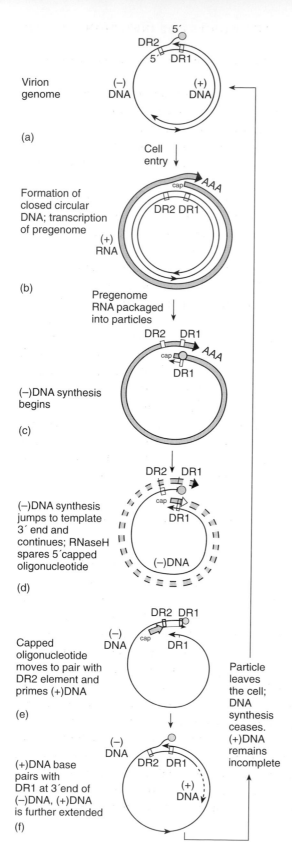

Virion genome

(–) DNA (+) DNA

(a)

Cell entry

Formation of closed circular DNA; transcription of pregenome

(+) RNA

(b)

Pregenome RNA packaged into particles

(−)DNA synthesis begins

(c)

(−)DNA synthesis jumps to template 3′ end and continues; RNaseH spares 5′capped oligonucleotide

(−)DNA

(d)

Capped oligonucleotide moves to pair with DR2 element and primes (+)DNA

(−) DNA

(e)

(+)DNA base pairs with DR1 at 3′end of (−)DNA, (+)DNA is further extended

(−) DNA

(+) DNA

(f)

Particle leaves the cell; DNA synthesis ceases. (+)DNA remains incomplete

Fig. 8.5 Replication of the hepadnavirus genome by reverse transcription (see text for details). Thin lines, DNA; shaded lines, RNA. DR1 and DR2 represent two copies of a short, directly repeated sequence in the genome. The viral polymerase (reverse transcriptase) attached to the genome 5′ end is shown as a grey circle. Filled arrowheads represent 3′ ends. Other arrows indicate the polarity (5′ → 3′) of nucleic acid strands. cap denotes the methylated cap structure on the RNA 5′ end and AAA its 3′-polyadenylate sequence.

derived from studies involving HBV and its relatives, woodchuck and duck hepatitis viruses (Fig. 8.5).

Upon entry into cells, the DNA genome is transported to the nucleus, where the attached protein is removed and both strands are completed and ligated to give a covalently closed circle (Fig. 8.5b). The negative strand provides a template for transcription by host RNA pol II, producing various mRNAs (see Chapter 9) that are exported to the cytoplasm. The longest mRNA, which has a terminal repetition because it extends over more than the full circumference of the circular template, encodes a multifunctional polymerase and also serves as the template for genome DNA synthesis. The polymerase interacts with a sequence close to the 5′ end of the RNA that encoded it and directs its packaging by core protein into particles (Fig. 8.5c). It is only once this has occurred that creation of the DNA genome begins.

The polymerase itself serves as a primer for the negative-sense DNA strand, explaining its presence attached to the 5′ end of this genome strand in the particle. Synthesis begins at the polymerase-binding site near the positive-sense RNA 5′ end (Fig. 8.5c). The short negative-sense DNA produced then moves to base pair with the second copy of its template sequence at the RNA 3′ end, and DNA synthesis continues, RNaseH activity degrading the template as it does so (Fig. 8.5d). A specific positive-sense RNA oligonucleotide from the most 5′ end of the RNA, containing the repeat sequence DR1, is spared this degradation. It transfers to base pair with DR2 near the 5′ end of the new negative strand, where it primes synthesis first to the most 5′ end of its template (Fig. 8.5e) and then, by base pairing instead to the second copy of DR1 at the other end of the template, onwards into the body of the genome (Fig. 8.5f). It is very unusual for this positive strand to be completed before the particle leaves the cell, hence the incompletely double-stranded genomes observed in virus preparations.

8.3 Comparing the retroviruses, hepadnaviruses and caulimoviruses

The retroviruses and hepadnaviruses are, in most senses, completely unrelated. However, some molecular aspects of their replication show considerable similarity. Reverse transcription in the two viruses follows a similar course; whereas this process occurs at the initiation of the interaction of a retrovirus with its host cell, it occurs at the end of the HBV infectious cycle. Also, in each type of virus, host RNA pol II is used to produce RNA that serves as genome or pregenome, respectively. These replication cycles therefore represent temporally permuted versions of each other.

The caulimoviruses, the only truly double-stranded DNA virus family in the plant kingdom, have been viewed with considerable interest as likely vectors for plant genetic manipulation. Investigation of the representative virus, cauliflower mosaic virus (CaMV), has shown that it has properties intermediate between those of retroviruses and hepadnaviruses. Similar to HBV, the CaMV genome is a double-stranded DNA circle, with a complete but gapped negative strand and an incomplete positive strand. However, similar to retroviruses, negative-strand DNA synthesis is primed by a host-cell tRNA which base pairs to the template RNA close to its 5′ end. All three viruses have reverse transcription mechanisms which involve shifts or jumps of the extending polymerase from the 5′ end to the 3′ end of a template molecule, mediated through sequences repeated at the two ends of the template. Only the retroviruses carry a specific integration function; although HBV and CaMV sequences can become integrated into the host genome, they do so inefficiently, presumably by non-homologous recombination.

8.4 Further reading and references

Cullen, B. R. (1991) Human immunodeficiency virus as a prototypic retrovirus. *Journal of Virology* **65**, 1053–1056.

Goff, S. P. (1992) Genetics of retroviral integration. *Annual Review of Genetics* **26**, 527–544.

Hohn, T., Hohn, B. & Pfeiffer, P. (1985) Reverse transcription in CaMV. *Trends in Biochemical Sciences* **10**, 205–209.

Hull, R. & Covey, S. N. (1986) Genome organisation and expression of reverse transcribing elements: variations and a theme. *Journal of General Virology* **67**, 1751–1758.

Levy, J. A. (ed.) (1992–5) *The Retroviridae* (Vols 1–4). New York: Plenum.

Linial, M. L. (1999) Foamy viruses are unconventional retroviruses. *Journal of Virology* **73**, 1747–1755.

Nassal, M. (1999) Hepatitis B virus replication: novel roles for virus–host interactions. *Intervirology* **42**, 100–116.

Nassal, M. & Schaller, H. (1993) Hepatitis B virus replication. *Trends in Microbiology* **1**, 221–228.

Seeger, C. & Mason, W. S. (1996) Reverse transcription and amplification of the hepatitis B virus genome. In: *DNA Replication in Eukaryotic Cells*, pp. 815–831. DePamphilis, M. (ed.). Cold Spring Harbor, NY: Cold Spring Harbor Laboratory Press.

Seeger, C. & Mason, W. S. (2000) Hepatitis B virus biology. *Microbiology and Molecular Biology Reviews* **64**, 51–68.

Whitcomb, J. M. & Hughes, S. H. (1992) Retroviral reverse transcription and integration: progress and problems. *Annual Review of Cell Biology* **8**, 275–306.

Also check Chapter 22 for references specific to each family of viruses.

Chapter 9

The process of infection: IIIA. Gene expression and its regulation in DNA viruses

9.1 Introduction

The process of gene expression by the various DNA viruses closely parallels that of their host organisms. The genome must be transcribed to form positive-sense RNA, and this must then be translated into polypeptide chains. In eukaryotes further steps intervene: the RNAs must be capped and polyadenylated, and may also need to be spliced, to give functional mRNA. The mRNA must then be moved to the cytoplasm to allow translation. Each stage in the pathway of gene expression, from transcription of RNA through to post-translational modification of protein, represents a potential point of control and DNA viruses exploit these possibilities in various ways so that each one achieves an organized programme of gene expression. In most cases, expression is phased, with early genes being expressed before DNA synthesis begins and late genes being activated only after this event. In some cases, further temporal divisions are also apparent. These patterns of expression have been characterized in considerable detail and, as well as furthering our understanding of virus growth cycles, such studies have increased greatly our understanding of the molecular biology of eukaryotic cells. This chapter, and the accompanying one on RNA virus gene expression (Chapter 10), demonstrates these points by examining gene expression and its control in a variety of virus systems.

9.2 The DNA viruses: Baltimore classes 1, 2, 6 and 7

All DNA viruses synthesize their mRNA by transcription from a double-stranded (ds) DNA molecule; those from Baltimore class 2 have to convert their genomes to double-stranded forms before transcription begins. The viruses of class 6 can also be considered here because they must create a dsDNA version of their genome RNA before transcription can take place. Among those infecting eukaryotes, all except the poxviruses carry out transcription in the cell nucleus. Thus the cell's own transcription machinery and RNA modification machinery are available and the viruses make use of these to produce their mRNA. These viruses are therefore good model systems for studying the synthesis of cellular mRNA, and several significant milestones in understanding eukaryotic gene expression have come from this work, e.g. transcription enhancers and transcription factors that regulate RNA polymerase II (RNA pol II) activity were first identified during studies of SV40 gene expression, and splicing was discovered by analysing adenovirus mRNA.

The amount of genetic information varies by more than 100-fold between different DNA virus families, as discussed in Chapter 4. Some have severe restrictions on their coding capacity and depend greatly on their host for essential replicative functions despite having evolved to maximize their use of the available genetic information, whereas others have had the opportunity to evolve or acquire many functions that are not essential for virus multiplication. The larger viruses are less dependent on the cell and can, for example, multiply when cells are not in S phase (synthesizing cellular DNA), whereas the smallest viruses are unable to do so. These larger viruses can also use inhibition of host macromolecular synthesis as a strategy to target all of the host's resources towards their own replication. Several of the viruses have the capacity to induce the host cell to enter S phase to promote their replication. Such functions may not be crucial in rapidly growing experimental cell cultures, but are probably essential during natural infections, when the virus infects cells that are not dividing. As discussed in Chapter 17, many of the DNA animal viruses are capable of transforming cells. Transformation results when the virus functions that regulate cell division become permanently expressed in cells that have not died as a result of virus infection.

9.3 Polyomaviruses

Polyomavirus (of mice) and SV40 (of simians) exemplify the polyomavirus genus of the papovavirus family; there are also related human viruses BK and JC. These viruses have double-stranded circular genomes of around 5.3 kilobase-pairs (kbp) in length which encode early and late genes on opposite strands of the template, each occupying about half the genome (Fig. 9.1). The genes are transcribed by cellular RNA pol

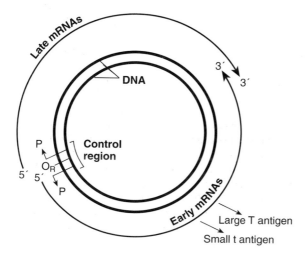

Fig. 9.1 Map of the genome of SV40 virus. P denotes the early and late promoters of transcription. O_R is the origin of replication. The mRNAs are shown in detail in Figs 9.2 and 9.3.

II to produce a single primary transcript in each case; these are then differentially spliced to form the various mRNAs. Differential splicing means that there is more than one acceptable splicing pattern for the RNA. Which pattern is used on a specific RNA molecule will depend on the relative affinity of the alternative splice sites for the splicing machinery. Altering the balance of use of alternative splice sites is a potential method for regulation of gene expression (see adenoviruses and influenza viruses, Section 9.5 and Section 10.8, respectively), but this mechanism has not been reported for these viruses.

SV40 has a major early protein, the large T (for tumour-specific) antigen, and a lesser protein called small t (Fig. 9.2a). Polyomavirus has three T antigens, called large T, middle T and small t (Fig. 9.2b). Each protein is made from a discrete spliced mRNA. In SV40, the largest early mRNA is translated to give the small t protein of 174 amino acids; an intron, which is removed during processing of the primary transcript for this

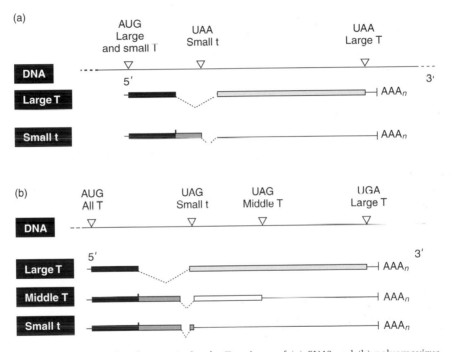

Fig. 9.2 Synthesis of early mRNAs for the T antigens of (a) SV40 and (b) polyomavirus. In SV40, splicing can remove the small t termination codon and fuse the N-terminal part of the small t sequence onto the large T reading frame downstream. In polyomavirus, the small t reading frame is formed by removal of a short intron close to its C ter minus; splicing from the same 5′ site to an alternative 3′ site downstream of the small t termination codon accesses the middle T reading frame. The large T reading frame is accessed by removal of a larger intron which fuses the N terminus of small t to the large T reading frame. The blocks indicate translated regions of each mRNA with fill patterns showing protein sequences that are identical within either (a) or (b). The dotted lines denote introns and thin lines untranslated RNA regions.

mRNA, lies downstream of this open reading frame (ORF). The other SV40 early mRNA is formed by removal of a larger intron, which includes the C-terminal half of the small t-antigen ORF. This alternative splicing event fuses together the first 82 codons of the small t-antigen ORF in frame with 626 codons from a much longer ORF located downstream. Although this reading frame is also present in the small t mRNA, it has no initiation codon and so is not accessed by ribosomes. Translation from this mRNA generates the large T antigen of 708 amino acids. Polyomavirus encodes its three T antigens in a similar fashion. These genes are good examples of how viruses maximize the coding potential of their genomes.

The SV40 early and late promoters lie close together in a region of the genome which also includes the origin of replication (see Fig. 9.1). The early promoter includes a TATA box and binding sites for the host cell transcription factor SP1. Its activity is also controlled by an enhancer, which binds several other host transcription factors. Large T antigen produced in the early phase of the infectious cycle inhibits its own synthesis by binding specifically to its promoter (this binding is also required to initiate replication; see Section 6.2). T antigen also activates late gene expression through effects on the activity of host-cell transcription factors and activates cell cycle progression, driving cells into S phase (small t also plays a role here). The effects of these proteins on cell cycle regulation are considered in Section 17.4. Late mRNAs encode three proteins—VP1, VP2 and VP3—which ultimately form the progeny capsid, using a strategy similar to that of early gene expression (Fig. 9.3). There is no viral protein produced to complex with the DNA; histones H2A, H2B, H3 and H4 from the host cell are used for this purpose.

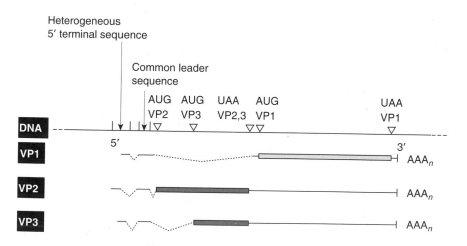

Fig. 9.3 Synthesis of late mRNAs for virion proteins (VP) of polyomavirus. The late mRNAs have heterogeneous 5′ ends because the late promoter lacks a TATA box, which normally serves to direct initiation to a specific location. SV40 is very similar but the VP1-coding region extends further 5′ to overlap with VP2/VP3. The blocks indicate translated regions of each mRNA, with fill patterns showing identical protein sequences; the dotted lines denote introns and thin lines untranslated RNA regions.

9.4 Papillomaviruses

Papillomaviruses are exemplified by bovine papillomavirus type 1 (BPV-1) and the many types of human papillomavirus (HPV). Details of the gene expression of these viruses have been much harder to work out than those of the polyomaviruses because of the difficulty of growing them in cell culture systems. A genome map of the widely studied BPV-1 is shown in Fig. 9.4. As with the polyomaviruses, all transcription occurs in the nucleus mediated by host RNA pol II; however, a striking difference from SV40 is that the genes are all organized in the same orientation, i.e. only one genome strand is transcribed. The slightly longer genome (8 kbp) probably encodes 11 proteins and mRNA for these is produced from seven different promoters. Transcription for the early and late genes is largely overlapping, covering most of the genome, but they are expressed from different promoters (six early and one late). Alternative splicing and polyadenylation are used to give multiple mRNAs from a primary transcript, especially in the case of the late transcripts where three different proteins are encoded (E4, L1 and L2).

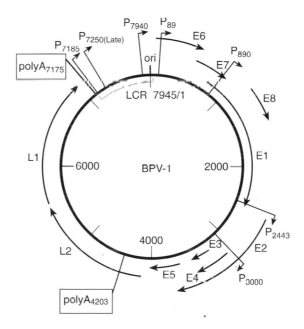

Fig. 9.4 The genome organization of bovine papillomavirus type 1 (BPV-1). E1–E8, L1 and L2 represent open reading frames (ORFs). Reading frames were originally designated E or L, based on the observation that a 69% fragment of the genome was sufficient for cell transformation *in vitro* and was hence defined as the early region by analogy with SV40. However, within the early region, the E4 gene is now known to be expressed only within the late temporal class of mRNA. Primary transcripts from each promoter undergo extensive differential splicing to produce multiple mRNAs (not shown). The region lacking ORFs is known as the long control region (LCR). P_{89}, etc., RNA pol II promoters; polyA, sites of mRNA polyadenylation; ori, origin of DNA replication.

Regulation of BPV transcription occurs through the E2 gene, whose full-length product, E2TA, dimerizes to form a DNA-binding transcription factor. There are two E2-dependent enhancer elements in the long control region, and also binding sites associated with most/all of the viral promoters. As well as E2TA, truncated E2 forms are produced, one initiated within the E2 reading frame (E2TR) and another an alternatively spliced form (E8E2); each contain only the C-terminal half of the E2TA protein. E2TR and E8E2 can form homo- and heterodimers, with each other or E2TA, which retain specific DNA-binding activity, but lack the transcription activation function of the E2TA dimer. Thus, the truncated forms of E2 modulate the level of E2-mediated transactivation in two ways—by inhibiting the formation of active E2TA through competition for dimerization and by competing with active dimers for DNA binding.

The biology of papillomavirus gene expression is particularly fascinating. These viruses cause warts, i.e. benign growths in epithelia, and expression of their early and late genes is separated in time and space. In infected epithelia, early genes are expressed in the dividing cells of the basal cell layer and, as the cells divide, the viral DNA is maintained at an approximately constant copy number per cell by limited replication. The E6 and E7 proteins act to alter cell cycle control within these cells, using mechanisms similar to those shown by SV40 large T antigen (see Section 17.4). The late events (accelerated replication and late protein synthesis) only begin once an infected cell has left this layer and is committed to terminal differentiation. Thus, the pattern of expression of the virus genes is determined by the differentiation state of the host cell. The factors that turn on this so-called vegetative phase in the growth cycle are not known. The human papillomaviruses are also of interest for their involvement in the development of certain human cancers (see Chapter 17).

9.5 Adenoviruses

Adenovirus gene expression has been studied best in the closely related human serotypes 2 and 5. In these viruses, transcription occurs from both strands of the 36 kbp linear genome. All mRNA is produced by host RNA pol II; there are also two short RNAs (the VA RNAs) produced by RNA pol III. Gene expression shows temporal regulation (Fig. 9.5a) and the terms 'early' (pre-DNA synthesis) and 'late' (post-DNA synthesis) were used originally to classify the genes as E or L. However, it is now appreciated that expression of the E1A gene commences before the other early genes, and so should be classified as immediate–early, and that the small IVa2 and IX protein genes form an intermediate class rather than being true late genes. The observation that defines E1A as 'immediate–early' is that, unlike the other early genes, it is transcribed in an infected cell when protein synthesis is blocked by a chemical inhibitor. This result also shows that E1A proteins are needed for expression of the remaining viral early genes.

Each of the five early (E) regions has its own promoter (Fig. 9.5b). Two of them (E2 and E3) have two alternative polyadenylation sites—A and B—leading to two families of mRNA in each case, whereas the major late region produces five families of mRNA (L1–L5), as a result of polyadenylation at any one of five possible sites. As

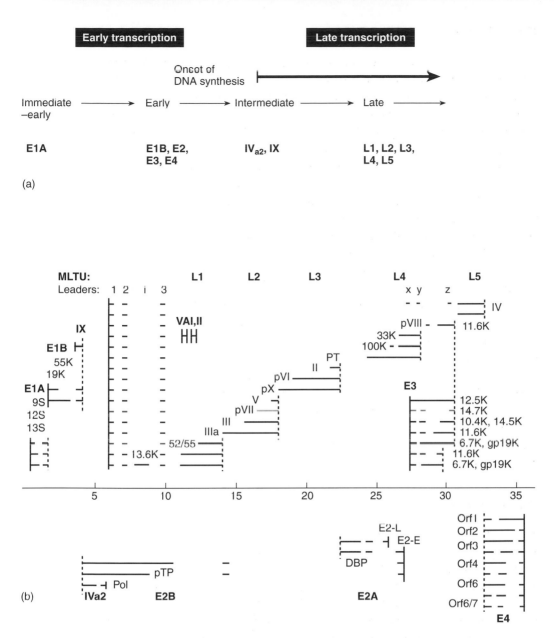

Fig. 9.5 Gene expression by adenovirus type 5. (a) The phases of gene expression: the numbers E1A, L1, etc. refer to regions of the viral genome from which transcription takes place. (b) A transcription map of the adenovirus 5 genome. The genome is represented at the centre of the diagram as a line scale, numbered in kilobase-pairs from the conventional left end, with rightward transcription shown above and leftward transcription below. Genes or gene regions are named in **bold**. Individual mRNA species are shown as solid lines, with introns indicated as gaps. Promoters of RNA polymerase II transcription are shown as solid vertical lines and polyadenylation sites as broken vertical lines. The RNA pol III transcripts VAI and VAII are delineated by paired vertical lines. The protein(s) translated from each mRNA is indicated adjacent to the RNA sequence encoding it. Structural proteins are shown by roman numerals (major proteins: II, hexon; III, penton; IV, fibre; pVII, core; see Chapter 11). PT, 23 M_r virion proteinase; DBP, 72 M_r DNA-binding protein; pTP, terminal protein precursor; Pol, DNA polymerase. (Reproduced with permission from Leppard, K. (1998) *Seminars in Virology* **8**, 301–307. Copyright © Academic Press.)

well as alternative polyadenylation, transcripts from adenovirus genes show complex patterns of differential splicing. Each of the early regions gives rise to several differently spliced mRNAs which, except for E1A where the principal products are closely related in sequence (see Section 17.4), generally encode distinct proteins. The E2 region, for example, encodes the three viral proteins that are directly involved in DNA replication (see Section 6.4), whereas splice variation within the five families of major late region mRNA allows synthesis of at least 13 different proteins, mostly involved in forming progeny particles.

RNA processing from the major late gene is particularly wasteful of the cell's resources; each late mRNA that is 1–5 kbp long is made from a precursor up to 28 kbp in length. There is no possibility of making multiple mRNAs from the same precursor because all the mRNAs include the same three RNA segments from the precursor's 5' end; these are spliced together to form the so-called tripartite leader sequence. Thus, a lot of newly synthesized viral RNA, from both the major late gene and other genes, is removed as intron sequence and degraded. However, the advantage of this system is that the virus can achieve expression of around 40 proteins using a minimum of genome space to direct transcription, and with coordinated transcriptional control.

Control of gene expression during adenovirus infection uses a diversity of mechanisms. First, transcription is regulated so that it takes place in an ordered sequence. The E1A proteins, especially the 13S mRNA product (see Fig. 17.3), activate expression of the remaining genes, but significant transcription from the intermediate and late genes awaits the onset of DNA synthesis; the molecular mechanism of this switch is not fully understood. Second, RNA processing is regulated so that the relative proportions of the possible mRNA products of a gene change as the infection proceeds; this involves regulated splice site and polyadenylation site usage, e.g. when the major late gene is first transcribed, processing produces mRNA for the L1 52/55 M_r protein almost exclusively, whereas, later on, mRNA is produced from all of the regions L1–L5 and the predominant L1 mRNA is that which encodes the IIIa protein. The mechanism by which this regulation is achieved is beyond the scope of this text. Third, the movement of mRNA within the cell is regulated so that viral mRNA reaches the cytoplasm in preference to cellular mRNA. Early proteins from E1B and E4 are required to achieve this. Fourth, translation is switched from a cap-dependent to a cap-independent mode during the late phase of infection through inactivation of cap recognition by the L4 100 M_r protein. This shuts off host protein synthesis while permitting continued viral late protein production because the late viral mRNAs, although capped, can recruit ribosomes in a cap-independent manner through sequences in their 5'-untranslated regions. Overall, this complexity of controls produces each of the proteins at the time, and in the amount, that is required.

9.6 Herpesviruses

Herpes simplex virus types 1 and 2

Herpesvirus genomes (130–230 kbp) are very much larger than those of adenoviruses and papovaviruses and show considerable diversity. Their gene expression has been studied principally using herpes simplex virus type 1 (HSV-1), which productively infects epithelial cells and establishes latency in sensory neurons *in vivo* (see Section 15.6) and which grows readily in standard cell cultures *in vitro*. HSV-2 has very similar properties. HSV-1 has around 70 genes interspersed on both strands of its 153 kbp genome. An immediate contrast with the smaller DNA viruses is that transcription of most HSV-1 genes is controlled separately by specific initiation and termination elements, and there is very little use of splicing during mRNA production. A further contrast is that about half of the genes can apparently be mutated without affecting the growth of the virus in cell culture. These dispensable genes are presumed to be important *in vivo*.

As with the other animal DNA viruses, HSV-1 gene expression is arranged into temporal phases. These are generally termed α, β and γ, corresponding to immediate early, early and late phases. α, β and γ genes are not grouped together but are found scattered throughout the genome. The α genes of HSV-1 produce various regulators of gene expression, the β genes produce proteins required for DNA replication, and the γ genes produce mainly structural proteins for progeny particle formation. Although expression of the γ genes accelerates with the onset of replication, it is in fact more or less independent of this process. Thus, when DNA synthesis is inhibited, both early and late transcripts are detected in the nucleus.

The different phases of HSV-1 gene expression are linked through the production of proteins in one phase that activate the next (Fig. 9.6). Thus, in the same way that adenovirus E1A proteins activate other viral genes, so an HSV-1 α-gene product,

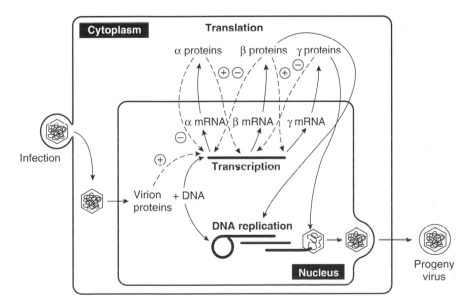

Fig. 9.6 Gene expression in herpes simplex virus type 1 infection. The three successive phases of gene expression are denoted α, β, γ (see text). Solid lines indicate the flow of material. Dashed lines indicate regulatory effects on gene expression; + indicates activation; − indicates repression.

known variously as ICP4 and IE175, turns on transcription of the β genes. One or more β gene product(s) then inhibit expression of α genes and induce γ gene expression, proteins from which inhibit expression of the β genes. This pattern of gene expression, with a series of genes controlled by activators and inhibitors, ensures that the relevant proteins are present when required and enhances the efficiency of the virus replication cycle. A particularly interesting feature of this cascade of regulated HSV-1 gene expression is that it is circular. A protein produced during the late phase of infection (γ-gene product) is packaged into the particle and serves as an activator of α-gene expression in the next round of infection. This protein is known as VP16 and is a potent transcriptional activator. Although VP16 has a transcriptional activation function, two host proteins, Oct1 and HCF, are required for it to upregulate α-gene expression. Oct1 is a DNA-binding transcription factor and VP16 must interact with HCF before it can bind to the Oct1–DNA complexes at α-gene promoters, thus bringing its own powerful activation domain to the promoters. The reason why the virus has evolved this indirect mechanism for activation by VP16 is uncertain, but it may be relevant to the establishment of, and reactivation from, latency (see below).

In contrast to this extensive and well-characterized pattern of lytic gene expression, the activity of the HSV-1 genome in a latently infected neuron is very limited. It is generally agreed that the primary reason why the lytic programme does not proceed in these cells is a failure to express the α proteins, but it is uncertain why this failure occurs. The only viral RNAs that can be detected are the latency-associated transcripts (LATs). These are actually stable introns, produced by splicing of longer RNAs, and they do not encode proteins. LATs are transcribed from the opposite strand to the α gene, ICP0, leading to the suggestion that they interfere with ICP0 synthesis by an antisense mechanism and so prevent the normal pattern of gene expression from proceeding. Certainly, ICP0 mutants reactivate poorly from latency. Alternatively, it has been proposed that the virion protein activator of the α genes, VP16, fails to enter the nucleus in infected neurons, resulting in a general failure of α-gene expression and the establishment of latency by default. One possible explanation for this failure may lie in the localization of the cellular cofactor, HCF, which is required to transport VP16 to the nucleus. HCF is a nuclear protein in most cell types, but is cytoplasmic in sensory neurons. Stimuli that cause HSV-1 reactivation from latency also cause relocalization of HCF to the nucleus, supporting this view.

Epstein–Barr virus

Similar to HSV-1, Epstein–Barr virus (EBV) can establish latency. However, it does so in B lymphocytes rather than sensory neurons (see Sections 13.3 and 15.6). Although poorly characterized, as productive infection is hard to achieve in cell culture, the molecular events of the lytic phase of the EBV life cycle are believed to resemble those defined for HSV-1. What has been possible, however, is to study the pattern of EBV gene expression during latency. Between one and nine proteins are expressed, depending on the growth state of the cells and whether or not latency is fully established. Transcription covers more than half of the 172 kbp genome, with extensive differential splicing to produce alternative mature mRNAs from a primary transcript. Unlike HSV-1 latency where DNA replication does not occur, EBV has a specific replication mechanism that is used during latency establishment and maintenance; this mechanism is completely distinct from that operating during a productive EBV or HSV-1 infection (see Section 6.3). Only one of the latency proteins, an origin-recognition DNA-binding protein EBNA-1, is required directly for DNA replication. The remainder of the latency proteins appear to be involved in modulating cell signalling and cell cycle control. The role of latent replication is to maintain the viral genome in an expanding pool of B lymphocytes, which are stimulated to divide upon infection. Subsequently, latency is maintained in a pool of resting, memory B cells.

EBV latency can be broken within individual cells by a variety of stimuli, with a resulting switch into a lytic pattern of gene expression and DNA replication. Expression of two virus-encoded transcriptional activators, BZLF-1 and BRLF-1, is thought to be the event that commits a latently infected cell to make this switch. In infected B-lymphocyte cell lines, a few cells are actively producing virus at any given time. Similar sporadic reactivation is also thought to occur *in vivo*.

9.7 Poxviruses

The poxviruses are the only DNA viruses of eukaryotes so far described that multiply in the cytoplasm. The molecular details of gene expression have been studied using vaccinia virus. Its linear double-stranded genome (190 kbp) encodes a large number of proteins, and these allow the virus to replicate with a considerable degree of autonomy from its host. It appears that viral gene expression is totally independent of the host nucleus because infection can proceed in experimentally enucleated cells. This cytoplasmic life cycle requires that the infecting virion carries the enzymes needed for mRNA synthesis, including a DNA-dependent RNA polymerase and enzymes that cap, methylate and polyadenylate the resulting mRNAs. Splicing activities are not known to be used by these viruses.

Using the enzymes from the infecting particle, viral transcription begins in cytoplasmic replication 'factories'. Only a subset of viral genes (the early genes) is transcribed initially. Early gene products include proteins needed for replication and to activate the intermediate genes, which then activate late genes. Thus, infection leads to a phased sequence of gene expression punctuated by viral DNA synthesis, much as is seen for the other DNA viruses. Independence from the cell's own transcription machinery allows the virus the opportunity to shut the cell nucleus down, so that all the metabolic resources of the cell are devoted to the virus. Thus, unlike the other families of DNA viruses, all of which can establish chronic, persistent or latent infections *in vivo*, vaccinia is exclusively cytolytic.

9.8 Parvoviruses

Adeno-associated virus (AAV) is the best-characterized parvovirus. Its linear, single-stranded (ss) genome, once converted to double-stranded form (see Section 6.6),

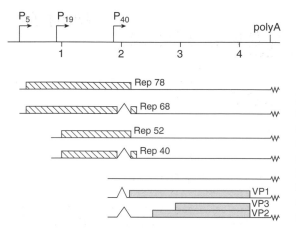

Fig. 9.7 Adeno-associated virus gene expression. The 4.6 kbp genome is represented as a line scale, with mRNAs beneath it with 5′ → 3′ polarity from left to right. Encoded proteins are shown as shaded boxes. (See text for further details.)

contains three promoters, each of which produces at least two related mRNAs by differential splicing (Fig. 9.7). The proteins produced by expression from P5 and P19 (the Rep proteins) are all sequence related and can be regarded as early proteins, because they activate the third promoter, P40. The unique functions of the Rep proteins have not yet been fully defined, but include the site-specific DNA-cleavage activity that is essential to genome replication (see Section 6.6). The P40 mRNAs code for the structural proteins, VP1, -2 and -3, from which new virions are assembled. These three proteins are all sequence related, the major capsid protein being VP3. This protein is produced by internal initiation at an AUG codon within the shortest mRNA. Read-through to this initiation point occurs in the majority of translation events because the alternative initiation event, specifying VP2, uses a non-canonical ACG codon. As noted in Sections 6.6 and 13.3, AAV requires a helper virus for growth under most circumstances and can establish latency when such help is not available. However, although transcriptional activators from the helper (such as adenovirus E1A—see Section 9.5) can upregulate AAV gene expression and other proteins may facilitate gene expression at a post-transcriptional level, there does not appear to be an obligatory requirement for any helper function to achieve AAV gene expression (see Section 6.6).

9.9 Retroviruses

Retroviral gene expression occurs exclusively from the provirus, a DNA copy of the viral RNA genome (typically 8–10 kbp), which is normally integrated into the host chromosome (see Section 8.1). The promoter for transcription lies in the upstream long terminal repeat (LTR) of the provirus, and this is regulated by a variety of cell transcription factors. In more complex retroviruses, such as HIV, there are additional virus-coded regulators of gene expression (see Chapter 19). The mechanism of mRNA production exactly mirrors the process for any host-cell gene, i.e. it involves host RNA pol II, capping, splicing and polyadenylation functions. Subsequently, mRNAs are transported to the cytoplasm where they are translated. Unlike the true DNA

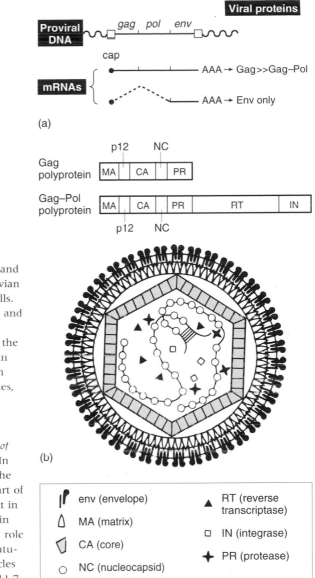

Fig. 9.8 (a) The mRNAs and proteins synthesized in avian leukosis virus-infected cells. (b) Processing of the Gag and Gag–Pol polyproteins of avian leukosis virus, and the location of proteins within the virion. (Adapted from Whitcomb, J. M. & Hughes, S. H. (1992) Retroviral reverse transcription and integration: progress and problems. *Annual Review of Cell Biology* **8**, 275–306.) In non-avian retroviruses, the protease is encoded as part of Pol and so is only present in the Gag–Pol fusion protein (see text for details). The role of the protease in the maturation of retrovirus particles is considered in Section 11.7.

viruses, simple retroviruses have no temporal phasing to their gene expression. However, the additional functions of complex retroviruses do allow them to control their gene expression in this way.

The mRNAs and protein-coding strategy for a typical simple retrovirus, avian leukosis virus (ALV), are shown in Fig. 9.8. ALV is the virus from which Rous sarcoma virus arose by recombination with host DNA sequences, adding an extra gene (see Section 8.1). The two ALV mRNAs, produced from a start site in the 5′-LTR, have identical 5′ and 3′ ends. One is equivalent to the full-length genome and the other has a section removed by splicing. Although both mRNAs include the *env* gene, only the spliced message is translated to produce the Env proteins because eukaryotic ribosomes normally treat mRNA as being monocistronic, translating only the 5'-proximal ORF. Spliced mRNAs are also used by complex retroviruses for expression of their extra genes which are located between *pol* and *env* and/or between *env* and the 3'-LTR.

The full-length ALV mRNA must encode both *gag* and *pol* gene products. This requires a special mechanism because translation in eukaryotes is normally initiated at the first AUG in the message, scanning from the 5' end, and such a mechanism would produce exclusively Gag protein. In fact no free Pol protein is produced by ALV and the protein is only synthesized fused to the C terminus of Gag. The *pol* gene is in the −1 reading frame with respect to *gag*. Ribosomes translating *gag* avoid its termination codon in about 5% of translation events by slipping backwards one nucleotide, an event known as a ribosome 'frame-shift'. Once this has occurred, the translation termination codon for the *gag* gene becomes invisible to the ribosome because it is no longer in the reading frame being used, so translation continues into the *pol* gene, producing a Gag–Pol fusion protein. The extremely high frequency of ribosome 'error' at this position at the 3' end of *gag* is caused by a specialized sequence context, comprising a slippery sequence on which the frameshift occurs and a pseudoknot downstream that pauses the ribosome on the slippery sequence to promote frameshifting, a mechanism that closely resembles that operating in coronaviruses (see Section 10.6). This mechanism achieves lower levels of *pol* gene expression than of *gag*, reflecting the relative amounts in which these proteins are needed to form progeny particles. Other retroviruses face the same problem in achieving expression of their *pol* genes, but they find a variety of solutions. In the murine equivalent of ALV, known as MLV, the *gag* and *pol* genes are in the same frame separated only by a stop codon. Nonsense suppression occurs to produce a Gag–Pol fusion, again with about 5% efficiency.

For all retroviruses, the *gag* and *pol* genes code for polyproteins and the protein encoded at the *gag–pol* boundary is a specific proteinase (Pro), which is required to process them. In ALV, Pro is coded at the C terminus of *gag*, whereas in MLV it is at the N terminus of *pol*. A further variant mechanism for *pol* expression then arises in other retroviruses, such as human T-cell lymphotropic virus (HTLV), where the proteinase is encoded in a reading frame distinct from both *gag* upstream and *pol* downstream. One frame shift takes the ribosome from the Gag reading frame into the Pro reading frame, and a second shift takes it on into the Pol reading frame; three different polyproteins, Gag, Gag–Pro and Gag–Pro–Pol, are therefore produced. Finally, in the least well-characterized retrovirus genus, the spumaviruses, there is evidence for a separate spliced mRNA being used to encode Pol, so that no Gag–Pol fusion protein is made.

The spumaviruses actually represent an interesting intermediate state between standard retroviruses and the hepadnaviruses (see Section 9.10), based on features of both their genome replication (see Section 8.1) and their gene expression. As well as expressing Pol protein independently of Gag (in the same way as hepadnavirus P and C proteins are separate), they also make use of a second promoter to express two proteins from reading frames that are 3' to *env*. Although other complex retroviruses have additional genes in this same position, they use splicing to express them from the same LTR promoter as used for all gene expression in other retroviruses.

9.10 Hepadnaviruses

The hepadnaviruses have very small circular genomes, e.g. hepatitis B virus, 3.2 kbp. The particular feature of note regarding the gene expression of this virus is the density at which information is compressed onto the genome, with every nucleotide coding for at least one protein, and two unrelated proteins being encoded in alternative reading frames over a significant part of its length (Fig. 9.9). There are four RNA pol II promoters which produce pre-genomic and three classes of subgenomic mRNA, respectively. None of these mRNAs is spliced and all end at the same polyadenylation site. Transcription is regulated by cell-type-specific factors and this is part of the basis for the tropism of this virus for the liver.

The mRNAs from the four promoters encode seven proteins (Fig. 9.9). All except X and pre-C are structural components of the virion. Pre-C is secreted from the cell (this is known as e antigen) and probably serves to modulate the immune response to the virus, whereas X is a transcriptional activator, which affects a wide range of viral and cellular promoters. The mRNAs from the genomic and S promoters show heterogeneous 5′ ends. This means that some molecules include the translation start sites for the pre C and M proteins, respectively, whereas the remainder do not. Translation from these latter mRNAs then begins at downstream AUG codons, encoding the C (core) and S (surface) proteins. Synthesis of the P protein presents a problem because there is no mRNA in which its reading frame is proximal to the 5′ end. Although the position of the P reading frame, overlapping the C terminus of the core protein sequence, is reminiscent of the retroviral arrangement of *gag* and *pol* genes, there is

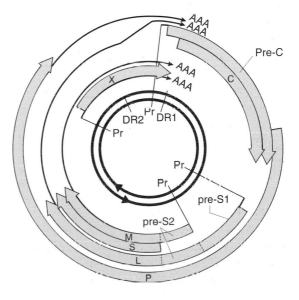

Fig. 9.9 Hepatitis B virus genome map. The closed circular DNA is transcribed from four promoters (Pr); arrowheads indicate 3′ ends. For details of protein expression, see text. Pre-S1, pre-S2 are the designations given to the upstream portions of the S reading frame which, when added to the N terminus of S, give rise to the longer L and M surface proteins. DR1, DR2 are repeated sequences shown to orientate the map (see Fig. 8.5).

no C–P fusion protein produced. Instead, a modified ribosome-scanning mechanism is believed to allow a subset of ribosomes, which load onto the mRNA at its 5′ end, to initiate translation at the start codon for the P protein. Again, this achieves a lower level of P expression relative to C, reflecting the differing requirements for these proteins.

9.11 DNA bacteriophages

Bacteriophage gene expression is controlled mainly at the level of transcription, because RNA processing is not a feature of prokaryotic gene expression, and the short half-life of prokaryotic mRNA makes transcriptional control particularly effective. Regulated transcription of phage genes might occur either through modification to the specificity of the host RNA polymerase or through the provision of a new phage-specified polymerase. Two examples of *Escherichia coli* bacteriophages that have dsDNA genomes, T7 and T4, are considered here briefly because they demonstrate strategies for achieving a controlled programme of gene expression not seen in the mammalian virus examples already described. Bacteriophage λ gene expression is covered separately in Chapter 12, focusing on lysogeny of this phage.

Bacteriophage T7

During T7 infection, mRNA molecules can be divided into two classes: early mRNA, which consists of four species, and late mRNA, consisting of eight or nine discrete species. Early mRNA is made without the need for virus-coded proteins (its transcription is therefore temporally equivalent to immediate–early gene expression by mammalian DNA viruses), whereas late mRNA, as it encodes both enzymes for genome replication and proteins for progeny particle construction, is temporally equivalent to the combined early and late phases of a mammalian virus. Studies of conditional lethal mutants show that the early gene 1 product is necessary for transcription of T7 late mRNA. This protein is a T7-specific RNA polymerase, which comprises a single polypeptide chain and is therefore much simpler than the corresponding host enzyme. The synthesis of a novel, phage promoter-specific, RNA polymerase allows T7 to inhibit host RNA synthesis and so direct the cell's resources exclusively towards progeny phage production. T7 RNA polymerase is frequently used in molecular biology research to produce RNA from cloned DNA *in vitro*.

Bacteriophage T4

The regulation of phage T4 transcription is more complex than that of T7. At least three classes of mRNA are found in T4-infected cells: immediate–early, delayed–early and late mRNA. In contrast to T7, all T4-specific RNA synthesis is sensitive to inhibitors of host DNA-dependent RNA polymerase, indicating the involvement of this enzyme in all phases of viral gene expression. However, the specificity and activity of the host RNA polymerase (subunit structure $\alpha_2\beta\beta'\omega\sigma$) are altered by covalent modification to generate this phased pattern of gene expression. By 2 min after infection, when delayed early RNA synthesis begins, 5′-adenylate is covalently added to the α subunit. By 10–15 min after infection, when late RNA synthesis starts, the mobility of the β′ subunit in electrophoretic gels has altered, suggesting that it too is modified. Furthermore, a newly synthesized polypeptide has been isolated from the core polymerase early in infection, the phage-specified ω factor. Thus, at different times, different core polymerases are present in the infected cell. These differ in their promoter specificity and/or their sensitivity to the chain termination factor (ρ) and so give rise to the programmed expression of the phage genome.

9.12 Further reading

This will be found in Chapter 22, listed under each family of viruses.

Chapter 10

The process of infection: IIIB.
Gene expression and its regulation
in RNA viruses

10.1 Introduction

The use of RNA as a genome means that the process of gene expression by RNA viruses differs from that of the host and of DNA viruses. However, an RNA genome must also be transcribed to form positive-sense mRNA, which is translated into proteins. All RNA viruses synthesize their mRNA from a negative-sense RNA molecule; those from Baltimore class 3 use the double-stranded (ds) genome as a template whereas class 4 viruses have to replicate their genomes to generate a negative-sense RNA template before transcription can occur. Class 5 viruses can use the genome RNA directly as a template for transcription. Host cells do not contain the enzymes capable of carrying out transcription from an RNA template and all RNA viruses must encode their own RNA-dependent RNA polymerases. Viruses of classes 3 and 5 carry the polymerase as an essential internal component of the virus particle and introduce it into the infected cell so that transcription can begin immediately. Class 4 virus particles do not carry a polymerase and the enzyme is generated by translation of the positive-sense RNA genome. The enzymes that carry out transcription of RNA viruses are frequently referred to as 'transcriptases' to differentiate them from the process of replication. However, as indicated in Chapter 7, both transcription and replication are carried out by the same enzyme for each virus, although they may be modified in different ways to function in the two separate processes.

10.2 The RNA viruses: Baltimore classes 3, 4 and 5

In general, all RNA virus genes are expressed simultaneously, although where several mRNAs are produced some may be more abundant than others. Thus, temporal control of gene expression, if present, is not usually as pronounced as it is for many DNA viruses. An exception is seen for the class 5 ambisense viruses in which one mRNA is produced only after genome replication has occurred. Unlike the DNA viruses, many RNA viruses produce mRNA which is either not capped or not polyadenylated, or neither capped nor polyadenylated. The lack of these structures, particularly the 5'-mRNA cap, requires the virus to use novel mechanisms to ensure translation of its mRNA.

Although most RNA viruses replicate in the cytoplasm of the host cell, many do so in the nucleus. Viruses cannot deviate from this 'choice'. As, unlike the DNA viruses,

RNA viruses do not require the host-cell DNA-dependent RNA polymerase, there are other reasons why some RNA viruses replicate in the nucleus, which are explained below. Despite the fundamental difference in the nature of the template nucleic acid, control of gene expression by RNA viruses frequently shows similarities with that seen in DNA viruses.

10.3 Reoviruses

Transcriptional regulation of gene expression

The genome of mammalian reoviruses is double-stranded RNA (dsRNA). The genome consists of 10 segments which are present in equimolar amounts, suggesting that a single copy of each is packaged in the virion (see Section 4.7). Each segment is perfectly base paired, with no overlapping single-stranded (ss) regions at either end of the molecule.

The mechanism by which reoviruses initiate the expression of their genome presented a problem to early investigators, because it was known that dsRNA could not function as mRNA. The first virus mRNAs in the cell must therefore result from transcription of the parental genome or by separation of the two strands. However, attempts to demonstrate strand separation were unsuccessful, and all cellular polymerases tested were unable to transcribe the dsRNA genome. This was not unexpected because reoviruses replicate exclusively in the cytoplasm whereas host-cell RNA polymerases are confined to the nucleus. Also, the synthesis of virus mRNA was insensitive to inhibitors of host-cell transcription such as actinomycin D. Together, this indicated that host enzymes were unlikely to be responsible for transcription of the virus genome.

Treatment of cells with inhibitors of translation such as cycloheximide before, and during, infection did not affect virus transcription, indicating that a pre-existing polymerase, carried by the virion, is involved and a dsRNA dependent RNA polymerase has been identified as an integral part of the virion particle. The polymerase is inactive in intact virions, but becomes activated at the second stage of uncoating. The activation step appears to be regulated by an outer capsid protein, called μ1c. This protein, which is modified by phosphorylation and glycosylation, is cleaved specifically during a second stage of uncoating, to leave a residual component, the ∂ polypeptide, attached to the cores. The polymerase is probably a multisubunit complex intimately associated with the structure of the core because transcription is conservative, i.e. single-stranded mRNA is copied using only one strand of RNA as template, and this mRNA is *exactly* the same length as the genome RNA. Added dsRNA is not recognized by the polymerase, suggesting that the genome is transcribed while still in the core. Newly synthesized mRNA leaves parental cores by way of channels through the external protein spikes (see Fig. 3.14). The newly synthesized mRNA from transcribing core particles has been visualized by electron microscopy (Fig. 10.1).

The virus mRNAs made at the beginning of the infection are modified at their 5′ end by the addition of a cap (see Chapter 4), but they do not contain the 3′ polyA

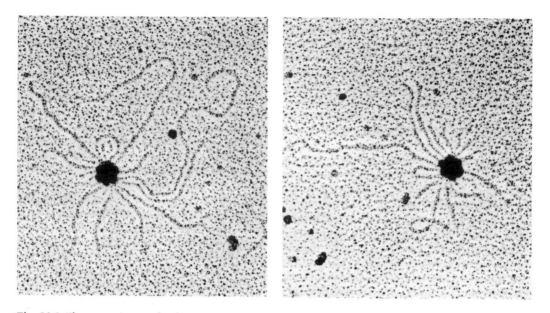

Fig. 10.1 Electron micrographs showing nascent mRNA leaving reovirus cores from the spike proteins. (From Bartlett *et al*. 1974.)

tail found on normal cellular mRNAs. The normal, cellular, capping enzymes are found only in the nucleus, but the reovirus particle contains several viral methylase activities that carry out this reaction. All of the reovirus mRNAs have the sequence GCUA at the 5' end.

After replication of the genome, as described in Chapter 7, new virus cores are generated in the cytoplasm and these also synthesize mRNA from all 10 genome segments. However, the mRNA produced from these progeny particles, in what is termed 'late transcription', does not contain a cap structure at the 5' end because the methylases in these immature virus particles are inactive. This has important consequences for translation of these mRNAs, as described below.

At 2–4 h after infection, all of the genome segments are transcribed, but some studies suggest that the mRNAs transcribed from segments 1, 6, 9 and 10 (L1, M3, S3 and S4) are predominant. Although these results remain the point of some debate, these are called *pre-early* mRNAs. However, by 6 h after infection, all of the genome segments are transcribed with equal efficiency. Thus, there appears to be some degree of temporal control of transcription, and transcription of each segment is independent of any other. The mechanism by which this control is achieved is not known, but is likely to be regulated through a host-cell repressor molecule which allows preferential transcription. One, or more, of the pre-early proteins would then cause derepression. The evidence for this is that, if cores expressing the pre-early genes preferentially are isolated from infected cells and allowed to continue transcription and translation *in vitro* in the absence of additional virus proteins, the pattern of transcription alters and all segments are transcribed efficiently; thus, the presence of a pre-

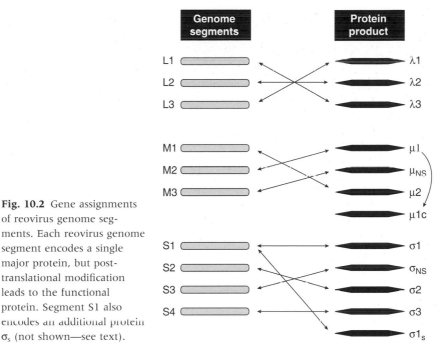

Fig. 10.2 Gene assignments of reovirus genome segments. Each reovirus genome segment encodes a single major protein, but post-translational modification leads to the functional protein. Segment S1 also encodes an additional protein σ_s (not shown—see text).

early protein is not required to activate transcription of the remaining segments. Similarly, if cores are generated *in vitro* by chymotrypsin treatment, they transcribe all segments with equal efficiency but when introduced into cells they exhibit the pre-early pattern of transcription, showing that a factor in the cells reduces transcription from some segments.

The reovirus proteins fall into three size classes, called λ, μ and σ by analogy with the L, M and S genome segments (see Section 4.7), and this implied that there might be a 1:1 relationship between each segment and a gene product. This was confirmed and any additional proteins were shown to be generated by post-translational modification of the precursor generated from each segment. The gene assignments and proposed functions of the proteins are summarized in Fig. 10.2 and Table 3.2.

Translational regulation of gene expression

An exception to the one segment–one protein rule for reovirus is seen with the S1 segment. The mRNA transcribed from the reovirus S1 segment contains two open reading frames (ORFs), each of which encodes a single protein, as described in Section 4.8 and Fig. 4.11. The relative level of the two proteins is determined by the efficiency of initiation of translation at the AUG initiation codons of the two ORFs, with the 5'-proximal AUG being used most to generate the $\sigma 1$ protein. This arrangement ensures that less of the protein encoded by the second ORF, $\sigma 1_s$, is produced.

When the 10 reovirus mRNAs are present in approximately equimolar amounts late in the infectious cycle, the proteins are made in greatly differing amounts, from very low levels of polymerase to high levels of other, major, structural proteins. This means that the various mRNAs must be translated with different efficiencies. How is this achieved? The sequence surrounding the AUG initiation codon for cellular mRNAs is important in determining the efficiency of translation. The consensus sequence, identified by M. Kozak, which is most favoured for initiation of translation, is G/ANNAUG(G), where N is any nucleotide, and the mRNAs bearing sequences closest to this are recognized most by ribosomes. In the case of the S1 segment, the Kozak rules appear to apply inasmuch as the sequence surrounding the second AUG, which initiates the $\sigma 1_s$ protein, agrees very closely with the proposed consensus, whereas the sequence surrounding the AUG for $\sigma 1$ (the first AUG) is not so good. This will encourage translation of the second ORF and enhance the production of $\sigma 1_s$, although the levels of production still do not come up to those of $\sigma 1$. However, the rate of translation of the other reovirus mRNAs does not correlate strongly with the Kozak sequence that each contains, and this alone is not sufficient to explain the different rates of translation.

In infected cell lysates, the level of translation of the virus mRNAs *in vitro* is equal. However, if the constituents of the lysate are altered such that one factor, e.g. an ion, becomes limiting, the situation becomes very much like that seen *in vivo*. It seems possible, therefore, that the level of translation of reovirus mRNAs is determined not only by the sequence surrounding the AUG, as pointed out by Kozak, but also by the ability of each mRNA to compete with each other for limiting factors. This ability is independent of the presence of a cap structure because it occurs both early and late in infection. The efficiency of competition may be determined by the sequence of the RNA, or its secondary structure, which is a consequence of the sequence. This is an example of control of gene expression at the level of translation.

In reovirus infected cells, there is no rapid shutoff of host protein synthesis and this is only gradually reduced, so that not until 10 h after infection do virus proteins form the majority of the translation products. Thus, for much of the infection, virus mRNAs are competing with existing host mRNAs—and winning. This results from a virus-induced alteration in the cellular translational process from dependence on the presence of a cap structure on the mRNA to cap-independent translation. Eventually, the uncapped viral mRNAs synthesized by the immature progeny particles are preferentially translated.

10.4 Picornaviruses

The picornavirus genome is a positive-sense ssRNA molecule of approximately 7500 nucleotides (nt.). It has a polyA tract at the 3′ end, resembling that seen in host cells, and the 5′ end has a small covalently attached protein, VPg, but no cap structure. The first event after uncoating of the picornavirus genome is translation using host-cell ribosomes. The presence of VPg and the absence of a cap mean that the picornavirus genome RNA, although resembling a normal host mRNA, cannot be translated directly on host ribosomes. During the replication cycle, the host translation system is altered so that it becomes cap independent and therefore unable to translate its own mRNA. However, this does not explain the ability of the genomic RNA to be translated immediately after uncoating. To overcome this potential barrier to gene expression, picornavirus genomes contain a region of RNA, 5′ to the AUG initiation codon, which adopts a specific three-dimensional conformation that directs ribosomes to initiate translation that is independent of a cap structure and internal to the mRNA. This region is referred to as the internal ribosome entry site (IRES). As a result of the virus-induced changes, host-cell protein synthesis is inhibited and only virus-specific synthesis takes place. Production of more mRNA is by replication of the genome, described in Section 7.3.

Control of gene expression by post-translational cleavage

The entire genome of picornaviruses is translated as a huge polyprotein that is approximately 2200 amino acids in length, which is cleaved in a series of ordered steps to form smaller functional proteins (Fig. 10.3), both structural and non-structural. Cleavage starts while the polyprotein is still being synthesized and is carried out by virus-encoded proteases, which also cleave themselves from the growing polypeptide chain. Only by inhibiting protease activity or altering the cleavage sites through the incorporation of amino acid analogues can the single polyprotein be isolated. Cleavage can

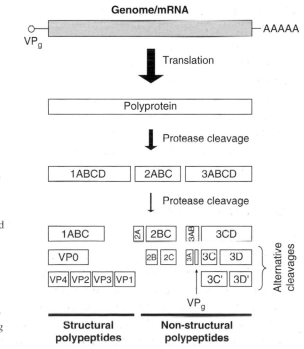

Fig. 10.3 Representation of the cleavages of the poliovirus polyprotein and the smaller products to yield mature viral proteins. Note that most poliovirus mRNA lacks the 5′ VPg protein present on virion RNA. 3C′ and 3D′ are produced by cleavage of 3CD at an alternative site to that producing 3C and 3D.

also be demonstrated by pulse–chase analysis. In theory, all picornavirus proteins should be present in equimolar proportions, but this is not found in practice, and it is the rate of cleavage that controls the amount of each protein produced. The rates of cleavage of the precursors vary markedly and this allows a great degree of control over the amounts of each protein produced. A differential is also introduced by degradation of some of the viral proteins, e.g. it is known that the virus-specified polymerase activity is unstable. The proteases 2A and 3C are obtained by autocatalytic excision when P2 and P3, respectively, achieve the required conformations.

10.5 Alphaviruses

Synthesis of a subgenomic mRNA

The genome of alphaviruses consists of one positive-sense ssRNA molecule of approximately 11 400 nt. with a sedimentation coefficient of 42S. In contrast to the situation with picornaviruses, *in vitro* translation of the positive-sense genomic RNA of alphaviruses, such as Semliki Forest virus, generates non-structural but not virion proteins. In virus-infected cells, two types of positive-sense RNA are produced: one of 42S identical to the genome RNA and one with a sedimentation coefficient of 26S. The 26S RNA is approximately one-third (3800 nt.) the length of the genomic RNA. Both of these positive-sense RNA molecules have a cap at the 5′ end, a polyA tail at the 3′ end and function as mRNAs. The 26S mRNA represents the 3′-terminal portion of the genomic RNA and is translated into the structural proteins. The primary product of translation of each mRNA is processed by proteolytic cleavage to produce functional proteins, in a manner similar to that described for picornaviruses. Pulse–chase experiments and tryptic peptide maps show clearly that the small functional proteins are derived from larger precursor molecules (Fig. 10.4).

As the sequence of the 26S mRNA is contained within the genomic RNA, and the latter does not direct the synthesis of the structural proteins, this implies that there is an internal initiation site for the synthesis of virion proteins within the genomic RNA which cannot be accessed by ribosomes.

The only negative-sense virus RNA found in infected cells sediments at 42S and this must be the template from which both the 42S and the 26S RNA are transcribed. To achieve this, the virus polymerase must be able to initiate transcription at both the 3′ terminus of the negative-sense template and also at a point within the template approximately one-third of the distance from the 5′ end (Fig. 10.4). These two mRNAs are co-terminal at the 3′ end.

It is not known how transcription is controlled, but the levels of 26S mRNA are very high, indicating that the internal initiation of transcription is efficient and ensuring that the structural proteins are abundant. As the 26S mRNA can be produced only after replication of the genome, this is a late mRNA and control of gene expression in this system is determined at the level of transcription. Synthesis of the 26S mRNA appears to be an adaptation for making large amounts of structural proteins without a similarly high level of non-structural proteins.

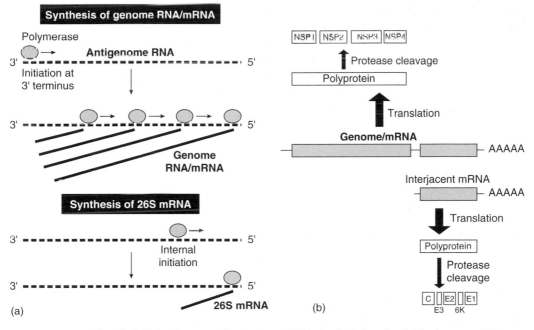

Fig. 10.4 Alphaviruses synthesize two mRNAs in the infected cell. The largest mRNA also acts as the genome and the smaller, 26S mRNA represents the 3' end of the genome. (a) The mechanism for production of the two mRNAs is shown. (b) The proteins encoded by each mRNA and the pattern of proteolytic cleavage that results in the generation of functional proteins.

10.6 Coronaviruses

Ribosomal frameshifting

Coronaviruses contain the largest RNA genomes described to date, with some reaching almost 30 000 nt. The single, linear, positive-sense, ssRNA molecule has a 5' cap and a 3' polyA tail. The first event after uncoating is translation of the virus genome RNA to produce the virus polymerase. This generates dsRNA using the plus strand as template, and then transcribes positive-sense RNA from a negative-sense RNA template in a replicative–intermediate (RI) replication complex (see Section 7.3).

The first translation event directs the synthesis of two proteins. The smallest, the 1a protein, is translated from the first ORF, which represents only a small proportion of the molecule. A second, larger protein is also produced, although in smaller amounts, and this is thought to contain the enzymatic component of the polymerase. The way in which the two proteins function together is not known. Nucleotide sequence analysis identified an ORF near the 5' end of the genome RNA that is capable of synthesizing the 1a protein, but no ORF that could synthesize the larger protein. Instead, a second ORF, called the 1b ORF, lacking an AUG initiation codon and overlapping with the 3' end of the 1a ORF, is present. The 1b ORF is in the −1 reading frame with respect to the 1a ORF. The large protein is generated by a proportion of ribosomes that initiate translation in the 1a ORF, switching reading frames in the region of the overlap of the 1a and 1b ORFs while continuing uninterrupted protein synthesis to generate a fusion, 1a–1b protein. A similar event occurs in the synthesis of certain retrovirus Gag–Pol fusion proteins (see Section 9.9). The amount of the fusion protein produced is determined by the frequency of the frameshifting event. This is an example of translational regulation of gene expression.

The frameshifting event in coronaviruses is determined by the presence of two structural features in the genomic RNA. The first is a 'slippery' sequence at which the frameshift occurs and the second is a three-dimensional structure called a pseudoknot, in which the RNA is folded into a tight conformation. These two features combine to effect the frameshift.

Functionally monocistronic subgenomic mRNAs

Coronaviruses carry the alphavirus strategy of producing a subgenomic mRNA to extremes, producing not two but several mRNAs. For mouse hepatitis virus (MHV), six mRNAs in addition to the genome are found in infected cells. The sizes of these mRNAs added together exceed that of the total genome. This observation was explained by sequence analysis, which showed that the sequences present in a small mRNA were also present in the larger mRNAs, and that all of the mRNAs shared a common 3' end with each other and the genome RNA. This is described as a nested set of mRNAs. In addition, each mRNA has a short sequence at the 5' end which is identical for each (Fig. 10.5). Each mRNA has a unique first AUG initiation codon, allowing translation of a unique protein. Although the largest (genome) mRNA contains all of the coding sequences for all of the proteins, the first ORF is used preferentially. For some MHV mRNAs, more than one protein is produced by each individual mRNA, although the second protein is present in low levels, as with certain reovirus and paramyxovirus mRNAs. The mRNAs are present in different quantities with respect to each other—some abundant, some less so—but the ratios of each do not alter during the infectious cycle. There is therefore no temporal control of gene expression.

The subgenomic mRNAs cannot be produced using the cell's splicing enzymes because these are located in the nucleus and coronaviruses replicate in the cytoplasm. The mechanism by which they are generated is novel, although there is some controversy over the details. Although it is possible that the negative-sense antigenome RNA is used as the template for synthesis of mRNA, most evidence points to the genome RNA as the primary template. The polymerase molecule begins transcription at the 3' end of the genome RNA, and at specific points the polymerase terminates and the complex of protein and negative-sense RNA dissociates, at least partially, from the template. The nascent RNA is then taken to a point near the 5' end of the genome RNA and synthesis continues to the end. This produces a series of negative-sense RNA molecules which contain sequences complementary to the 3' end of the genome and to a small region from the 5' end. The transfer mechanism is not understood and may be related to the secondary structure of the template RNA. The frequency with which

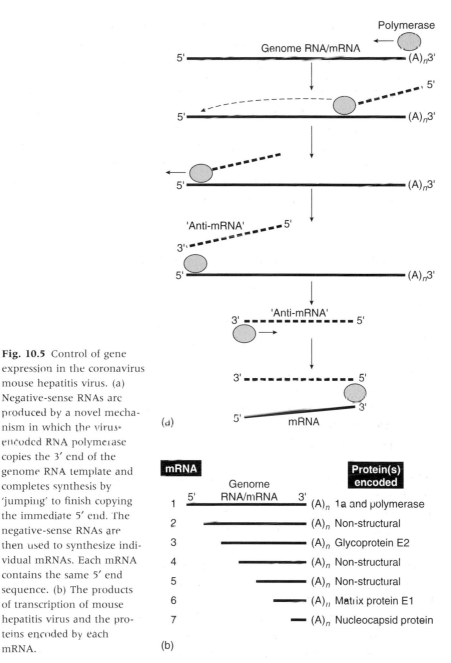

Fig. 10.5 Control of gene expression in the coronavirus mouse hepatitis virus. (a) Negative-sense RNAs are produced by a novel mechanism in which the virus-encoded RNA polymerase copies the 3′ end of the genome RNA template and completes synthesis by 'jumping' to finish copying the immediate 5′ end. The negative-sense RNAs are then used to synthesize individual mRNAs. Each mRNA contains the same 5′ end sequence. (b) The products of transcription of mouse hepatitis virus and the proteins encoded by each mRNA.

the polymerase terminates within the template determines the abundance of each RNA. The negative-sense RNA molecules are then used as templates to make faithful positive-sense copies that act as mRNAs (see Fig. 10.5). In effect, each mRNA is the product of the replication of a subgenomic negative-sense RNA. Each mRNA is then transcribed and translated into the appropriate protein. Coronavirus gene expression is therefore controlled primarily at the level of transcription.

Family	No. of segments/genome
Orthomyxoviridae	8
Bunyaviridae	3
Arenaviridae	2
Rhabdoviridae	1
Paramyxoviridae	1
Bornaviridae	1
Filoviridae	1

Table 10.1 The segmented class 5 viruses.

10.7 Negative-sense RNA viruses with segmented genomes

Class 5 viruses have a variable number of ssRNA segments making up their genome (Table 10.1), although gene expression is predominantly, but not exclusively (see Section 4.8), from monocistronic mRNAs. For viruses with segmented genomes, each segment must be transcribed into mRNA offering the opportunity for transcriptional control of gene expression. Normal cells do not contain enzymes capable of generating mRNA from an RNA template, and all viruses with non-segmented negative-sense RNA genomes synthesize mRNA using virus-encoded transcription machinery, which is carried into the cell with the infecting genome. Almost all of the negative-sense RNA viruses replicate in the cytoplasm. The exceptions are the orthomyxoviruses and Borna disease virus.

10.8 Orthomyxoviruses

Transcriptional regulation of gene expression

Of the orthomyxoviruses, most is known about the type A influenza viruses of humans and other animals. An unusual feature of influenza virus replication is that it occurs in the nucleus of the infected cell. Immediately after infection, the uncoated virus particle is transported to the nucleus, where viral mRNA is synthesized, and, later in infection, certain of the newly synthesized proteins migrate from their site of synthesis in the cytoplasm into the nucleus. As passage of molecules across the nuclear membrane is a highly selective process, this constitutes a level of control that is unique to eukaryotic cells.

Influenza viruses can replicate only in cells with a functional, cellular, DNA-dependent, RNA polymerase II (RNA pol II), although this enzyme cannot transcribe the virus genome. Two types of positive-sense RNA are generated in infected cells—mRNA and template for replication (antigenome RNA). These two types of RNA can be differentiated from each other by a number of criteria (Table 10.2).

The production of mRNA from the genome initiating the infection, termed 'primary transcription', uses the virion-associated polymerase and it is this step that requires active pol II. Each of the eight genomic RNA segments is used as a template for tran-

Table 10.2 Differences between the influenza virus mRNA and antigenome positive-sense RNA.

mRNA	Antigenome RNA
Shorter than the template genome segment	Exact copy of the template genome segment
Contains 3′ polyA tail	No 3′ polyA tail
Contains 5′ cap	No 5′ cap
Synthesis is insensitive to inhibitors of protein synthesis	Synthesis requires continuous virus protein synthesis

scription of a monocistronic mRNA, using a virus-encoded polymerase complex. Each genomic RNA segment contains its own signals to initiate transcription. The first 13 bases at the 5′ end of every segment are identical, and similarly the last 12 bases at the 3′ end of every segment are identical, but different from bases 1–13 at the 5′ end. The signal for initiation of transcription of the genome segments is contained in the 3′-terminal sequences.

The method of influenza virus transcription is unusual and explains the reliance on host pol II. The ribonucleoprotein (RNP) complex that carries out the transcription process has no capping or methylation activities associated with it, but the virus mRNAs have a cap 1 structure at the 5′ end. Protein PB2, a component of the RNP complex, binds to the capped 5′ end region of newly synthesized host-cell mRNAs. The host mRNA is then cleaved 10–13 bases from the cap, preferably after an A residue, although occasionally after a G. This small segment of RNA is then used as a primer for influenza virus mRNA synthesis. The proteins PB1 and PA of the RNP complex initiate transcription and extend the primer, respectively. Transcription is terminated at a specific point on the template at a homopolymer run of uracil residues, 17–22 residues from its 5′ end. These are copied into A in the mRNA and the virus polymerase continues to add A residues, in a reiterative fashion, to generate a polyA tail before the mRNA dissociates from the template. Synthesis of the antigenome positive-sense RNA does not use a primer and generates an exact copy of the template RNA. Influenza virus transcription is shown in Fig. 10.6.

Initially, similar amounts of each influenza virus mRNA are produced (as are the proteins) but, within an hour, the levels of mRNAs encoding the nucleoprotein (NP) and NS1 (a non-structural protein) increase greatly with respect to the others. Later, after genome replication has started, the levels of mRNAs encoding the haemagglutinin (HA), neuraminidase (NA) and matrix (M1) proteins predominate. Thus, although all of the genes are expressed throughout infection, there is some temporal control of transcription determined by the relative rates of transcription of some segments.

Control of gene expression by mRNA splicing

Superimposed on the basic pattern of transcription of a single mRNA from each influenza virus genome segment is a system that permits the synthesis of two additional proteins from influenza virus segments 7 and 8 by the process of splicing. Transcription

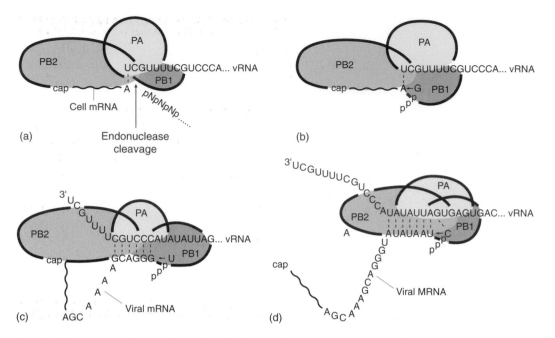

Fig. 10.6 The process of transcription from influenza virus genome segments. The three virus proteins, PB2, PB1 and PA, act together to remove cap structures, together with 10–13 nucleotides from the 5′ end of host cell mRNA, and use these as primers to initiate transcription.

of each of these segments produces an mRNA that encodes a specific protein—M1 and NS1, respectively. An internal section is removed from a proportion of the primary transcripts and the two remaining molecules are ligated (Fig. 10.7). In both cases the splicing events delete a substantial portion of the first ORF, leaving the AUG initiation codon in place. The result is that an alternative ORF, in a different reading frame and not previously accessed by ribosomes, is fused to the first few codons of the M1 or NS1 ORFs. The newly generated ORF directs the synthesis of novel proteins, called M2 and NS2. This is similar to the situation for the polyomaviruses (see Section 9.3). An additional, rare, mRNA is generated from the segment 7 primary transcript by an alternative splicing process. The splice removes the first AUG codon in the primary transcript, which is used to synthesize the M1 and M2 proteins. The first ORF in the alternatively spliced segment 7 mRNA contains only nine codons. The putative protein produced by translation of this mRNA is called M3, but it has not been detected in infected cells and its role in the infectious cycle is not known. The levels of expression of NS2 and M2 are determined by the frequency of the post-transcriptional splicing event.

Post-translational protein cleavage

Although some viruses, such as the picornaviruses, require proteolytic cleavage to produce functional proteins, most viruses do not encode proteases specifically to

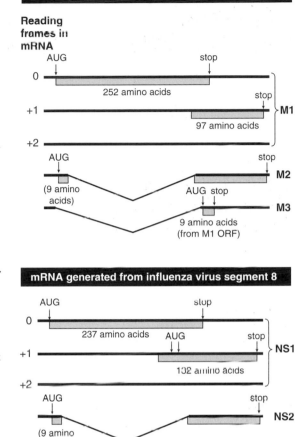

Fig. 10.7 Splicing in influenza virus: synthesis of mRNAs encoding M1 and M2 proteins from segment 7 and NS1 and NS2 proteins from segment 8 of influenza virus type A. The shaded areas represent the coding regions. The reading frames of the unspliced mRNAs are shown but only the first open reading frames (ORFs), encoding M1 and NS1 from segments 7 and 8, respectively, are used in these mRNAs. Note that the two major products from each pair of mRNAs share a short common amino-terminal amino acid sequence but differ in the majority of the sequence. The M3 protein has not been detected in infected cells.

cleave their proteins, and it might be expected that post-translational cleavage would not be involved. However, post-translational cleavage is frequently the final maturation event for viruses, e.g. after its synthesis, the influenza virus HA glycoprotein undergoes a single protease cleavage (Fig. 10.8; see Section 11.7). Uncleaved molecules occur in cells that are deficient in protease activity, but are still incorporated into virus particles that are morphologically normal and can agglutinate red blood cells. However, such virus particles are non-infective until the HA protein is cleaved. The cleaved HA changes conformation and the fusion peptide is now free at the N terminus of HA2 (see Fig. 3.17). The protease is a host-cell enzyme located in the Golgi network, through which the HA is translocated. The restriction of such enzymes to specific tissues such as the respiratory tract determines the site of productive virus replication. Some viruses, such as Sendai virus, require cleavage activation of proteins which can occur after the virus has left the host cell rather than inside the cell as for influenza virus.

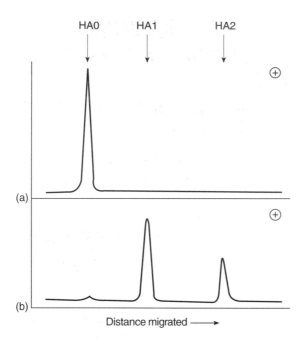

Fig. 10.8 Cleavage of influenza virus haemagglutinin (HA). (a) Electrophoresis of an uncleaved molecule from non-infectious virus grown in cells deficient in protease activity. (b) Haemagglutinin molecules isolated from the same virus incubated with 100 µg/ml trypsin for 30 min at 37°C. After incubation with trypsin the virus becomes infectious.

10.9 Arenaviruses

Ambisense coding strategy

Although arenaviruses appear similar to conventional segmented negative-strand viruses with just two (L and S) genomic RNAs, these have an unusual gene organization. A virus-encoded enzyme transcribes only the 3′ part of the genome RNAs into a capped, polyadenylated subgenomic mRNA. In the case of the S RNA this mRNA encodes the nucleocapsid (N) protein; for the L RNA it encodes the large polymerase protein. Following replication the 3′ end of the antigenome copy of each RNA segment is used as a template for transcription. This generates a subgenomic mRNA from the S antigenome RNA which encodes a protein called GPC, the precursor to the virus structural glycoproteins. A protein called Z is encoded by the mRNA copied from the L antigenome RNA. Because the GPC and Z mRNAs can only be synthesized after replication they are, by definition, late mRNAs. This mechanism means that both the genome and antigenome act as negative-sense RNA templates for transcription and are termed 'ambisense' RNA (Fig. 10.9). Some bunyaviruses such as Punta Toro virus also use an ambisense strategy for their S genome segment.

10.10 Negative-sense RNA viruses with non-segmented, linear, single-stranded genomes—rhabdoviruses and paramyxoviruses

Transcriptional regulation of gene expression

Of the rhabdoviruses, most is known about vesicular stomatitis virus (VSV). The proposed model for VSV transcription is used as a paradigm for all other viruses with

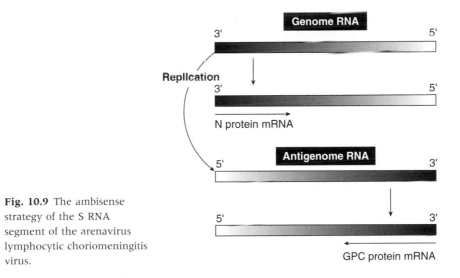

Fig. 10.9 The ambisense strategy of the S RNA segment of the arenavirus lymphocytic choriomeningitis virus.

non-segmented negative-sense RNA genomes, although the precise molecular details of each virus in terms of regulatory sequences, etc. differ. Transcription of the VSV genome yields five separate monocistronic mRNAs and gene expression is controlled at the level of transcription, with the abundance of mRNAs determining the abundance of proteins. A similar pattern of transcription is seen with the paramyxoviruses, although these may encode 6, 7 or 10 mRNAs, depending on the virus.

The ribonucleoprotein (RNP) complex that carries out the transcription (and replication—see Section 7.4) process of VSV consists of the genome RNA in association with large amounts of the virus nucleoprotein (N or NP), lesser amounts of the virus phosphoprotein (P) and a few molecules of a large (L) protein. The L protein is thought to contain the catalytic sites for RNA synthesis, and possibly also capping of the mRNA, and is often referred to as the polymerase. The RNP complex can initiate transcription only at the 3′ end of the genomic template RNA. For VSV, transcription begins with the production of a small, 49 nt., uncapped, non-polyadenylated RNA which is an exact copy of the terminus. This is the leader RNA for which no function is known. Each of the remaining five transcription units on the genome RNA is flanked by consensus sequences that direct the polymerase first to initiate and subsequently to terminate transcription. As the polymerase moves along the genome it meets a transcription initiation signal and begins mRNA synthesis.

At some point during this process, a cap is added, presumably by a component of the RNP complex because no capping enzymes are found in the cytoplasm. At the consensus transcription termination signal, which includes a homopolymer uracil tract, the polymerase adds a polyA tail as described for influenza virus. At this point, the mRNA dissociates from the template and is removed for translation. The polymerase then does one of two things. A proportion, possibly 50%, of the polymerase molecules also dissociate from the template and, having done so, can rebind only at the 3′ terminus where the transcription process begins again. The remaining polymerases move along the genome without transcribing, until they encounter the next

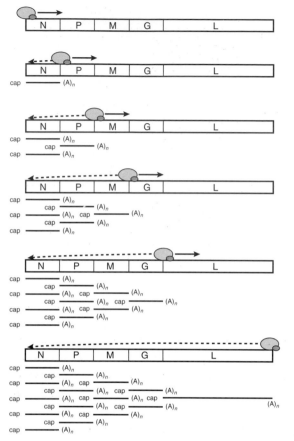

Fig. 10.10 Diagrammatic representation of the sequential transcription process of the rhabdovirus vesicular stomatitis virus (VSV). Transcription occurs in a sequential, start–stop process described in the text. The mRNAs are separated by non-transcribed regions, the lengths of which are virus specific.

consensus transcription initiation signal where they begin to transcribe a region of the genome. At the end of this transcription unit, the polymerase exercises the same two options of either dissociation or translocation followed by reinitiation. The result of this process is that the mRNAs are synthesized sequentially in decreasing proportions as the polymerase moves towards the 5′ terminus of the genome (Fig. 10.10). The relative abundances of the proteins reflect those of the mRNAs. A similar strategy is used by paramyxoviruses such as Sendai virus.

Translational regulation of gene expression

The mRNA encoding the phosphoprotein (P protein) of several, but not all, paramyxoviruses contains two ORFs. The 5′-proximal AUG codon directs translation of a large ORF encoding the P protein (Fig. 10.11). The second ORF, initiated by the next AUG, directs the synthesis of a protein called C. The amino acid sequences of the P and C proteins are completely different, similar to the situation for the reovirus σ1 and σ1$_S$ proteins. However, analysis of the proteins produced by translation of the Sendai virus P mRNA indicates that several other proteins are also generated. Initia-

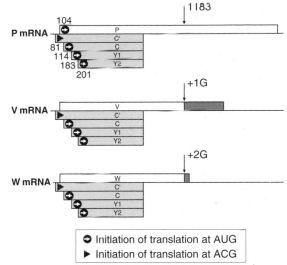

Fig. 10.11 Arrangement of the overlapping open reading frames (ORFs) of the Sendai virus P gene mRNA and the various potential protein products. The relative locations and protein products of the ORFs are indicated. The arrow shows the insertion site for non-templated G residues. Initiation of translation at AUG and ACG codons is indicated. The numbers refer to the nucleotide positions of the first nucleotide of the relevant codon.

tion of translation at the second and third AUG codons in the C protein ORF generates proteins called Y1 and Y2, respectively (Fig. 10.11). As these proteins are translated from the same reading frame as that used for the C protein, the Y1 and Y2 proteins are identical in sequence to the carboxyl terminus of the C protein. However, the differences at the amino termini of the three proteins mean that they are likely to have different functions in the virus replication cycle. This is similar to the papillomavirus E2 proteins described in Section 9.4.

The Sendai virus P mRNA also encodes a protein that is initiated at a codon ACG upstream of the first AUG codon, which initiates synthesis of the P protein. Initiation of translation at this point uses the same reading frame as that for the C protein and the protein produced, called the C′ protein, contains additional amino acids at the amino terminus when compared with the C protein. As with the Y1 and Y2 proteins, this difference may be sufficient to give the C′ protein a unique function. Use of a codon other than AUG for initiation of translation is rare, but is also seen in certain retroviruses.

Non-templated insertion of nucleotides during transcription

Nucleotide sequence analysis of mRNAs transcribed from the Sendai virus P gene identified a novel strategy for gene expression, which is also used by many other paramyxoviruses. During transcription of any gene it is essential that the polymerase makes a faithful copy of the template to prevent the introduction of mutations. However, RNA synthetic enzymes have no proofreading capacity. When the Sendai virus P gene is being transcribed, most of the mRNA produced is a faithful copy of the template. However, in approximately 30% of molecules an additional G nucleotide is inserted at a precise point in the ORF encoding the P protein. The mRNA with the additional G residue, when translated, directs the synthesis of a novel protein which has the same amino-terminal sequences as the P protein but a unique carboxyl terminus (see Fig. 10.11). The novel protein is called V. Strangely, some paramyxoviruses encode the V protein from the mRNA faithfully copied from the genome template and the P protein from the mRNA with the additional base(s). The frequency of the non-templated insertion event determines the relative abundance of the novel protein. The function of the V protein is not yet known, but it is thought to be involved in the disease process of the virus. Addition of two G residues produces a novel protein, W, with a unique carboxyl terminus.

Post-translational protein cleavage

In paramyxoviruses, two glycoproteins—the attachment and fusion (F) proteins—are inserted into the lipid bilayer. The F protein is inserted with the carboxyl terminus on the external surface of the cell and, eventually, the virus (a type I membrane protein), whereas the amino terminus of the attachment protein is external (a type II membrane protein). The F protein is synthesized as a precursor, F_0, which has an amino-terminal hydrophobic signal sequence cleaved off by a host-cell protease during insertion into the cell membrane. A further cleavage event, also carried out by a host-cell protease, this time in the Golgi network, is required for the F protein to become functional. The two components of the F protein, called F_1 and F_2, generated by this cleavage activation event are covalently attached to each other by disulphide bonds. The attachment protein is not cleaved.

10.11 RNA bacteriophages

The genomes of the small, positive-sense, ssRNA bacteriophages, such as MS2 and R17, code for only three proteins: phage coat protein, maturation protein (A protein) and the enzyme RNA synthetase in the order 5'–maturation protein–coat protein–synthetase–3'. It might be thought that, with such limited genetic potential, these phages might reproduce without the use of any control mechanisms. This is not so. Analysis of radioactively labelled phage proteins from infected cells shows that the coat protein is synthesized in great excess over the other two proteins. In fact, the three proteins are produced in the ratio of 20 coat proteins : 5 synthetase molecules : 1 maturation protein.

Translational regulation of gene expression

It now appears that differences in the frequency of initiation of synthesis of the three proteins is the basis for regulation of RNA phage development. The relative amounts of coat protein, synthetase and maturation protein initiated in *Escherichia coli* extracts depend on the integrity and conformation of the viral RNA. With native RNA isolated from phage particles, synthesis of only coat protein is initiated in *in vitro* translation assays. However, alteration of the conformation of the viral RNA by formaldehyde treatment or fragmentation of the RNA allows all three proteins to be initiated. The similar effects of these two treatments strongly suggest that specific conformational features in native RNA restrict initiation at the other two sites.

Determination of the nucleotide sequence of an extensive stretch of the coat protein and adjacent synthetase cistrons has revealed a probable region of hydrogen bonding, 21 nt. long, between codons 24 and 32 of the coat cistron and the synthetase initiation site. Hydrogen bonding in this region could account for the inability of ribosomes to bind to the synthetase initiation site. Furthermore, an amber mutation in any of the first 24 codons of the coat cistron would prevent ribosomes opening up this double-stranded region, thus accounting for the polar effect of some coat protein mutants on initiation of synthetase.

In contrast to synthetase formation, initiation of the maturation protein is not affected by conformational changes that occur during translation of the coat gene. Thus, amber mutations in the coat cistron have no effect on production of maturation protein *in vitro*. As described for picornaviruses (see Section 7.3), much of the RNA in the infected cell exists as RI, and this consists of an intact negative-sense template, with one or more nascent single-stranded positive-sense RNA molecules extending from it. *In vitro*, RI directs synthesis of about five times as much maturation protein, relative to the total protein initiated, as does ssRNA isolated from phage particles. Presumably, the conformation of the 5' end of the ssRNA chains in the RI is different from that of completed RNA molecules, allowing ribosomes access to the initiation site for the maturation protein on the nascent strands. Only a fraction of the maturation protein molecules initiated on the RI *in vitro* are actually completed, in contrast with coat protein, which is the major product directed by RI. This failure to produce complete molecules of maturation protein could be explained by assuming that only the shortest nascent strands have open initiation sites for the maturation protein, and that these strands are not long enough to contain the entire cistron. According to this hypothesis, as each nascent RNA chain is synthesized in the infected cell, a ribosome attaches to the maturation protein initiation site in the short 5' tail; as the RNA chain is elongated, that ribosome proceeds to translate the maturation protein cistron, but folding of the elongated RNA strand (Fig. 10.12) prevents additional ribosomes from initiating at the initiation site.

It appears that conformation of the phage RNA is not the only factor influencing the independent translation of the three genes. There is some evidence indicating that the specificity of the ribosomes and initiation factors is also important. As the ribosome-binding sites differ in nucleotide sequence for each phage cistron, it is possible to imagine that the same ribosome can recognize all three initiation sites but bind with different affinity to each.

None of the three control systems so far described accounts for the observed cessation of synthesis of the synthetase protein late in infection. This is explicable by a fourth mechanism of translational control—the specific binding of phage proteins to the messenger. When *E. coli* is infected with phage-carrying mutations in the coat protein, there is an enhancement of synthetase formation, suggesting that coat protein acts as a translational repressor. Furthermore, when coat protein is incubated with phage RNA and the RNA is then used as a messenger *in vitro*, the formation of syn-

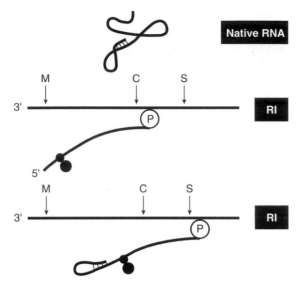

Fig. 10.12 Model to account for the linear rate of production of maturation protein. In native RNA the initiation site for maturation protein synthesis (M) is not available to the ribosome because of hydrogen bonding. During replication, an RNA tail is produced at the 5′ end in which the initiation site is available to the ribosome. As the nascent RNA chain grows in length, hydrogen bonding again prevents ribosome binding. P, polymerase; C, coat protein; S, synthetase.

thetase is specifically inhibited. By way of contrast, the presence of coat protein does not affect the initiation or elongation of either the coat protein or the maturation protein. Thus, the role of coat protein as a translational repressor of synthetase formation has been established.

10.12 Further reading and references

Bishop, D. H. L. (1986) Ambisense RNA genomes of arenaviruses and phleboviruses. *Advances in Virus Research* **31**, 1–51.

Conzelmann, K.-K. (1998) Nonsegmented negative-stranded RNA viruses: genetics and manipulation of viral genomes. *Annual Review of Genetics* **32**, 123–162.

Curran, J. & Kolakofsky, D. (1999) Replication of paramyxoviruses. *Advances in Virus Research* **54**, 403–422.

Curran, J., Latorre, P. & Kolakofsky, D. (1998) Translational gymnastics on the Sendai virus P/C mRNA. *Seminars in Virology* **8**, 351–357.

Kormelink, R., de Haan, P., Meurs, C., Peters, D. & Goldbach, R. (1992) The nucleotide-sequence of the M RNA segment of tomato spotted wilt virus, a bunyavirus with 2 ambisense RNA segments. *Journal of General Virology* **74**, 790.

Lai, M. M. C. & Cavanagh, D. (1997) The molecular biology of coronaviruses. *Advances in Virus Research* **48**, 1–100.

Lamb, R. A. & Horvath, C. M. (1991) Diversity of coding strategies in influenza-viruses. *Trends in Genetics* **7**, 261–266.

Portela, A., Zurcher, T., Nieto, A. & Ortin, J. (1999) Replication of orthomyxoviruses. *Advances in Virus Research* **54**, 319–348.

Strauss, J. H. & Strauss, E. G. (1994) The alphaviruses—gene-expression, replication, and evolution. *Microbiological Reviews* **58**, 491–562.

Also check Chapter 22 for references specific to each family of viruses.

Chapter 11

The process of infection: IV.
The assembly of viruses

11.1 Introduction

In infected cells, virus-encoded proteins and nucleic acid are synthesized separately and must be brought together to produce infectious progeny virus particles (virions) in a process referred to as assembly or morphogenesis. For some viruses, there is an additional aspect when assembly takes place in the nucleus and proteins synthesized in the cytoplasm must be moved to that location. There is little information available about assembly for most viruses. Despite the diversity of virus structure (see Chapter 3), there appear to be only three ways in which virus assembly occurs. First, the various components may spontaneously combine to form particles in a process of self-assembly. In principle this is similar to crystallization, in which the final product represents a minimum achievable energy state. Second, the viral genome may encode certain proteins which, although they do not ultimately form a structural part of the virus, are required for normal assembly. Finally, a particle may be assembled from precursor proteins, which are then modified, usually by proteolytic cleavage, to form the infectious virion. In the last two cases, it is not possible to dissociate and then reassemble a mature particle from its constituent parts. The presence of a lipid envelope in some viruses, surrounding a nucleocapsid core, introduces an additional aspect to assembly of an infectious particle because this must be acquired separately from the main assembly process.

Introduction of the virus genome into the progeny virus particle is a critical step in the assembly of viruses. There are two processes by which this can be achieved. Reconstitution and other studies have shown that, for some viruses, the genome which forms part of the infectious particle plays an integral role in assembly, acting as an initiating factor with the particle components forming around the nucleic acid. For other viruses, the particle, or a precursor of the particle, is formed first and the nucleic acid is then introduced at a late stage in the assembly process. The first process obligatorily involves the interaction of virus-encoded structural proteins with specific nucleotide sequences in the genome, referred to as packaging signals. Although the second process may require packaging signals to be present in the genome, this is not always the case.

The structure of virus particles, features of which are considered in Chapter 3, determines, to some extent, the process of assembly. Recent advances in the determination of the three-dimensional structures of virus particles by cryoelectron microscopy and X-ray crystallography have shed more light on the precise nature of interactions between virion components. This has led to suggestions of how some of these components may come together during the assembly process.

11.2 Self-assembly from mature virion components

Absolute proof of self-assembly requires that purified viral nucleic acid and purified structural proteins, but no other proteins, are able to combine *in vitro* to generate particles that resemble the original virus in shape, size and stability, and are infectious. A critical step in demonstrating assembly *in vitro* is the process used to effect disassembly. Disassembly should release the subunits in such a way that they retain the ability to reassociate in a specific manner in order to assemble the virus particle. Ideally, the disassembled constituent monomers of the virion should not be denatured by the disassembly process. This has been demonstrated for very few viruses to date.

For many viruses, it is not possible to demonstrate spontaneous self-assembly even though it is suspected that it is an integral aspect of the process, e.g. the assembly of a virus may be spontaneous but acquisition of infectivity then requires a maturation event which modifies one of the structural proteins after assembly has occurred. After dissociation of purified infectious particles, the modified protein may not be able to interact spontaneously with the other components to re-form the virion. Similarly, where a virus is enclosed in a lipid envelope, the disassembly process is likely to destroy this structure so an infectious particle cannot be regenerated.

11.3 Assembly of viruses with a helical structure

The best studied example of self-assembly is the *in vitro* reconstitution of tobacco mosaic virus (TMV), which has been used as a paradigm for the assembly of many other viruses with helical symmetry.

Assembly of TMV

The TMV particle consists of a single molecule of positive-sense single-stranded (ss) RNA, embedded in a framework of small, identical, protein molecules (A protein), arranged in a right-handed helix (see Chapter 3), with each protein binding to three nucleotides of RNA. As already outlined in Chapter 1, TMV can be disassembled to yield protein and RNA components, which can be reassembled *in vitro* to yield active virus. However, the isolated protein, free from any RNA, can also be polymerized into a helical structure, indicating that bonding between the subunits is a specific property of the protein. Although the most likely model for the assembly of the virus would be for the protein molecules to arrange themselves like steps in a spiral staircase, enclosing the RNA as a corkscrew-like thread, research has indicated that the assembly of TMV is a much more complicated process in which the genome RNA plays an essential role.

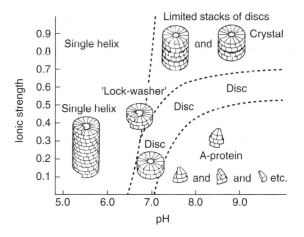

Fig. 11.1 Effect of pH and ionic strength on the formation of aggregates of tobacco mosaic virus (TMV) A protein.

In solution, TMV A protein forms several distinct kinds of complex, depending on the environmental conditions, particularly ionic strength and pH (Fig. 11.1). The complexes differ in the number of individual proteins that constitute them. Of these complexes, the disc structure is considered the most important, because it is the dominant one found under physiological conditions. The discs contain 17 subunits per ring, which is close to the 16.34 per turn of the helix in the virus particle, so the bonding between the subunits is probably very similar. A rod-like particle built of a succession of stacked rings could arise as a variant of the normal helical structure in which the protein subunits would have a similar bonding pattern to that in the virus, but in the absence of RNA–protein interactions there could be small local differences in packing so that turns of the helix are transformed into closed rings. However, this does not explain why the discs consist of only two rings rather than generating long filamentous structures. The absence of filaments suggests that the discs do not readily extend.

Careful analysis of electron micrographs reveals that there is a slight axial alteration at the edges of the subunits, such that a double disc presents a slightly different region to the underside of any further disc adding to it (see Fig. 11.4), which may prevent the addition of further discs at this pH. When discs are taken to a lower pH, aggregation progressively occurs, with the appearance of short rods made up of imperfectly meshed sections of two helical turns; after many hours, these can combine to give the regular virus-like structure (Fig. 11.1). This is thought to result from the lower pH reducing the repulsion between the carboxylic acid side chains of two adjacent amino acids, leading to a conversion of the disc into a 'lock-washer'. During the in vitro reconstitution of the virus, the conversion to the helical mode could be brought about by interaction of the protein with RNA, which provides the additional energy necessary for the stabilization of the helical form.

When A-protein subunits and small aggregates are mixed with TMV RNA, polymerization is slow and formation of virus requires about 6 h. However, when discs are mixed with RNA under the same conditions, polymerization is rapid and mature virus forms within 5 min. Addition of small aggregates as well as discs to the RNA does not

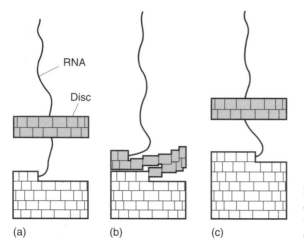

(a)   (b)   (c)

Fig. 11.2 Simple model for assembly of tobacco mosaic virus (TMV). (See text for details.)

increase the rate of polymerization. These results strongly suggest that discs are the normal precursor for the assembly of virus. A possible model for the assembly of TMV is shown in Fig. 11.2 (but note that an alternative model is presented later). A disc is added to the growing helix (Fig. 11.2a) and this is converted to the lock-washer form (Fig. 11.2b), as a result of neutralization of the adjacent carboxyl groups by interaction with the viral RNA. After conversion to the lock-washer form, the subunits progressively unroll, entrapping the RNA in the groove between successive turns of the helix until the structure is ready to receive another disc (Fig. 11.2c). In this way the RNA is contained within the helical structure.

When purified TMV RNA is mixed with limited amounts of viral coat protein in the form of discs, the RNA is incompletely encapsidated. Treatment of the incomplete structure with nucleases identified a unique region of the RNA which is resistant to digestion. The protected RNA consists of fragments up to 500 nucleotides (nt.) long. The shortest fragments define a core about 100 nt. long common to all the fragments, whereas larger ones are extended by up to 400 nt. in one direction and up to 30 nt. in the other. These data are interpreted as showing that assembly is initiated at a unique internal packaging site on the RNA, and that growth occurs bidirectionally but at greatly unequal rates. The major direction of assembly would be $3' \rightarrow 5'$, and consistent with this view is the finding, from sequence analysis of the protected fragments, that the packaging site is close to the 3' end of TMV RNA. Computer-assisted sequence analysis of the packaging site suggests strongly that it exhibits a hairpin configuration (Fig. 11.3).

Although the model described above is compatible with the available data, an alternative explanation is also possible. This is the 'travelling loop' model, which suggests that the hairpin structure at the packaging site in the TMV genome RNA inserts itself through the central hole of the disc into the groove between the rings of subunits. The nucleotides in the double-stranded stem then unpair and more of the RNA is bound within the groove. As a consequence of this interaction, the disc becomes con-

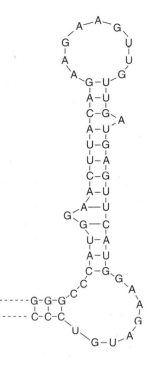

Fig. 11.3 The packaging site of tobacco mosaic virus (TMV) RNA. The loop probably binds to the first protein disc to begin assembly. The fact that guanine is present in every third position in the loop and adjacent stem may be important in this respect.

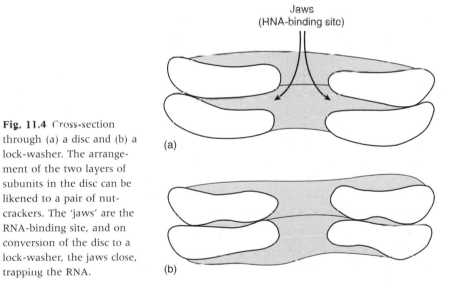

Jaws
(RNA-binding site)

(a)

(b)

Fig. 11.4 Cross-section through (a) a disc and (b) a lock-washer. The arrangement of the two layers of subunits in the disc can be likened to a pair of nut-crackers. The 'jaws' are the RNA-binding site, and on conversion of the disc to a lock-washer, the jaws close, trapping the RNA.

verted to a lock-washer structure, trapping the RNA (Fig. 11.4). The special configuration generated by the insertion of the RNA into the central hole of the initiating disc could subsequently be repeated during the addition of further discs on top of the growing helix; the loop could be perpetuated by drawing more of the longer tail of the RNA up through the central hole of the growing virus particle. Hence, the parti-

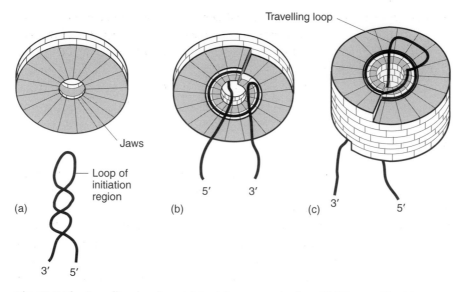

Fig. 11.5 The 'travelling loop' model for tobacco mosaic virus (TMV) assembly. (a) Nucleation begins with the insertion of the hairpin loop of the packaging region of TMV RNA into the central hole of the first protein disc. The loop intercalates between the two layers of subunits and binds around the first turn of the disc. (b) On conversion to the lock-washer, the RNA is trapped. (c) As a result of the mode of initiation, the longer RNA tail is doubled back through the central hole of the rod, forming a travelling loop to which additional discs are added rapidly.

cle could elongate by a mechanism similar to initiation of packaging, only now instead of the specific packaging loop there would be a 'travelling loop' of RNA at the main growing end of the virus particle (Fig. 11.5). This loop would insert itself into the central hole of the next incoming disc, causing its conversion to the lock-washer form and continuing the growth of the virus particle.

A feature of the first model shown in Fig. 11.2 is that discs have to be threaded on to the RNA chain and this would obviously be the rate-limiting step. However, the 'travelling loop' model shown in Fig. 11.5 overcomes this problem as far as growth in the 5' direction is concerned, because incoming discs would add directly on to the growing protein rod. Discs would still have to be threaded on to the 3' end of the RNA, and thus elongation in this direction would be much slower, as has been observed experimentally. One prediction of the 'travelling loop' model is that both the 5' and 3' tails of the RNA should protrude from one end of partially assembled TMV particles. Electron micrographs of such structures have been observed. Currently, it is not known which model is correct.

11.4 Assembly of viruses with an isometric structure

As indicated in Chapter 3, the particles of all isometric viruses are icosahedral, with 20 identical faces. The process of triangulation describes the mechanism used by isometric viruses to increase the size of their capsid, with a concomitant increase in the number of capsomere subunits. As indicated above, reconstitution experiments have had only limited success and the most detailed study so far reported on the self-assembly and reconstitution of an isometric virus is that of cowpea chlorotic mottle virus (CCMV)—a plant virus (a bromovirus of Baltimore class 4). The formation of infectious CCMV particles from a stoichiometric mixture of initially separated CCMV RNA and protein affords proof of the ability of this virus to self-assemble.

Assembly of picornaviruses

Over recent years the three-dimensional structures of a number of viruses have been determined by X-ray crystallography. Many of these have been picornaviruses, including poliovirus, which have icosahedral particles, of approximately 30 nm in diameter. Poliovirus particles consist of 60 copies of each of the four structural proteins VP1–VP4. The proteins associate together in a complex and these complexes are arranged in groups of three on each of the 20 faces of the icosahedron (see Fig. 3.12). Much information is available about the sequence of events in the assembly of poliovirus which illustrates how an icosahedral virus particle can be generated.

The entire genome of poliovirus is translated as a single giant polypeptide, which is cleaved as translation proceeds into smaller polypeptides (see Section 10.4). The first cleavage generates a polypeptide called P1, which is the precursor to all four virion proteins. Synthesis of P1 is directed by the 5' end of the genome and it is synthesized completely before being cleaved, suggesting that folding is necessary for cleavage. Subsequent cleavages of P1 give rise to proteins called VP0, VP1 and VP3. These three proteins associate with each other in infected cells to produce a complex with a sedimentation coefficient of 5S. Five of the 5S complexes come together to form a 14S pentamer complex. Twelve of the 14S complexes in turn aggregate together to form an empty 73S capsid. At this point, the genome positive-sense ssRNA is added. The mechanism by which the genome is recognized is not known. The VPg protein covalently attached to the 5' end of the genome RNA may be involved in the genome acquisition, but this cannot be the only factor because the antigenome RNA produced during replication (see Section 8.3) also has VPg at the 5' end and it is not encapsidated. Consequently, it is believed that a packaging signal sequence must be present in the genome RNA. After the RNA is encapsidated, VP0 is cleaved to yield the virion proteins VP2 and VP4 (see Section 11.7) with an alteration in the sedimentation coefficient of the particle. VP4 is located inside the particle with the other three proteins on the particle surface. As a result of this late cleavage event, dissociation and reassembly to form infectious particles *in vitro* are not possible. These steps, summarized in Fig. 11.6, clearly show that the poliovirus particle can spontaneously assemble in the cell and that the genome RNA is added to an immature particle rather than being an integral component in the assembly process such as is seen for TMV.

Assembly of adenoviruses

Adenovirus particles range in size from 70 to 90 nm in diameter, depending on the strain being studied, and appear as icosahedra in the electron microscope (see Fig.

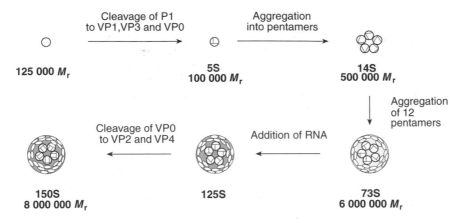

Fig. 11.6 Summary of the steps involved in the assembly of poliovirus.

3.13). The particle is more complex than that of the picornaviruses, containing at least 10 proteins. The 252 capsomeres that form the external surface consist of 240 proteins arranged such that they have sixfold symmetry (called the hexon capsomeres) and the remaining 12 are arranged at the vertices of the icosahedron with fivefold symmetry (penton proteins). Fibre structures project from the pentons. Although many details of the assembly of adenoviruses remain to be established, those available indicate that, unlike the picornavirus assembly process, the individual components of the adenovirus particle are assembled independently of each other and are brought together in a directed fashion (Fig. 11.7).

Assembly of adenovirus takes place in the nucleus of the infected cell where the virus genome is replicated, and the proteins must be translocated there at an early stage. The proteins that form the fibre, the base of the penton capsomere and the hexon capsomere are synthesized independently of each other in the cytoplasm. The fibre and hexon proteins come together to form independent trimer intermediates whereas the penton monomers form a pentameric penton base. The fibre trimers and penton base then associate to give a penton capsomere. The formation of the hexon trimer unit requires the presence of an additional adenovirus protein with a relative molecular mass (M_r) of 100 000 which interacts directly with the hexon proteins. This protein with an M_r of 100 000 is not found in the mature virus particle and is termed a 'scaffolding' protein.

The remaining steps in the adenovirus assembly process have been inferred from detection and analysis of putative intermediate structures. A key aspect is the formation of immature virus particles that do not contain the genome double-stranded (ds) DNA or core proteins, but which do contain at least three proteins that are not found in the infectious particle. These three scaffolding proteins may be removed, in part, by proteolytic degradation. The hexon capsomeres come together in a nonamer complex and subsequently 20 of the nonamer complexes interact to produce an icosahedral cage-like lattice. This structure then acquires the virus DNA, core proteins and

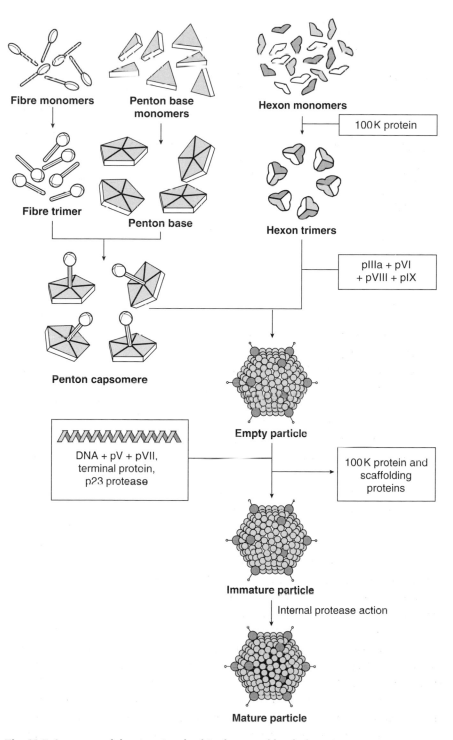

Fibre monomers   **Penton base monomers**   **Hexon monomers**

100 K protein

Fibre trimer   **Penton base**   **Hexon trimers**

pIIIa + pVI + pVIII + pIX

Penton capsomere

DNA + pV + pVII, terminal protein, p23 protease

Empty particle

100 K protein and scaffolding proteins

Immature particle

Internal protease action

Mature particle

Fig. 11.7 Summary of the steps involved in the assembly of adenovirus.

remaining structural components, with concomitant loss of the scaffolding proteins. It is not known whether the DNA and the core proteins enter the immature particle together in a complex or in rapid succession. Finally, a protease in the particle cleaves various components to create an infectious virion.

Adenovirus DNA contains a protein covalently attached to the termini (see Section 4.6), but this is not involved in packaging of the genome into particles. Analysis of DNA packaged into adenovirus defective–interfering (DI) particles (see Section 8.1) and generation of deletion mutants have identified a region essential for packaging. Approximately 400 base-pairs (bp) at one end of the DNA, adjacent to the E1A gene (see Fig. 9.5b), must be present for the DNA to enter the immature particle. The conclusion that this is a packaging signal responsible for the specific acquisition of DNA by the particle is supported by evidence that the genome enters the particle in a polar fashion, one end first, and that this polarity is lost when the sequence is duplicated at the other end of the DNA.

Sequence-dependent and -independent packaging of virus DNA in virus particles

The adenovirus assembly process indicates clearly that packaging of genome DNA is sequence dependent. Many other viruses also rely on the presence of specific sequences to select the virus DNA or RNA, although the process may be different. In the case of herpes simplex virus, the assembly process is coupled to the excision of the genome from the concatemeric replication intermediate (see Section 6.3).

The herpes simplex virus particle is covered by a lipid envelope, which is acquired in a process described in Section 11.6, but the internal component of the particle is formed first in a process similar to that described for adenovirus. The herpes simplex virus assembly process generates an immature particle lacking the virus DNA. As indicated in Section 4.4, the HSV genome contains direct repeats (the 'a' sequence) at the ends of the DNA. Each 'a' sequence contains within it two short regions, called $pac1$ and $pac2$, which are required for packaging. During assembly, the immature virus particle recognizes the pac sequences and cleaves the DNA at the precise termini, while at the same time inserting it into the interior of the particle. The presence of the 'a' sequence alone cannot be sufficient to specify cleavage and ensure that the entire genome is taken into the particle, because an additional copy of the 'a' sequence is located within the genome at the boundary of the internal long and short repeat elements (IR_L and IR_S). The most likely explanation is that the length of the DNA also plays a role in assembly, with constraints on the minimum and maximum permissible lengths of the DNA accepted into a particle.

A similar process occurs with bacteriophage λ, which also generates concatemeric DNA from a circular intermediate. A protein within the phage head recognizes the cos sequence in the concatemeric DNA and cleaves at that point, to produce the single-stranded overhang described in Section 4.5. The DNA is then brought into the empty head until it is full as described for phage T4 in Section 11.5. The correct length of DNA within the phage head ensures that the next cos sequence in the concatemer is readily accessible to the cleavage enzyme and the DNA is cut to leave an overhang. If two *cos* sequences are brought close together by a deletion of the genome, no *cos* sequence is available to the enzyme when the phage head is full and particle assembly aborts. Similarly, if an insertion occurs to increase the distance between *cos* sequences, the head becomes full before the next cleavage site is reached and again the assembly process aborts. In this way, phage λ regulates the size of the genomic DNA that can be packaged and ensures that all the genes are present. However, the phage λ packaging system tolerates some flexibility in genome length, which permits the virus to carry additional DNA derived from the host during the process of specialized transduction (see Section 12.4).

For a long time, it was considered essential that dsDNA animal viruses had controls on the assembly of particles to ensure that only virus DNA could be packaged. For those viruses that have a packaging signal, this is the only control present and the rest of the sequence is irrelevant. However, recently it has been shown that polyomavirus (a papovavirus) can, under certain conditions, package foreign DNA. This indicates that the control mechanisms for some viruses may not be as precise as was originally thought, and it may be possible to exploit this in the development of new ways to transfer DNA into cells for gene therapy.

11.5 Assembly of viruses with complex structures

The particle morphologies of many viruses do not conform to the relatively simple geometric organizations seen with helical and isometric viruses. For some of these viruses, the particle may consist of components that individually have elements of symmetry with which we are familiar, whereas others do not appear to have any recognizable pattern to their particle structure. For most of these viruses, little is known about the process of assembly. Exceptions are the T even bacteriophages and, to a lesser extent, bacteriophage λ. The assembly of bacteriophage T4 has been worked out in detail and demonstrates elements in common with some animal viruses, such as assembly of helical and isometric components, the involvement of scaffolding proteins, and acquisition of genome DNA by immature particles.

Assembly of bacteriophage T4

As described in Chapter 3, the structure of the tailed bacteriophages is considerably more complex than that of the viruses considered above. Several conditional lethal mutants of T4 are available which are unable to produce infectious particles when bacteria are infected under non-permissive conditions. Electron microscopic examination of extracts from these bacteria revealed the presence of structures readily recognizable as components of phage particles. Although some of the mutants were defective in one or other of the genes that encoded the structural proteins, many of these mutants were not. This indicated that additional, non-structural, gene products were involved in the assembly process. The ability of the mutants to synthesize some recognizable structural components of the phage T4 particle, such as the head, tail and tail fibres, also showed that these are assembled independently of each other, similar to the situation for adenovirus capsid components (see Section 11.4).

A significant advance came with the discovery that the morphogenesis of phage T4 from partially assembled components could be made to occur *in vitro*. In one experiment, purified fibreless phage isolated from cells infected with a tail fibre mutant were mixed with an extract from cells infected with a mutant that could not synthesize heads. Infectivity rapidly increased by several orders of magnitude, indicating that the 'headless' mutant extract was acting as a tail fibre donor, whereas the other extract supplied the heads (Fig. 11.8). This type of experiment is analogous to a genetic complementation test but is different in that it occurs *in vitro*.

To determine the exact sequence of events in phage T4 assembly, the precursor structures were isolated before the complementation tests, e.g. free base plates were isolated from bacteria infected with T4 mutants that have a defect in one of gene X, gene Y or gene Z, and after isolation they retained their *in vitro* complementation activity. By determining the ability of each to complement the other two unfractionated extracts, it was possible to establish the sequence in which the three gene products interact with the base plate. If isolated gene X-defective base plates complement a gene Z-defective extract, the gene Z product must have carried out its function by interacting with the base plate in the absence of the gene X product, i.e. the gene Z protein acts before the gene X protein. If, in the converse experiment, isolated gene Z-defective base plates do not complement a gene X-defective extract, the gene X product cannot act before the gene Z product and the sequence Z, X is established. Using these techniques in conjunction with analysis of polypeptides in polyacrylamide gels, it has been possible to determine the way in which heads, tails and tail fibres are constructed before their assembly into mature virions.

Bacteriophage T4 head assembly and acquisition of genome DNA

The order in which the phage T4 head proteins are introduced into the growing structure is shown in Fig. 11.9. Several proteins come together to form an immature prohead structure which contains all of the elements of icosahedral symmetry. The immature head then undergoes a maturation process, during which it acquires the genome DNA together with several additional proteins and loses a scaffolding protein present in the immature structure. The time course of phage head maturation can be followed by briefly labelling the nascent proteins in infected bacteria with radioactive amino acids, and determining the nature of the labelled head structures present at various time points afterwards. This has shown that the first particles formed (prohead I) have a sedimentation coefficient of 400S. These are converted into prohead II (350S), followed by prohead III (550S) and finally into mature heads, sedimenting at 1100S. The conversion of prohead I to prohead II is achieved by the proteolytic cleavage of a single protein called P23. Prohead III is approximately half full of DNA and appears to represent an intermediate between prohead II and the mature head. Its isolation suggests that head filling may take place in two stages: first, half of the DNA is inserted rapidly and then the second half is inserted more slowly, or after some specific enabling event has occurred. These stages are shown in Fig. 11.9.

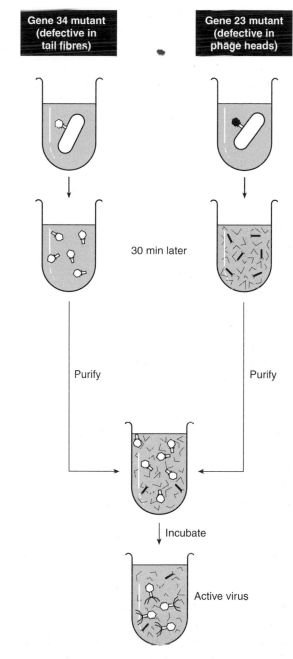

Fig. 11.8 *In vitro* complementation between two mutants that are blocked at different stages in morphogenesis.

As indicated in Section 4.5, bacteriophage T4 DNA is circularly permutated. The immature phage head excises a defined length of DNA from the concatemeric replication intermediate but, unlike adenovirus and herpes simplex virus, does not respond to a specific packaging sequence at the terminus of the genome. The cleavage event in DNA acquisition by the phage T4 head is sequence independent, although the head

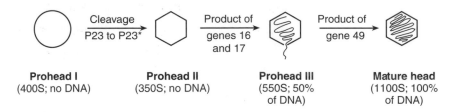

Prohead I	Prohead II	Prohead III	Mature head
(400S; no DNA)	(350S; no DNA)	(550S; 50% of DNA)	(1100S; 100% of DNA)

Fig. 11.9 Assembly and maturation of the head of bacteriophage T4.

does confer specificity on the length of the DNA accepted. The dimensions of the T4 head (100 nm×65 nm) correspond to an internal volume of $2.5 \times 10^4 \, \mu m^3$, assuming that the head proteins themselves occupy no space. As the DNA genome of T4 is 56 μm long, it probably occupies a minimum volume of approximately $1.8 \times 10^4 \, \mu m^3$. Clearly, the T4 genome must be compactly packed into the T4 head. Exactly how this is achieved is not clear, but it is the size of the head that determines the length of the encapsidated DNA. When abnormal phage T4 particles with grossly extended heads are produced, the length of the DNA molecules cleaved from the concatemer and packaged increases in direct proportion to the increase in head size.

Assembly of the bacteriophage T4 tail and tail fibres

As indicated, the phage T4 tail and tail fibre components assemble independently of each other. The tail takes the form of a helical structure which includes the head–tail connector, the core to which it is attached, the surrounding sheath and the base plate and tail fibres. The generation of a functional structure requires that each of the proteins is added in a strictly controlled manner. For both the tail and tail fibres, scaffolding proteins are involved in assembly. The pathway of tail morphogenesis is shown in Fig. 11.10.

The overall bacteriophage T4 assembly process

The pathway by which the different components of T4 virion are assembled into infectious particles is shown in Fig. 11.11. After a modification process, carried out by two non-structural proteins, heads spontaneously join to tails. In contrast, the tail fibres do not spontaneously join to the base plate but require the active participation of the product of gene 63. The way in which a head vertex with fivefold symmetry stably and spontaneously unites with a tail with sixfold symmetry is not clear. Nor is it clear why tails attach only to one vertex. It is possible that one end of the DNA protrudes through a vertex, disturbing the fivefold symmetry, and that this promotes tail addition. Indeed, protrusion of the DNA a short way into the tail may be a necessary structural feature for successful injection after contact of the phage with a susceptible bacterium.

11.6 Assembly of enveloped viruses

A large number of viruses, particularly viruses infecting animals, have a lipid envelope as an integral part of their structure. These include herpes-, filo-, retro-,

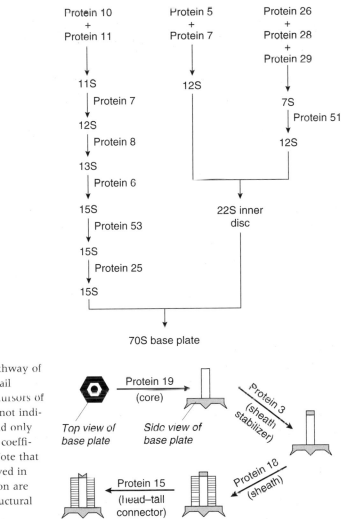

Protein 10
+
Protein 11

Protein 5
+
Protein 7

Protein 26
+
Protein 28
+
Protein 29

11S   12S
   │ Protein 7   7S
   ↓   │ Protein 51
  12S   12S
   │ Protein 8
  13S
   │ Protein 6
  15S   22S inner
   │ Protein 53   disc
  15S
   │ Protein 25
  15S

70S base plate

Fig. 11.10 The pathway of bacteriophage T4 tail assembly. The precursors of the base plate are not individually named and only the sedimentation coefficients are given. Note that the proteins involved in base plate formation are not necessarily structural proteins.

Top view of base plate   Side view of base plate

Protein 19 (core)

Protein 3 (sheath stabilizer)

Protein 18 (sheath)

Protein 15 (head–tail connector)

orthomyxo-, paramyxo-, corona-, arena-, pox- and iridoviruses, the tomato spotted wilt virus group, and rhabdoviruses of plants and animals (see Chapter 22). For each virus, the component held within the envelope—the nucleocapsid—is of a predetermined morphology which may be helical, isometric or of a more complex nature. For most enveloped viruses the nucleocapsid is formed in its entirety before acquisition of the lipid envelope.

Assembly of helical nucleocapsids

The assembly process of TMV is usually considered to be a model for the generation of all helical virus structures. However, the structures of nucleocapsids of enveloped viruses differ in many ways. This is particularly true of the nucleocapsids of the negative-sense ssRNA viruses such as filo-, paramyxo-, rhabdo- and orthomyxo-

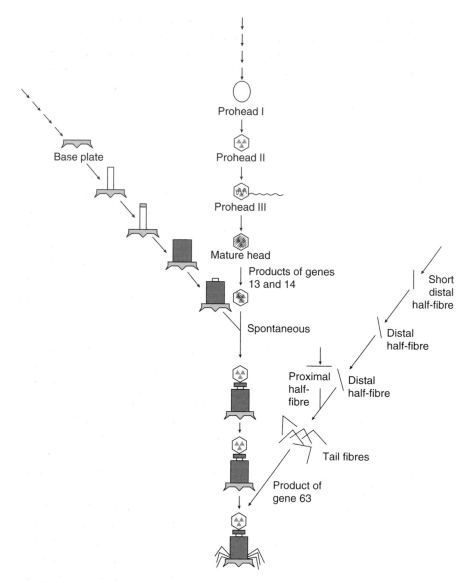

Fig. 11.11 Assembly of bacteriophage T4.

viruses. For members of some of these virus groups, the RNA genome in the nucleo-capsid is protected from degradation by nucleases *in vitro*, whereas for the others it is not. This indicates that the structures must be arranged differently. For all Baltimore class 5 viruses except the orthomyxoviruses, the basic structure of the nucleocapsid consists of the genome RNA encapsidated by a nucleoprotein, called N or NP, in association with smaller numbers of a phosphoprotein and a few molecules of a large protein. This complex carries out the processes of replication (see Section 7.4) and transcription (see Section 10.8). For orthomyxoviruses, the nucleoprotein associates with three proteins—PA, PB1 and PB2 (see Section 7.1), which carry out RNA synthesis. Although some of the steps in the assembly process for nucleocapsids of specific viruses are understood, no precise details are available to describe all aspects.

Analysis of genomes of negative-sense ssRNA DI viruses (see Chapter 10) and the generation of synthetic genomes for reverse genetics studies have shown that the termini of the genomes are essential for the encapsidation process. It is thought that specific sequences are located in the terminal regions of the genome (and antigenome), which initiate the generation of nucleocapsid structures and ensure that virus RNA is packaged. Expression of the measles virus nucleoprotein in bacteria, in the absence of virus genome RNA, results in the production of short nucleocapsid-like RNA–protein complexes, indicating that the nucleoprotein has the inherent ability to form nucleocapsids around any RNA. Presumably the signal in the genome termini gives a significant advantage to the genome in competition with any other RNA. For the paramyxovirus Sendai virus, the phosphoprotein that is part of the nucleocapsid associates strongly with the nucleoprotein before the interaction with RNA. This association prevents the nucleoprotein from interacting non-specifically with RNA and gives specificity to the encapsidation process. This may be a general feature of many class 5 viruses.

The molecular details of the interaction between the nucleoprotein and genome RNA have not yet been established for any helical nucleocapsid complex. For several paramyxoviruses, such as Sendai virus and measles virus, the genome RNA must always have a number of nucleotides that is divisible by six. This 'rule of six' is interpreted to be the result of the nucleoprotein binding to groups of six nucleotides in the RNA. If the genome does not conform to the rule of six, unencapsidated nucleotides will be present at one, or both, termini and this will prevent replication and hence propagation of the genome. However, for most viruses there is no such strict length requirement for the genome and the nature of the interaction between the nucleoprotein and the genome is not known.

A significant difference between the structure of the TMV capsid and the nucleocapsid of some negative-sense ssRNA viruses lies in the location of the genome RNA. In the TMV particle, the RNA is located entirely within the capsid. For influenza virus and the rhabdovirus vesicular stomatitis virus (VSV), biochemical analysis suggests that the RNA is wound around the outside of the nucleocapsid complex with the nucleotide bases exposed. This leaves the nucleotides available to be used as templates during replication and transcription. It is not yet known whether this is a common feature of the class 5 viruses, but the difference between this structure and that of the TMV capsid reflects their different roles: in TMV the capsid serves to protect the genome RNA because the particle does not contain the extra, lipid layer found in the class 5 viruses.

Assembly of isometric nucleocapsids

In the absence of detailed information, it is assumed that the assembly of isometric nucleocapsids is similar to the processes described for non-enveloped viruses. Thus, either the various components are assembled around the virus genome in response to a packaging signal, as is the case for the togaviruses, or an immature particle is formed and the genome is added subsequently as seen for herpes simplex virus (see Section 11.4).

Acquisition of the lipid envelope

Cells contain a large quantity of lipid bilayer membranes. These membranes define boundaries between cellular compartments, such as the nucleus and cytoplasm, as well as making up the external surface of the cell itself. Most enveloped viruses acquire their envelope by budding from the plasma membrane of the infected cell (Fig. 11.12). Four events leading to budding have been identified. First, the nucleocapsids form in the cytoplasm. Second, viral glycoproteins, which are transmembrane proteins, accumulate in patches of cellular membrane. Third, the cytoplasmic tails of the glycoproteins that protrude from the membrane interact with the nucleocapsid. This interaction may be direct, or indirect via an intermediary matrix protein which becomes aligned along the inner surface of the modified membrane. Finally, once these interactions have taken place, the virion is formed by budding. A simple way of envisioning the process is that the glycoproteins in the membrane progressively inter-

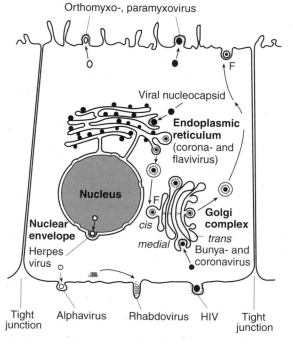

Apical (outer) surface of plasma membrane
Orthomyxo-, paramyxovirus

Viral nucleocapsid

Endoplasmic reticulum
(corona- and flavivirus)

Nucleus

Golgi complex

Nuclear envelope
cis

medial   *trans*

Herpes virus
Bunya- and coronavirus

Tight junction   Alphavirus   Rhabdovirus   HIV   Tight junction

Basal surface of plasma membrane

Fig. 11.12 Sites of maturation of various enveloped viruses. F, fusion of a vesicle with a membrane.

act with the nucleocapsid, or the matrix protein and nucleocapsid where appropriate. As the number of interactions increases, the membrane with the glycoproteins inserted in it is 'pulled' around the nucleocapsid. When the nucleocapsid is completely enclosed, the lipid bilayer pinches off to release the virus particle. As a result of this, viruses that bud from the plasma membrane are automatically released when the budding process is complete. An unusual feature of the budding process is that host membrane proteins are excluded entirely from viral particles, except for the retroviruses. The mechanism by which this exclusion is achieved is not fully understood, but it is likely that the virus glycoproteins are inserted into membranes in specific areas where host proteins are not readily found. Retrovirus envelopes contain specifically selected host proteins.

Some viruses, such as corona- and bunyaviruses, bud into the endoplasmic reticulum (ER), acquiring cytoplasmic membranes as their envelopes; they are then released to the exterior via the Golgi complex. First, the virus particle buds into the ER, acquiring an envelope in a process similar to that described for budding with the plasma membrane. The enveloped virus then exits the ER within a vesicle, following the normal mechanism and route for trafficking of material from the ER to the Golgi complex. The vesicle containing the enveloped virus moves to and fuses with the *cis*-Golgi complex. The virus then moves through the Golgi apparatus and exits in another vesicle which buds off from the concave *trans*-Golgi network surface. This vesicle is transported to the plasma membrane where it fuses to release the enveloped particle to the exterior of the cell. The virus particle does not fuse with any of the vesicle membranes and retains its original envelope throughout this process.

Herpesviruses assemble in the cell nucleus, with many of the viral proteins synthesized in the cytoplasm being transported into the nucleus. After entry of the virus genomic DNA into the nucleocapsid, the virus must acquire an envelope. The source from which, and process by which the envelope is obtained are still not clear. Currently, there are three possibilities. The first is that the virus buds through the inner nuclear membrane and thus becomes enveloped. The lumen of the ER is continuous with the space between two nuclear membranes and the enveloped virus particle can leave the ER in a vesicle in the same way as described for the bunya- and coronaviruses. Before the budding process, it would be necessary for the membrane to be modified by incorporation of virus-specified glycoproteins which are found in the virus envelope. The second possibility is that the herpesvirus nucleocapsid goes through a process of envelopment by budding through the inner nuclear membrane followed by fusion with the outer nuclear membrane. The fusion event would be similar to that occurring during entry by enveloped viruses (see Chapter 5) and as a result the envelope would be lost. The nucleocapsid can then migrate to the plasma membrane where the virus glycoproteins are located and leave by budding, thus acquiring an envelope in the process. The third possibility is that the nucleocapsid escapes from the nucleus into the cytoplasm by an unknown mechanism, possibly through the pores in the nuclear membrane. Once in the cytoplasm, it migrates to the modified plasma membrane and exits by budding. Evidence to support elements of all three possible mechanisms has been described.

For other viruses, the process by which the lipid envelope is acquired by the nucleocapsid is unclear. Very little is known about the assembly of lipid-containing phages and the iridoviruses, but it appears that the envelope is not incorporated by a budding process. In this respect, they resemble poxviruses, the morphogenesis of which has been studied extensively by electron microscopy of infected cells. In thin sections, poxvirus particles initially appear as crescent-shaped objects within specific areas of cytoplasm, called 'factories', and even at this stage they appear to contain the trilaminar membrane that forms the envelope. The crescents are then completed into spherical structures. DNA is added, and the external surface undergoes a number of modifications to yield mature virions. The origin of the lipid in the virus is not known because it does not appear to be physically connected to any pre-existing cell membrane.

One complication is that many differentiated cells *in vivo* are polarized, meaning that they carry out different functions with their outer (apical) surface and their inner (basal) surface. For instance, cells lining kidney tubules are responsible for regulating Na^+ ion concentration. They transport Na^+ ions via their basal surface from the endothelial cells lining blood vessels to the cytoplasm and then expel them from the apical surface into urine. Some cell lines, such as Madin–Darby canine kidney (MDCK) cells, retain this property in culture, but this can be demonstrated only when they form confluent monolayers and tight junctions between cells. The latter serve to separate and define properties of the apical and basal surfaces. These properties reside in cellular proteins that have migrated directionally to one or other surface. This directional migration and insertion into only one of the cell surfaces also occurs with some viral proteins. The lipid composition of a polarized cell is also distributed asymmetrically between apical and basal surfaces, e.g. if a cell is dually infected with a rhabdovirus and an orthomyxovirus, electron microscopy shows the rhabdovirus budding from the basal surface, which is in contact with the substratum, and the orthomyxovirus from the apical surface (see Fig. 11.12). This property is determined by a molecular signal on the viral envelope glycoproteins which directs them to one surface or the other; tight junctions are needed or proteins diffuse laterally and viruses can then bud from either surface.

11.7 Maturation of virus particles

For most animal and bacterial viruses, formation of infectious virions requires the cleavage of precursor protein molecules into functional proteins. These cleavages may occur before or after the precursors have been assimilated into the virus particle. Examples of cleavage of a precursor before assembly into the virion are seen with the glycoproteins of orthomyxo- and paramyxoviruses, as well as many others. For the orthomyxoviruses, such as influenza virus, the haemagglutinin (HA) protein is synthesized as an inactive precursor (HA0). Cleavage of the HA0 protein into HA1 and HA2 is carried out by a host protease immediately before insertion into the plasma membrane of the infected cell. The HA protein is acquired by the virus during budding when the envelope is added. The cleavage of HA protein is essential because uncleaved protein cannot fuse

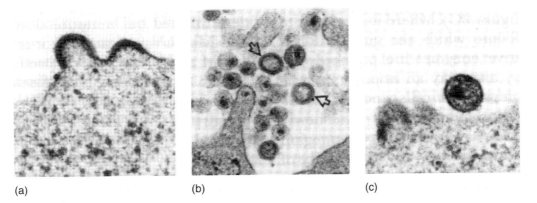

(a)   (b)   (c)

Fig. 11.13 Morphogenic changes seen in human immunodeficiency virus (HIV) particles after budding from the surface of the host cell. (a) Budding of an HIV capsid. (b) Mixture of both immature (arrowed) and mature HIV particles. (c) Mature HIV virion with characteristic cone-shaped core. (From Hunter 1994.)

with the host cell membrane to initiate the next round of infection. A similar situation is seen with the fusion proteins of the paramyxoviruses. For Sendai virus and the economically important avian paramyxovirus, Newcastle disease virus, the protease responsible for the cleavage of the fusion protein is restricted only to cells of the respiratory tract and this distribution prevents the viruses replicating elsewhere in the body. It is for this reason that they cause only respiratory disease in their hosts. An example of a maturation cleavage occurring after particle formation is seen with poliovirus, where the polypeptide VP0, which is assembled into a particle (see Section 11.4), is cleaved to form VP2 and VP4. Without this cleavage the particle is not infectious.

A more dramatic example of morphogenic alterations in a virus particle after assembly is seen with human immunodeficiency virus (HIV). The HIV virion consists of a nucleocapsid surrounded by a lipid envelope. The envelope contains two glycoproteins, gp41 and gp120, which are cleaved from a precursor by a cellular protease similar to the influenza virus HA protein and the paramyxovirus fusion proteins. This cleavage is essential for infectivity of the progeny particle. The HIV nucleocapsid in newly budded particles does not show the distinctive structure that is seen with infectious, mature particles. Electron microscopic examination of particles at various times after budding has shown that the core of the particle undergoes considerable alteration in structure before achieving the final, infectious form. This process is shown in Fig. 11.13. Its basis is the action of a virion protease which cleaves the assembled gag precursor proteins (see Section 9.9) into their functional products—the matrix, capsid and nucleocapsid proteins—after which the morphology of these particle components can mature.

11.8 Further reading and references

Butler, P. J. G. (1984) The current picture of the structure and assembly of tobacco mosaic virus. *Journal of General Virology* **65**, 253–279.

Compans, R. W. (1991) Protein traffic in eukaryotic cells. *Current Topics in Microbiology and Immunology* **170**, 1–186.

Earnshaw, W. C. & Casjens, S. R. (1980) DNA packaging by the double-stranded DNA bacteriophages. *Cell* **21**, 319–331.

Fujisawa, H. & Hearing, P. (1994) Structure, function packaging signals and specificity of the DNA in double-stranded DNA viruses. *Seminars in Virology* **5**, 5–13.

Hendrix, R. W. & Garcea, R. L. (1994) Capsid assembly of dsDNA viruses. *Seminars in Virology* **5**, 15–26.

Homa, F. L. & Brown, J. C. (1997) Capsid assembly and DNA packaging in herpes simplex virus. *Reviews in Medical Virology* **7**, 107–122.

Hunter, E. (1994) Macromolecular interactions in the assembly of HIV and other retroviruses. *Seminars in Virology* **5**, 71–83.

Jones, I. M. & Morikawa, Y. (1998) The molecular basis of HIV capsid assembly. *Reviews in Medical Virology* **8**, 87–95.

Strauss, J. H., Strauss, E. G. & Kuhn, R. J. (1995) Budding of alphaviruses. *Trends in Microbiology* **3**, 346–350.

Tucker, S. P. & Compans, R. W. (1993) Virus infection of polarised epithelial cells. *Advances in Virus Research* **42**, 187–247.

Also check Chapter 22 for references specific to each family of viruses.

Chapter 12
Lytic and lysogenic replication of bacteriophage λ

12.1 Introduction

A common perception of virus infection is that the only possible outcome is an immediate and rapid production of progeny virus particles. Although this is the most common result of infection, it is not the only one and certain, but not all, viruses can enter a different type of replicative cycle in which the host cell survives, either for a fixed period or indefinitely. The appreciation that some viruses have the capacity to establish a non-lytic infection has significant implications for our understanding of virus-associated diseases and potential therapies. It also has relevance to the potential use of animal viruses, such as the defective parvovirus adeno-associated virus, as vectors for gene therapy (see Chapter 13).

The first indication that two alternative outcomes of infection are possible was with bacteriophage λ and can be seen after infection of a culture of *Escherichia coli*. Shortly after infection with phage λ, most of the bacteria in the culture lyse, releasing more infectious phages. However, a small number of bacteria survive the infection and can be propagated. The surviving bacteria are resistant to infection by phage λ and also by related phages. However, if grown in culture, the resistant bacteria spontaneously, and continuously, produce very low levels of infectious phages. When cultures of the resistant bacteria are subjected to ultraviolet irradiation at levels sufficient to stop the culture growing, lysis occurs with release of large amounts of phage λ. These observations indicated that the phage λ introduced initially was present in the resistant bacteria and could be propagated from one generation to the next. However, the phage must be present in a non-infectious form that can spontaneously revert to produce virus. Several studies have clarified this and have shown that the phage DNA in the resistant bacteria is inserted in the genome of the host where it is replicated with the host chromosome. In this state the phage genome is called a prophage. The bacteria carrying the phage DNA are called lysogens. Bacteriophages that are able to adopt either a lytic or a lysogenic replication cycle are called temperate phages. The study of lysogeny has led to an understanding of mechanisms of control of gene expression which has had a significant impact on many areas of research, including control of gene expression in humans.

When phage λ enters a bacterium, it must initiate either a lytic or a lysogenic cycle. For the latter to occur, the phage DNA must insert itself into the host chromosome and be maintained, in contrast to the rolling circle replication of a lytic infection (see Section 6.3). To understand fully the mechanism by which lysogeny is established and maintained, and how the virus regains the lytic replication cycle, it is necessary first to consider the genes involved in a lytic infection.

12.2 Gene expression in the lytic cycle of bacteriophage λ

After attachment, the phage λ linear genomic DNA is introduced by an injection mechanism (see Section 5.3) into the *E. coli*, and is immediately converted into a covalently closed double-stranded DNA (dsDNA) circle by host enzymes. Circularization is possible because of the single-stranded *cos* sequences at the ends of the genome (see Section 4.5). Phage λ genes are then switched on and off in a tightly regulated, coordinated, manner which determines the outcome of the infection.

The phage λ genome contains two principal promoters which are recognized by the host-cell, DNA-dependent, RNA polymerase; this immediately begins transcription of mRNA. One of the promoters (P_L—Fig. 12.1) directs transcription in a leftwards direction and the mRNA terminates at the end of the gene encoding the N protein. The other promoter (P_R) directs transcription in a rightward direction to encode the protein Cro. However, termination of transcription at the end of the *cro* gene is not complete and some mRNAs extend through the *cII*, *O* and *P* genes, and these proteins are trans-

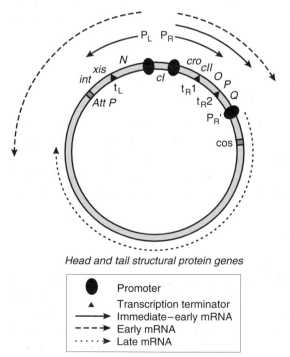

Head and tail structural protein genes

- ● Promoter
- ▲ Transcription terminator
- ——➤ Immediate–early mRNA
- – – –➤ Early mRNA
- · · · · ·➤ Late mRNA

Fig. 12.1 The circular map of the bacteriophage λ genome. The genes involved in the lytic cycle are indicated. The positions of the phage promoters and transcriptional terminators involved in the lytic cycle are shown, together with the mRNAs produced at immediate–early, early and late times.

lated from the polycistronic mRNA. The N protein causes the RNA polymerase to transcribe through the regions of DNA at the ends of the *N*, *cro* and *P* genes, where it had previously stopped, to generate polycistronic mRNAs encoding several proteins. The mRNA from P$_L$ extends through the *N*, *cIII*, *xis* and *int* genes, and mRNA from P$_R$ extends through the *cro*, *cII*, *O*, *P* and *Q* genes. Thus, N protein acts as a transcriptional anti-terminator and allows expression of additional genes. These events occur before phage λ DNA synthesis and are referred to as immediate–early (*N* and *cro* gene expression) and early (*cIII*, *xis* and *int* expression from P$_L$, and the *Q* gene from P$_R$).

The Cro and Q proteins are important for the next phase of gene expression. This occurs after phage λ DNA synthesis and is therefore, by definition, a late event in the replication cycle. Immediately on infection, and at the same time as P$_L$ and P$_R$ are used, host-cell RNA polymerase recognizes a third promoter region in phage λ DNA, P$_R'$, located immediately after the *Q* gene. However, transcription is terminated just downstream to synthesize a very short mRNA, even in the presence of the N protein. This short mRNA does not encode a protein. The Q protein is an anti-terminator that causes the RNA polymerase molecules initiating transcription at P$_R'$ to ignore the termination signal and to continue mRNA synthesis. The resulting mRNA is extremely long and extends through the genes encoding the structural proteins, which make up the phage head and tail (Fig. 12.1). At the same time as the Q protein is exerting its activity, the Cro protein is also at work. It binds to the phage λ DNA at the operator elements (O$_L$ and O$_R$, respectively) of the promoters P$_L$ and P$_R$. By doing this Cro inhibits transcription from these promoters, stopping production of the early mRNAs. Sufficient Q protein is present to ensure that P$_R'$, which is not affected by Cro protein, remains active. The result is that the only mRNA found at late times encodes the structural proteins which can package the newly synthesized phage λ DNA into progeny virions using the *cos* site (see Section 11.4).

12.3 The establishment and maintenance of bacteriophage λ lysogeny

Phage λ is able to diverge from the pattern of gene expression seen in the lytic cycle, to insert its DNA into the chromosome of the host and to maintain this state. Subsequently, the phage DNA can be excised from the chromosome and lytic replication initiated to produce infectious progeny. These various components of the lysogenic process are controlled by a fine balance between several inhibitors of gene expression.

Establishment of lysogeny

The initial events in the process of lysogeny are identical to those seen in a lytic infection. The circularized phage DNA is transcribed from the two major promoters, P$_L$ and P$_R$, and also from P$_R'$. As transcription from P$_R'$ makes only a small mRNA that does not encode a protein and it is not involved in the lysogeny process, it will not be considered further. Transcription from P$_L$ and P$_R$ makes the mRNAs encoding the N and Cro proteins, with small amounts of the cII, O and P proteins. Subsequently, the anti-terminator action of the N protein results in the synthesis of cIII, Xis, Int and Q proteins and larger quantities of cII, O and P, as described above (see Section 12.2). The cII protein acts as a gene activator, directing the host RNA polymerase to begin transcription at two promoters which would otherwise be inactive. These are P_{RE} (promoter for repressor expression) and P_{int} (Fig. 12.2a). The mRNA initiated at P_{int} directs the synthesis of the Int protein which is responsible for integrating the phage λ DNA into the host chromosome (see below). The use of the P_{int} promoter ensures expression of larger quantities of Int protein than can be generated from the polycistronic mRNA transcribed from P_L.

P_{RE} directs transcription of a single gene, *cI*. The cI protein, usually referred to as the cI, or lambda, repressor, is critical in lysogeny and phage with mutations in the *cI* gene can only replicate lytically. The cI protein binds to O_L and O_R, inhibiting transcription from P_L and P_R and preventing production of the early proteins. By inhibiting synthesis of the early proteins, the cI protein prevents the subsequent appearance of the late, structural proteins and, consequently, of infectious particles. To ensure that its own synthesis is not prevented by an absence of cII protein, the cI protein directs RNA polymerase to an additional promoter, P_{RM} (promoter for repressor maintenance), which initiates transcription of the *cI* gene alone (Fig. 12.2b).

The action of the cII protein in activating expression from P_{RE} has an additional consequence. The mRNA transcribed from P_{RE} contains some sequences that are antisense to those of the mRNA from P_R encoding the Cro protein. Hybridization of these will prevent translation of the *cro* gene mRNA and reduce the level of Cro protein in the bacterium.

The final result of the process described above is that the phage λ DNA is integrated into the host chromosome and the only gene that is active is the one encoding the cI protein. The cI protein, by inhibiting P_L and P_R, prevents the expression of any other phage genes. By activating P_{RM}, the cI protein ensures transcription of its own gene and therefore its own continued synthesis (Fig. 12.2c).

The choice between the lytic and lysogenic pathways

As described above, the initial steps in both the lytic and lysogenic pathways are the same. Although the lytic process occurs much more frequently, it is clear that some circumstances must favour lysogeny. Several aspects of what determines the choice between lysis and lysogeny remain unclear, but a key factor is the balance between the various repressors and activators produced and, among these, the role of the cII protein is critical. If cII is very active, cI protein will be produced in large amounts. This will efficiently inhibit synthesis of all genes except its own, and lysogeny will result. On the other hand, if the cII protein is poorly expressed or has low activity, very low levels of cI protein will be present. The Cro protein will inhibit the activity of P_L and P_R and the synthesis of the cII protein, among others, will be significantly reduced. Without sufficient cII protein, P_{RE} will not be strongly activated and synthesis of the cI protein will, in turn, be further reduced. In this case the lytic pathway will follow.

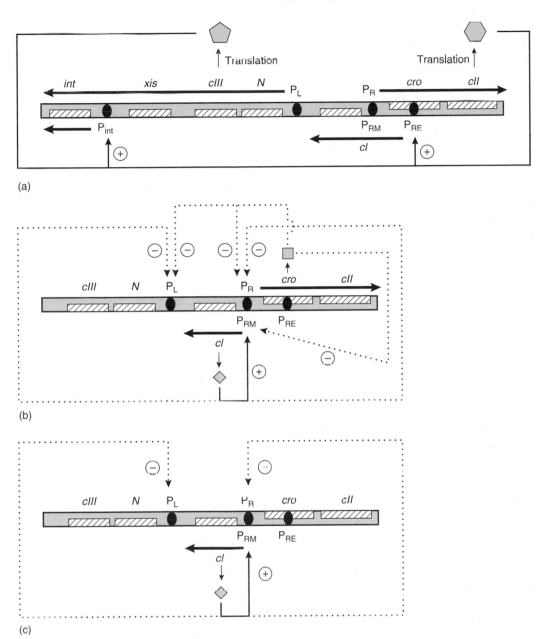

Fig. 12.2 The region of the bacteriophage λ genome encoding the genes responsible for lysogeny. (a) The cII protein activates the P_{RE} and P_{int} promoters to produce mRNAs for the cI and Int proteins, respectively. Note that the mRNA from P_{RE} contains sequences antisense to the *cro* gene. (b) The cI protein binds O_L and O_R, inactivating P_L and P_R, respectively, while activating P_{RM}, ensuring its own continued synthesis throughout lysogeny. The Cro protein also inhibits P_L and P_R activity. (c) The steady state of *cI* gene expression during lysogeny.

The activity of the cII protein is determined, at least in part, by environmental factors. The cII protein is susceptible to proteolytic degradation by host enzymes. The levels of host proteases are affected by many factors, especially growth conditions. Healthy bacteria grown in rich medium contain high levels of proteases and when infected are more likely to support lytic replication. In contrast, poorly growing bacteria have lower levels of proteases and so will encourage establishment of lysogeny. It should be clear that it is to the advantage of the phage to undertake a lytic infection in healthy bacteria, which will contain high levels of ATP and are equipped to synthesize the virus proteins. When bacteria are nutritionally deficient, it is to the advantage of the phage to establish itself as a lysogen and wait until conditions improve.

The cIII protein also plays a role in the choice between lysis and lysogeny. The function of cIII is to protect cII from proteolytic degradation. This protection is not complete because cII can be destroyed even in the presence of cIII. However, if cIII is absent or not functional, cII is almost always degraded and the phage can only undergo a lytic lifestyle.

Integration of bacteriophage λ DNA during lysogeny

The integration of phage λ DNA into the host chromosome is carried out by the Int protein. As indicated above, the cII protein activates the P_{int} promoter upstream of the *int* gene. The mRNA transcribed from this promoter is translated into Int protein and the action of cII ensures that it is present in large quantities.

The integrated DNA is referred to as a prophage and genetic mapping studies showed that there is only one site of integration, adjacent to the galactose (*gal*) operon in the *E. coli* genome at a position referred to as the lambda attachment (*att* λ) locus. In 1962, J. Campbell suggested that crossing over between the circularized phage DNA and the host genome results in insertion of the entire phage genome into that of the bacterium. Thus, the phage genome is inserted as a linear structure into the host genome by reciprocal recombination (Fig. 12.3a). The position of the recombination site in the phage DNA lies downstream of the *int* gene and the region is referred to as the *att P* site. The bacterial and phage attachment sites, *att* λ and *att P*, contain identical tracts of 15 base-pairs (bp), indicating that homologous recombination occurs during the integration event. The actual site of the crossover for both integration and excision must take place within, or at the boundaries of, this common sequence (Fig. 12.3b).

12.4 Induction and excision of the bacteriophage λ lysogen DNA

Induction and excision of integrated DNA

Early observations showed that ultraviolet irradiation of a lysogenic culture stimulates the phage to relinquish the lysogenic state and to enter a lytic mode of replication. This is termed 'induction of the lysogen'. Several other stimuli, such as treatment with potent mutagens, also induce the phage and a common feature of these treatments is

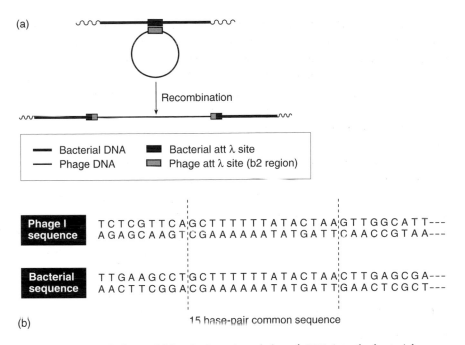

Fig. 12.3 (a) Campbell's model for the insertion of phage λ DNA into the bacterial genome. (b) The common sequence in the bacterial and λ attachment sites.

that they cause significant changes in the host *E. coli*, and in particular expression from an array of genes not previously active. This is referred to as the *E. coli* SOS response and the role of the new gene products is to protect the bacterium against the effects of the stimulus. Just as lysogeny may allow the phage to wait for the host to provide an optimum environment for replication, induction may be seen as a means by which such a phage can escape from a host the survival of which is threatened.

A critical component in the SOS response is the Rec A protein. The function of this protein is to mediate recombination between DNA molecules. However, when the bacterium is subjected to stress, such as irradiation with ultraviolet light, the Rec A protein alters its activity to become a specific protease. The primary target of the proteolytic Rec A protein is a repressor called Lex A. The Lex A protein represses expression from a range of genes and its cleavage removes this repression, giving expression of the genes. Rec A protein also cleaves the phage λ cI repressor protein and, when this occurs, the promoters P_L and P_R become functional, transcribing the *N*, *cro* and other genes described above. However, the activity of the altered Rec A protein ensures that no functional cI protein can be made and so only lytic replication can occur even if cII protein is active.

Reversal of the recombination event between the phage and host genomes, which generated the prophage, results in the regeneration of the circular phage genome. Although the integration event requires only the Int protein, for excision to occur the *xis* and *int* genes must both be transcribed because both proteins are required to act together. The mRNA encoding the Xis and Int proteins is transcribed from the reactivated P_L promoter.

Specialized transduction

The process of excision is usually very precise but occasional errors can occur, such that the region excised includes a small amount of bacterial DNA, with a corresponding piece of phage DNA deleted. If the portion of bacterial DNA is located to the left of the prophage, it may include some or all of the genes of the *gal* operon, with a compensatory loss of some of the genes at the right end of the prophage. This incomplete phage chromosome then serves as the template for replication, such that essentially all of the phage progeny issuing from that bacterium carry the genes for galactose use and have lost some of the phage genes. Infection of a Gal⁻ bacterium with such a phage, under conditions allowing lysogenization, can confer the ability to ferment galactose, because the necessary genes become inserted into the bacterial chromosome. The process of transferring genes from one host to another is called transduction and, as phage λ is able to transfer only the genes immediately adjacent to the *att* λ region of the bacterial chromosome, it is referred to as a specialized transducing phage. Many transducing phages are defective in replication, because they may lack some essential phage genes, and these can be propagated only if a normal ('helper') phage is present to supply the missing gene functions.

12.5 Immunity to superinfection

An unusual feature of lysogeny is that a bacterial lysogen is immune to superinfection by a second phage of the same type that it carries, but it is usually not immune to other, independently isolated, temperate phages. Phages that induce immunity to one another are termed 'homoimmune' and, if they do not, they are called 'heteroimmune'. The immunity that the lysogens display is very different from the immunity seen in animals. Analysis of the genetic factors involved in the specificity of immunity showed that the *cI* gene is responsible.

Lysogens continuously express the *cI* gene and contain significant levels of functional cI repressor until induced (see Section 12.4). The cI repressor acts by binding to the specific O_L and O_R sequences adjacent to promoters P_L and P_R and preventing the host DNA-dependent RNA polymerase from initiating transcription from either promoter. When the DNA of a superinfecting, homoimmune phage enters the lysogenic bacterium, the cI repressor binds to the O_L and O_R sequences and prevents phage gene expression, aborting the infection. If the superinfecting DNA is from a heteroimmune phage, the cI repressor of the lysogen cannot bind the incoming DNA and the second phage is able to initiate an infection.

12.6 Benefits of lysogeny

The fact that temperate phages carry so many genes involved in lysogeny indicates that lysogeny has some advantages for them. A significant advantage would be to provide a method for a phage that has infected a host low in energy and synthetic capacity to persist in the bacterium until conditions improve. The phage can then 'reappear' when conditions have improved.

It is important that in the persistent, lysogenic state, the phage does not represent a significant disadvantage for the survival of the host. In fact, lysogeny often carries a strong selective value for the bacterium, because temperate phages frequently confer new characteristics on the host. This phenomenon manifests itself in many ways and is referred to as lysogenic conversion. For λ, the only known lysogenic conversion is the capacity of λ lysogens to block the multiplication of a particular class of mutant (the rII mutant) of bacteriophage T4. This block involves the product of the *rex* gene and perhaps also the cI repressor. As rII mutants may not be very common in the natural environment of λ, the *rex* gene probably has other, more significant roles to play in the phage λ replication cycle.

An interesting example of lysogenic conversion is that observed in *Corynebacterium diphtheriae*. Strains of this bacterium that cause the serious childhood disease diphtheria carry the diphtheria toxin (and are called toxigenic). Infection of *Corynebacterium* sp. with phage β isolated from virulent, toxigenic bacilli of the same species produces lysogens that acquire the ability to synthesize toxin. Loss of the prophage results in loss of toxin production because the structural gene for toxin is carried by the phage itself.

12.7 Further reading and references

Johnson, A. D., Poteete, A. R., Lauer, G., Sauer, R. T., Ackers, G. K. & Ptashne, M. (1981) λ repressor and *cro*-components of an efficient molecular switch. *Nature (London)* **294**, 217–223.

Landy, A. & Ross, W. (1977) Viral integration and excision: structure of the lambda *att* sites. *Science* **197**, 1147–1160.

Ptashne, M. (1986) *A Genetic Switch*. Palo Alto: Blackwell Scientific Publications & Cell Press (a distillation of the principles of gene regulation using the phage λ system).

Ptashne, M., Jeffrey, A., Johnson, A. D., Maurer, R., Meyer, B. J., Pabo, C. O., Roberts, T. M. & Sauer, R. T. (1980) How the λ repressor and *cro* work. *Cell* **19**, 1–11.

Also check Chapter 22 for references specific to each family of viruses.

Chapter 13

Interactions between animal viruses and cells

In this chapter we are concerned with the various types of interactions that occur between animal viruses and cells in culture. We do not discuss plant viruses, because plant cell culture is technically more difficult and still not widely used. Virus–cell interactions are classified here into acutely cytopathogenic, persistent, latent, transforming, abortive and null infections. They have all been studied in the laboratory to determine the molecular events involved and to pave the way for our eventual understanding of infection of the whole organism (see Chapter 15). Two points should be borne in mind: first, that a prerequisite for any of these types of infection is the initial interaction between a virus and its receptor on the surface of the host cell, and hence any cell lacking the receptor is automatically resistant to infection; and, second, that both virus and cell play a vital role in establishing the interactions described below, and a virus may exhibit, for example, an acutely cytopathogenic infection in one cell type and latency in another.

13.1 Acutely cytopathogenic infections

Acutely cytopathogenic infections are those that result in cell death. Elsewhere these are called 'lytic' infections, but this term is not entirely accurate, because in some infections cells die without being lysed, i.e. by apoptosis or programmed cell death (see below). The replication of viruses that cause acutely cytopathogenic infections are the ones most commonly studied in the laboratory, because cell killing is an easy effect to observe, production of infectious progeny can usually be monitored without difficulty and the time scale is usually measured in hours. The one-step growth curve (see Section 1.2) describes the essential features of any eukaryotic or prokaryotic virus–host interaction that results in cell death, and Fig. 13.1 shows the successive appearance of cell-associated virus infectivity, infectivity that has been released from the cell into the tissue culture fluid and the cytopathic effect (CPE). The pathological effects are always last to appear, just as in human infections (see Chapter 15). These are often first seen under the microscope as a change from a spread-out, flattened morphology to a rounded-up structure. Quite early on in infection, viruses often inhibit the synthesis of cellular proteins, DNA or RNA, but frequently cell death occurs sooner than can be explained by these events. Exactly how an animal cell is killed in most cases is still not certain (see Section 13.7), but it is nothing to do with lysis by

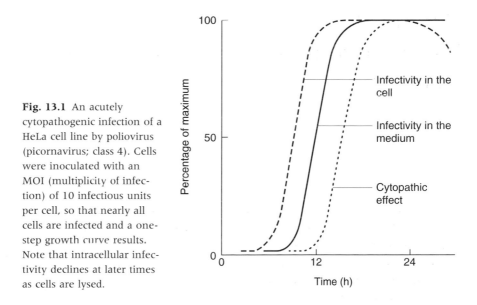

Fig. 13.1 An acutely cytopathogenic infection of a HeLa cell line by poliovirus (picornavirus; class 4). Cells were inoculated with an MOI (multiplicity of infection) of 10 infectious units per cell, so that nearly all cells are infected and a one-step growth curve results. Note that intracellular infectivity declines at later times as cells are lysed.

lysozyme, which is restricted to a minority of bacteriophages and bacteria. Lysis is one of the ways that a nucleocapsid virus such as poliovirus (Fig. 13.1) is liberated from the infected cell. If the cell is not lysed, the virus may enter a cytoplasmic vesicle, which then releases its contents into the tissue culture fluid by fusion with the plasma membrane. All membrane-bound viruses bud from cell membranes (see Section 11.6).

Apoptosis or programmed cell death of the infected cell

Apoptosis is the way in which cell numbers are regulated during normal development. Familiar manifestestions are the separation of webbed digits in the human embryo, the loss of the tail of the amphibian tadpole and the removal of self-reactive T cells. The uniqueness of apoptosis is that the dying cell remains intact and its contents stay within the plasma membrane. This contrasts with cell lysis (or *necrosis*) where the cell disintegrates and its contents are released. *In vivo*, these necrotic products are inflammatory and have to be tidied up by scavenger cells whereas the apoptotic cell is not inflammatory. During apoptosis the cell undergoes profound internal changes, which include fragmentation of its chromosomal DNA. These processes follow a clear, well-regulated pattern. In the end, the cell rounds up and is disposed of by being engulfed by a phagocyte, wherein it is hydrolysed. It seems likely that apoptosis is the cell's response to infection rather than being induced by the virus because, by committing suicide, the cell prevents the production of virus progeny. Some viruses have proteins that trigger the apoptosis pathway as an inevitable consequence of their interactions with the cell, but then make other proteins that inhibit apoptosis while their replication proceeds. This is relevant to the mechanisms by which viruses transform cells (see Section 17.4).

13.2 Persistent infections

Persistent infections result in the continuous production of infectious virus and this is achieved (as described below) either by the survival of the infected cell or by a situation in which a minority of cells are initially infected and the spread of virus is limited, so that cell death is counterbalanced by new cells produced by division, i.e. no net loss. Persistent infections result from a balance struck between the virus and its host by the following mechanism: (i) the interaction of virus and cells alone; (ii) the interaction of virus and cells, with antibody or interferon to limit virus production; (iii) the interaction of virus and cells, with the production of defective–interfering (DI) virus to limit virus production (see Section 7.1); or (iv) a combination of these events.

Persistent infections resulting from the virus–cell interaction

Simian virus 5 causes an acutely cytopathogenic infection with cell death in the baby hamster kidney (BHK) cell line (Fig. 13.2a) but, by changing the host cell to a mono-layer of monkey kidney (MK) cells, it gives an outstanding example of a persistent infection. The virus multiplies at the same rate in both cell types, and multiplies with a classic one-step growth curve in MK cells (Fig. 13.2b), but the MK cells show no CPE, remain healthy and produce progeny virus continuously (Fig. 13.2c). Infection by simian virus 5 does not damage the MK cell, in the sense that it does not perturb the synthesis of cellular protein, RNA or DNA, and cell division continues as normal. Calculations show that this virus infection makes little demand on the host's resources, e.g. total viral RNA synthesis is < 1% of cellular RNA synthesis (even though each cell is producing about 150 000 particles/day). Thus, in monkey cells, simian virus 5 causes

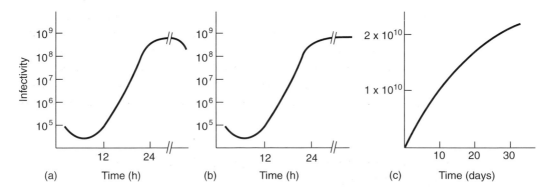

Fig. 13.2 Different types of infection caused by simian virus 5 in baby hamster kidney (BHK) cells and monkey kidney (MK) cells. (a) The acutely cytopathogenic infection in BHK cells. Note that virus yield drops after 24 h. (b) The initial one-step growth curve in MK cells which kills no cells and becomes persistent. (c) The cumulative yield from cells infected in (b) over 30 days. Cells grow normally during this infection and have to be subcultured at intervals of approximately 4 days.

a harmless persistent infection, whereas in BHK cells it causes an acutely cytopathogenic infection from which all cells die. The cell is thus all important in determining
the outcome of this relationship.

Persistent infections can also arise when viruses are able to inhibit the apoptotic
response that normally gives rise to an acutely cytopathogenic event, e.g. Sindbis virus
(a togavirus) kills normal cell lines, but persists in primary cultures of neurons. In the
neurons, Sindbis virus activates the host-cell gene, *bcl*-2, which prevents apoptosis.
Other viruses are known to have specific gene products (e.g. from adenovirus E1B)
that inhibit apoptosis. As the persistent infection allows the virus to multiply for a
long period of time, it is to the advantage of the virus to have evolved the means to
suppress apoptosis.

In thinking ahead to infection of the whole animal, the theory is that viruses evolve
to a state of peaceful coexistence with their host. In other words, the virus gains no
advantage in killing its host—rather the reverse because the virus depends absolutely
on the host for its very survival. Thus, the benign persistent infection of monkey cells
by this monkey virus may reflect a long evolutionary association. When a virus is
lethal to its host, we suspect that this is a new virus–host relationship which in time
will evolve so that fewer cells die.

Persistent infections resulting from interactions between viruses, cells and antibody, or viruses, cells and interferon

This is a situation in which the infection would otherwise be acutely cytopathogenic.
Usually, only a few cells are infected initially, and the addition of a small amount of
specific neutralizing antibody (often low-avidity antibody is most effective) decreases
the amount of progeny virus available to reinfect cells. Alternatively, the addition of
a critical amount of the natural antiviral, interferon (see Section 14.5) depresses the
extent of overall virus multiplication. The result is that the rate of infection and hence
cell death are exceeded by the division of non-infected cells, so that on balance the
cell population survives. This is a question of establishing a dynamic equilibrium tilted
in favour of the cell. Of course, the overall net production of cells is less than in an
uninfected culture but, in the animal, cell division would be upregulated by the
normal homeostatic mechanisms that control cell numbers. Indeed, this situation is
thought to mimic certain sorts of persistent infections in the whole animal (see
Chapter 15).

Persistent infections resulting from interactions of viruses, cells and DI virus

Defective–interfering genomes are produced by all viruses as a result of errors in replication, which delete a large part of the viral genome (see Section 7.1) making them
defective for replication. The DI genome retains the sequences that are needed for recognition by viral polymerases and for the packaging of the genome into a virus particle,
but needs nothing else. Thus, the DI genome is replicated only in a cell that has been

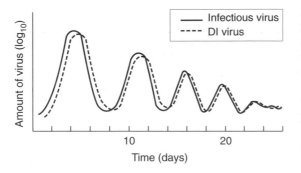

Fig. 13.3 A persistent infection established between a virus that normally causes an acutely cytopathogenic infection and defective–interfering (DI) virus. The dynamic cycles of production of infectious and DI virus (see text) eventually give way to a low-level, steady-state, persistent infection.

infected with infectious virus of the type from which the DI genome was generated, because this is needed to supply replicative enzymes and structural proteins. In this sense the DI genome is parasitic on infectious virus. The DI virus particles that result from this collaboration are usually indistinguishable from infectious particles. *Interference* comes about because more copies of the DI genome can be made in unit time than those of the full-length genome, e.g. if the DI genome represents one-tenth of the infectious genome, then, for every full-length genome synthesized, there will be 10 copies made of the DI genome. As viral polymerases are made in modest amounts, the large number of DI genomes produced will eventually sequester all the polymerase and synthesis of the infectious genome will cease. At this point, production of DI genomes declines as a consequence. In practice the situation is more complex, because the generation of DI genomes is very dependent on the type of cell infected, so that both cell and virus contribute to a balanced situation.

Persistent infections result when there is an equilibrium among the multiplication of infectious virus, the multiplication of DI virus and cell division. In the beginning, DI genomes are generated *de novo* and the increase of DI genomes follows that of infectious virus, upon which it is dependent (Fig. 13.3). Initially there is sufficient polymerase to allow replication of infectious and DI genomes, so there is no interference. Interference is apparent only when the number of DI genomes is so great that they sequester most of the polymerase proteins. At this point, the multiplication of infectious virus inevitably declines, which in turn results in a concomitant decrease in the dependent DI virus. When the amount of infectious virus reaches a low point, cell numbers recover. As they do so, infectivity increases in the relative absence of DI virus, but then the cycle of events is repeated. In this very dynamic way infectious virus persists under conditions where, without DI virus, it would produce a short-lived, acutely cytopathogenic infection. In some systems, the cycles become progressively smaller, until there is only a low level production of infectious virus and DI virus, and a steady-state persistent infection results (Fig. 13.3).

13.3 Latent infections

The term 'latent' is defined as existing, but not exhibited. Thus, in the context of a cell infected with a virus, this means that the viral genome is present but no infec-

Table 13.1 Examples of latent infections.

Virus	State of virus genome	Synthesis of		
		At least one transcribed RNA*	Viral protein(s)	Infectious progeny
Herpes simplex virus	Episomal	+	−	−
Epstein–Barr virus	Episomal	+	+	−
Adeno-associated virus	Integrated	+	+	−

*The amount and nature of gene expression is limited but varies between virus systems, and is strictly controlled.

tious progeny are produced. However, latency is an active infection and some virus-coded products are always expressed (Table 13.1). Lysogeny by temperate phages (see Chapter 12) is clearly a latent infection, and in animal cells *in vitro*, for example, herpesviruses and adeno-associated virus (AAV) exhibit latency. Understanding of the molecular control processes is still, however, at an early stage compared with phage λ. Bacterial viruses and AAV (see below), which achieve latency, do so by integrating a DNA copy of their genome into the host's genome. This ensures that the viral genome will be replicated, together with host chromosomal DNA, and transmitted to daughter cells; it will be protected from degradation by nucleases. On the other hand, the DNA of herpesviruses is not integrated, but remains episomal, although the normally linear molecule is circularized.

It is in fact inaccurate to call a virus latent, because latent infections always start and finish as acutely cytopathogenic infections. The initial, acutely cytopathogenic infection is converted into a latent infection, and then delicate molecular controls operate to maintain the latent state. Eventually, latency is broken by certain external stimuli, and there is full expression of the virus genome, production of infectious virus and re-establishment of the acutely cytopathogenic infection, giving the virus the opportunity of infecting new hosts. Thus, latency can be seen as an evolutionary strategy to remain in a host for a long period of time. In humans, latent infections with herpesviruses last for a lifetime, interspersed with periodic reverses into an acutely cytopathogenic infection. It is apparent that bacteriophage latency is fundamentally different from that of animal viruses, because the former is maintained primarily by virally encoded repressors of lytic replication (see Chapter 12), whereas the latter is controlled by host factors that are necessary for the expression of early virus gene products (see Section 15.6). The presence of a latent virus can sometimes be shown by using labelled antibody specific for viral proteins or, in all cases, by polymerase chain reaction (PCR) amplification of virus genome sequences.

Epstein–Barr virus (EBV), a member of the herpesvirus family, becomes latent when it infects B lymphocytes (non-dividing, non-antibody-producing cells that become antibody-synthesizing cells when activated by antigen) *in vitro*. Its very large 172 kilobase-pair (kbp) linear double-stranded DNA (dsDNA) is circularized and not integrated. As a result of infection the B cell is stimulated into continuous cell division. The viral DNA replicates semiconservatively, once each cell cycle, and each molecule segregates to a daughter cell like a host chromosome. During latency, up to nine of the 100 viral genes are expressed (see Section 9.6). However, about one in 10^3–10^6 cells convert spontaneously to the acute phase of infection, where the full set of genes is expressed, and infectious virus is synthesized and released. EBV is also associated with various malignancies (see Section 17.5).

Adeno-associated virus has a small single-stranded (ss) linear DNA genome of 4680 nucleotides (nt.). This is an unusual virus in that it is usually dependent on co-infection of the same cells with an adenovirus or a herpesvirus for its replication (see Section 6.6). In the absence of a helper virus, AAV becomes latent with the help of the Rep protein, integrating into chromosome 19 of the host genome, and is replicated as part of the host genome. There is minimal gene expression during latency. On superinfection with helper virus, and in the presence of Rep, latency is broken. Viral DNA is excised from the host genome and the production of infectious AAV commences.

13.4 Transforming infections

As a result of infection with a variety of DNA viruses or some retroviruses, a cell may undergo more rapid multiplication than its fellows, concomitant with a change in a wide variety of other properties, i.e. it is transformed. This is often preceded by integration of at least part of the viral genome with that of the host. Chapter 17 deals in detail with transformation and other aspects of tumour viruses. However, the term 'tumour virus' is a misnomer, because transformation is a rare event resulting from a virus that usually causes acutely cytopathogenic or persistent infections. Nevertheless, transformation is a significant event, because one immortalized cell can take over the whole population. For DNA viruses, transformation has no evident evolutionary significance, because the transformed cell contains only a fragment of the genome and cannot give rise to any progeny (Table 13.2).

13.5 Abortive infections

Any cell that possesses the appropriate receptors will be infected, but different cells do not replicate viruses with equal efficiency. There may be a reduction in the total

Table 13.2 Viruses that (very rarely) cause transformation of the infected cell.

Family	Genome	Proportion of genome integrated	Progeny
Retrovirus	RNA	Whole	+
Papovavirus	DNA	Part	−
Adenovirus	DNA	Part	−
Herpesvirus	DNA	None	−
Hepadnavirus	DNA	Part	−

yield of virus particles (sometimes to zero), or the quality of the progeny may be deficient as shown by their particle:infectivity ratio. Both of these reflect a defect in the production or processing of some components necessary for multiplication, be it DNA, RNA or protein. One example is avian influenza virus growing in the mouse L cell line, in which both the amount of progeny and its specific infectivity are reduced, probably as a result of the synthesis of insufficient virion RNA. Another is infection of other non-permissive cells with human influenza viruses, which gives rise to normal yields of virions that are non-infectious. This is because these cells lack the type of protease required to cleave the haemagglutinin precursor protein into HA1 and HA2 (see Section 11.7). It can be reversed by adding small amounts of trypsin to the culture or even to released virions. Abortive infections present difficulties to virologists trying to propagate viruses, but have been used to advantage when research into the nature of the defect has furthered understanding of productive infections. In natural infections, abortiveness contributes to the character of the infection, by effectively restricting virus to only those cells in which a productive infection takes place.

13.6 Null infections

This category represents the majority of cells which, lacking the appropriate receptors, cannot be infected. However, for many cells this is the sole block to infection, because, when they are transfected with infectious viral nucleic acids, they produce progeny virions showing that they are permissive to infection.

13.7 How do animal viruses kill cells?

The recognition that animal viruses often kill the cells in which they multiply gives the simplest criterion of infectivity, that of viral CPEs. However, such toxicity is not a property of the virus, but of a specific virus–cell interaction; as exemplified by simian virus 5 above, a virus does not necessarily kill the cell in which it multiplies (see below). Surprisingly, it is still by no means clear exactly how a cell is killed, except that in all cases viruses need to undergo at least part of the multiplication cycle. Thus, it seems that a toxic product is produced by the virus and that viruses with different replication strategies are likely to invoke different mechanisms of toxicity. One problem in these studies is in distinguishing between an effect on a cell function that operates early enough to be responsible for toxicity and those that appear late on and are a consequence of the toxic effects. Viruses can inhibit synthesis of host proteins, RNA and DNA, but here we deal only with the proteins, about which slightly more is known (Table 13.3).

Studies of the inhibition of host-cell protein synthesis by poliovirus suggest that the inhibition results from inactivation of those translation initiation factors that are responsible for recognizing capped messenger RNAs (mRNAs). Poliovirus mRNA itself is not capped and relies on a special mechanism of initiation. Its translation is therefore unaffected (see Section 10.4). In Semliki forest virus-infected cells, there is evidence that the virus inhibits host protein synthesis by affecting the plasma membrane

Table 13.3 Some suggested mechanisms of viral cytopathology.

Mechanism	Virus
Loss of ability to initiate translation of cellular mRNA	Poliovirus, reovirus, influenza virus, adenovirus
Imbalance in intracellular ion concentrations	Semliki forest virus, rotavirus
Competing out of cellular mRNA by excess viral mRNA	Vesicular stomatitis virus, Semliki forest virus
Degradation of cellular mRNA	Influenza virus
Failure to transport mRNA out of the nucleus	Adenovirus
Apoptosis	Sindbis virus, Semliki forest virus, influenza virus A, B, C, HIV-1, adenovirus, measles virus

HIV-1, human immunodeficiency virus type 1.

Na^+/K^+ pump, which controls ion balance. As a result, intracellular Na^+ concentration increases to a level where viral but not cellular mRNA is translated. The rotavirus non-structural protein, nsP4, has been shown in a number of studies to affect the ion balance in infected cells by altering the permeability of the plasma membrane and other cell membranes, and is toxic even when expressed in a cell by itself. It has thus been claimed as the first known viral endotoxin. In cells infected with vesicular stomatitis virus (VSV), a vast excess of viral mRNA succeeds in competing with cellular mRNA for translational machinery. However, there is evidence against this hypothesis, because a DI VSV, which is able only to undergo primary transcription, also kills cells. Here, it is thought that an early virus product is toxic. Recently it has become apparent that cells infected by some viruses die through apoptosis (see Section 13.1). There is no cell lysis during apoptosis.

Evidently, viruses do not kill cells by any one simple process, and we are far from understanding the complex mechanisms involved. However, cells would not die immediately their macromolecular synthesis is switched off, unless there was rapid turnover of some vital molecule. Thus, the mechanisms discussed above, with the exception of the upset in Na^+/K^+ balance and apoptosis, seem more akin to death by slow starvation than to acute poisoning. Lastly, it is by no means clear what advantage, if any, the virus accrues in killing its host cell, because most progeny viruses leave the cell by exocytosis or by budding out of the cell. However, there is a distinction to be made between cell death that leads to the death of the animal host, and death of cells (e.g. in the gut) that can readily be replaced by the host, often without the host realizing that they are infected. The latter does no harm, but the former suggests that the infection represents a very new virus–host interaction which is poorly evolved in terms of survival of virus and host (e.g. HIV-1—see Chapter 19; myxoma-virus—see Chapter 18.4) or the invasion of the 'wrong' type of cell (as in dengue haemorrhagic fever).

13.8 Recapitulation

• All virus–cell interactions can be classified under just six headings.

• Many viruses can be classified into more than one category, depending on the type of cell infected.

• Any definition of the type of infection caused by a virus should thus specify the circumstances of the infection.

13.9 Further reading

Carrasco, L. (1995) Modification of membrane permeability by animal viruses. *Advances in Virus Research* **45**, 61–112.

Holland, J. J. (1990) Generation and replication of defective viral genomes. In: *Virology* (2nd edn), pp. 77–99. Fields, B. N. & Knipe, D. M. (eds). New York: Raven Press.

Holland, J. J. (1990) Defective viral genomes. In: *Virology* (2nd edn), pp. 151–165. Fields, B. N. & Knipe, D. M. (eds). New York: Raven Press.

Jones, C. (1999) Alphaherpes virus latency: its role in disease and survival of the virus in nature. *Advances in Virus Research* **51**, 81–133.

Levine, B., Huang, Q., Isaacs, J. T., Reed, J. C., Griffin, D. E. & Hardwick, J. M. (1993) Conversion of lytic to persistent alphavirus infection by the *bcl*-2 cellular oncogene. *Nature (London)* **361**, 739–742.

Miller, L. K. & White, E. (1998) Apoptosis in virus infection. *Seminars in Virology* **8**, 443–523 (several articles).

Mims, C. A. & White, D. O. (1984). *Viral Pathogenesis and Immunology*. Oxford: Blackwell Scientific Publications.

Mims, C. A., Nash, A. & Stephen, J. (2000) *Mims' Pathogenesis of Infectious Disease* (5th edn). London: Academic Press.

O'Brien, V. (1998) Viruses and apoptosis. *Journal of General Virology* **79**, 1833–1845.

Oldstone, M. B. A. (1991) Molecular anatomy of viral persistence. *Journal of Virology* **65**, 6381–6386.

Roulston, A., Marcellus, R. C. & Branton, P. E. (1999) Viruses and apoptosis. *Annual Review of Microbiology* **53**, 577–628.

Shatkin, A. J. (1983) Molecular mechanisms of virus-mediated cytopathology. *Philosophical Transactions of the Royal Society of London, Series B* **303**, 167–176.

de la Torre, J. C. & Oldstone, M. B. A. (1996) Anatomy of viral persistence: mechanisms of persistence and associated disease. *Advances in Virus Research* **47**, 303–343.

Also check Chapter 22 for references specific to each family of viruses.

Chapter 14

The immune system, virus neutralization and interferon

Infections of multicellular animals are complicated by the variety of cell types present in an individual and the possession, in higher animals, of an elaborate defence against infection by any foreign invaders. It may be helpful to summarize the responses of the latter to virus infections. Broadly speaking, there are two types of host response: *innate immunity* and *adaptive immunity*. This chapter is intended as a refresher course in immunology as it relates to viruses and virus infections. Readers are referred to the specialized immunology texts for a more comprehensive treatment.

14.1 Introduction

Innate immunity

Innate immunity is the first line of defence and is possessed in some form by all animals. It is made up of soluble chemical components (e.g. interferon-α and -β and complement), and cellular components (macrophages, polymorphonuclear leukocytes (PMNLs) and natural killer (NK) cells). These do not have virus-specific recognition systems and thus act non-specifically, but they are present constitutively and there is no delay in mobilizing them. Their activity is not permanently increased as a result of experiencing an infection, although it may rise transiently.

Adaptive immunity

Adaptive immunity is specific for any foreign molecule, in our case a virus product (see below). It is mediated by the activation of B and T lymphocytes. These are inactive, non-dividing (resting) cells, which do not exert any immune response and are present in very low numbers. On their surface are epitope-specific receptors (protein molecules called BCRs—B-cell receptors—or TCRs—T-cell receptors). The epitope is defined as that part of the antigen that the BCR or TCR binds to (Fig. 14.1). Each cell synthesizes a receptor that is specific for just one epitope. When a receptor binds to its cognate epitope, the cell is stimulated to undergo a large number of divisions. The daughter cells then differentiate into *effector* cells, which mount immune responses, or *memory* cells (see below). In effect such virus-specific cells adapt to the ongoing infection and can deal with it. However, all this takes a few days, during which time the

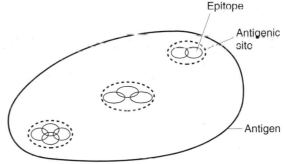

Fig. 14.1 Schematic inter-relationship of the epitope (or antigenic determinant), antigenic site (a collection of overlapping epitopes) and antigen. However, a T-cell receptor (TCR) reacts only with an epitope on a peptide derived from a protease-processed protein (see text).

virus can multiply and cause clinical disease. T- and B-effector cells are non-dividing, and short-lived. A population of T- and B-effector cells is maintained in the body only by the continued stimulation of the production of new effector cells by antigen. Detection of long-lived T-effector cells, or more usually B-effector cells making antibody, indicates that long-lived antigen is present. This is not necessarily in the form of infectious virus and it is not known where exactly it is located. After the virus is vanquished, the population of T- and B-effector cells falls, but the population of new cells called memory cells, which have the same receptor specificity, remains. On reinfection by the same virus, there is now a sufficiently large population of memory cells to provide effector cells that repulse the infection before it can cause disease. These cells constitute *immunological memory*. Memory cells closely resemble lymphocytes in morphology and lack of effector functions. They require stimulation by the cognate epitope (one that reacts with a particular antigen receptor) just like lymphocytes, but respond at a lower receptor: virus concentration.

T cells were the first type of lymphocyte to evolve, and their activity enhanced that of the innate immune system. As discussed above, stimulation by the cognate virus results in a cloned population of T-effector cells that is responsible for *cell-mediated immunity*. B lymphocytes evolved more recently and, as will be seen, provided the immune system with another type of weapon to counter infection. They first appeared in primitive fish. B lymphocytes develop in much the same way as T cells; their receptors bind the cognate virus epitope, and then they divide and differentiate into memory cells and B-effector cells that synthesize antibody (immunoglobulin). This is called humoral or antibody-based immunity.

Antibody made by a B cell is identical in sequence to that of BCR except that the latter has an extra domain that anchors it in the plasma membrane. In effect, an antibody is a soluble BCR. Each T or B cell has in its plasma membrane many copies of the TCR or BCR, respectively. Both receptors recognize not the whole virus particle, but a small region (the epitope) of a foreign molecule (the antigen). Antibody can recognize an epitope that is protein, lipid, carbohydrate or nucleic acid, provided that it is 'foreign' (see below). Embryologically, we develop with a random set of TCRs and BCRs but, early in life, cells bearing receptors that can react with our own epitopes are eliminated, so that, by definition, all remaining BCRs and TCRs recognize only foreign epitopes. This is the immunological doctrine of self and non-self. In respect to proteins, it can be said that a BCR binds to the whole molecule, but it actually recognizes and interacts with a small region of approximately 17 amino acid residues, which constitutes its epitope. The TCR is much more restricted and recognizes only peptides. These are 8–22 amino acids in length and are derived by intracellular proteolysis of proteins, as described below. The peptides are always found complexed with cellular plasma membrane proteins, called major histocompatibility complex (MHC) proteins. A TCR recognizes only its cognate peptide in a complex with an MHC protein. Any one B cell or T cell and its clonal descendants have epitope receptors of only one sequence, and so recognize only one epitope.

Although immune cells circulate as single cells, after meeting their cognate antigen they migrate to one of the lymphoid centres. There are many of these loosely organized aggregations of cells, and the best known are the lymph nodes, located at strategic sites around the body (e.g. at the junction of limbs and torso, torso and head), and Peyer's patches in the gut wall. Here their maturation into effector cells takes place. Activated B cells stay there and secrete antibody, but activated T cells return to patrol the body.

The immune system is usefully thought of as consisting of two parts: the systemic immune system (which looks after the body as a whole) and the mucosal immune system (which looks after the mucosal surfaces) (Fig. 14.2). The latter comprise the surfaces of the respiratory tract, the intestinal tract, the urogenital tract and the conjunctiva of the eye. These are important in virology because most viruses (and indeed other micro-organisms) infect through these sites. Mucosal surfaces are particularly vulnerable because they are very extensive in area, and their physiological activities require them to be composed of naked epithelium, unlike the skin which is largely impermeable to viruses. The problem is that the cells and antibodies that constitute systemic immunity do not get through in sufficient concentration to protect the mucosae. However, the mucosae have their own lymphocytes, and these are stimulated when antigen is directly administered to the mucosal surface. This is particularly

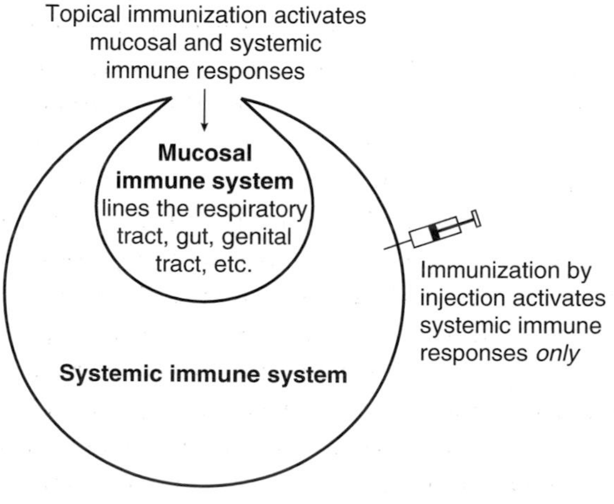

Fig. 14.2 Schematic relationship of the compartmentalization of the systemic and mucosal immune systems.

relevant to the use of non-replicating vaccines, which have been traditionally administered by subcutaneous or intramuscular injection. However, if the mucosa itself is stimulated with antigen, this stimulates immunity in both the mucosal and the systemic immune systems. Furthermore, the mucosal areas are connected, so that immunity raised at one mucosal site will also be found at other mucosal sites. This is mediated by the migration stimulated T and B cells.

All innate and adaptive immune responses act in an interlocking and orchestrated fashion. It will be no comfort to the student to learn that for each infection there is one specific response that is the key to overcoming it, even though all responses are stimulated. Only the key response is effective, the other responses being ineffective. To compound this difficulty, there is no way of predicting which element of the immune system is needed to overcome a particular virus infection. The concept of key immune responses has clear and important implications for vaccines (see Sections 16.1–16.7).

Figure 14.3 summarizes the essential parts of the very complex action of the immune system against virus infection. The cells of the immune system that patrol the body are single and mobile. Many of these, the T and B lymphocytes, are purely sentinels and cannot carry out offensive action until they have clonally expanded and differentiated. There are two targets: virus particles and the factories that make them—virus-infected cells—and the immune response acts in the three ways described in Sections 14.2 and 14.3.

14.2 Innate immunity (Fig. 14.3, lower left)

There are two types of cells that carry out phagocytosis as their main function (these are sometimes called professional phagocytes to distinguish them from other cells that all have endocytic activity). There is a strict division of territory, with the PMNLs in the blood and the macrophages actually moving within the tissues. Both types of cell are constitutively capable of phagocytosing virus particles, although their activity can be transiently increased by chemical messengers. Phagocytes destroy virus particles in their lysosomal vesicles by means of a battery of powerful enzymes. However, some viruses can avoid being killed and turn the tables by infecting macrophages, sometimes with dire consequences. This happens with human immunodeficiency virus type 1 (HIV-1; see Chapter 19) and with Dengue fever virus complexed with non-neutralizing antibody (see Section 5.1). In the latter case the Fc region of the antibody binds Fc receptor proteins in the plasma membrane, and these then act as surrogate virus receptors to permit infection of otherwise uninfectable macrophages.

If innate immunity is effective the infection is aborted with no trace of it ever happening in terms of B- or T-cell responses. However, if infection is initiated, other parts of the innate system are called in—the first being the cellular proteins interferon-α and interferon-β. These are so similar in structure and function that they often called interferon-α/β. These molecules have two functions: they can directly inhibit virus replication without killing the cell, as described in Section 14.6, and they can upregu-

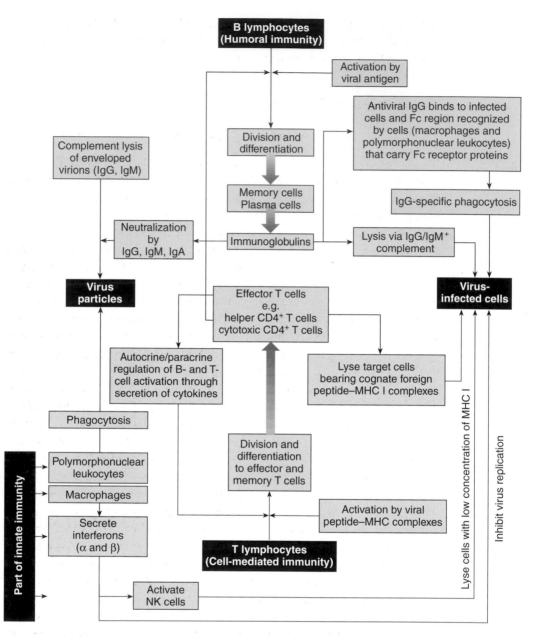

Fig. 14.3 Summary of some of the responses of the immune system to viruses and virus-infected cells. MHC, major histocompatibility complex; NK, natural killer.

late expression of MHC I proteins, which enhances the antiviral activity of T cells by increasing the concentration of foreign peptides presented on the cell surface (see Section 14.3). The last major cell of the innate system is the NK cell, which detects the presence of abnormally low concentrations of MHC proteins on the surface of cells. This is useful as it is often a sign of virus infection. The NK cell is able to lyse and destroy such cells. Thus, innate immunity can destroy both virus particles and infected cells, as well as inhibiting virus replication.

14.3 Adaptive immunity

T-cell-mediated or cell-mediated immunity (Fig. 14.3, lower centre)

The second arm of the immune system comprises the T cells. Any one T cell has many copies of an identical, epitope-specific receptor, the TCRs (see above), which recognizes a cognate foreign peptide complexed with an MHC protein on the surface of the cell. (In humans these are referred to as HLA[1] proteins. HLA stands for human leukocyte-associated antigen, an historical term, which is misleading because these proteins are present on all cells except red blood cells.) All cells except red blood cells express MHC class I proteins, but a few types of cell, notably dendritic cells and B lymphocytes, also express MHC class II proteins. There are three pairs of genes encoding MHC I proteins and another three pairs encoding MHC II proteins. The genes that code for MHC proteins are highly polymorphic, and expressed co-dominantly, so that in an outbred population a cell expresses six different MHC I and six different MHC II proteins. Unless complexed with a peptide, MHC proteins do not mature. Any one MHC protein combines specifically with its peptide through recognition of two terminally situated amino acid residues which anchor the peptide in a groove on the MHC protein. In the cell, peptides with the appropriate specificity associate with MHC at random, so in the non-infected cell these will all be self-peptides and, in the virus-infected cell, a proportion will be viral peptides. These peptides are formed by intracellular proteolytic processing by proteosomes, and complex with MHC proteins rather like an egg in an egg cup. Class I MHC proteins complex with peptides of 8–10 amino acids and class II MHC proteins with peptides of 17–22 amino acids. The complex is transported to the plasma membrane, where the peptide is displayed on the outside of the cell (antigen presentation). The epitope is the peptide sequence situated between the anchor residues. Thus, any one MHC protein can present an infinite number of epitopes, provided that they all have the appropriate anchor residues.

The TCR of a T lymphocyte recognizes an infected cell through a foreign peptide held by an MHC molecule, which results in its activation (clonal expansion and differentiation) to become a functional T-effector cell. T lymphocytes fall into two categories, which express either CD8 or CD4 proteins in their plasma membranes and are hence referred to as CD8[+] T cells or CD4[+] T cells. CD8 is a ligand for MHC I proteins and CD4 for MHC II proteins. This recognition is additional to that of the TCR for the foreign peptide, but equally important. All cells of the body have MHC I proteins and hence are under constant immune surveillance by CD8[+] T cells. There is a quantitative element to the recognition of MHC–peptide complexes by a T cell, and the sensitivity of this process is proportional to the MHC concentration—hence the importance of the interferons that upregulate MHC I protein expression in facilitating an antiviral immune response. Conversely, many viruses have evolved functions that

Table 14.1 Some functions of activated T cells.

	Cytotoxicity	DTH	Help	Suppression
CD4$^+$ T cell	+	++	+++	+
CD8$^+$ T cell	+++	++	−	++

+++ Indicates a major activity.
DTH, delayed-type hypersensitivity.

downregulate expression of MHC proteins but, when this reaches a level that T cells cannot recognize, the cell can be identified and killed by NK cells (see Section 14.2). One of the major functions of activated CD8$^+$ T-effector cells is to destroy MHC-I-positive cells that present foreign, i.e. viral, peptides (Table 14.1). However, in some infections (measles) CD4$^+$ T-effector cells predominate, although these kill only cells that express MHC II proteins. T cells carrying out this activity are also known as cytotoxic T cells or activated cytotoxic T lymphocytes (CTLs).

The major function of activated CD4$^+$ T cells is help and only these T cells can provide it. Help is a positive regulatory function which is essential for the activation of all T and B lymphocytes by their cognate foreign epitope. Thus, the entire immune response depends on such helper T cells—this is the main type of cell infected by HIV and explains why their destruction is so devastating (see Chapter 19). One specific helper function of CD4$^+$ T cells is to provide the cytokines that control immunoglobulin heavy chain gene switching, and drive an IgM-synthesizing cell to change to the synthesis of one of the other immunoglobulins (IgA, -G, -D or -E—see below) while retaining epitope specificity.

Help and many other functions of T cells are mediated by cytokines, which are soluble proteins through which cells of the immune system communicate with each other. Cytokines act exactly like hormones except that they operate over a very short distance, and their effects are said to be either paracrine, affecting neighbouring cells, or autocrine, affecting the cytokine-secreting cell itself. The effects can be either positive (i.e. stimulatory) or negative. In this way the fine regulation of the immune system is achieved. There are a large number of cytokines (>100), and their effects may be overlapping, additive or synergistic. Understanding them is the key to being able to manipulate the immune system in terms of making more effective vaccines, and to treating the excesses of the immune response which can manifest itself in pathological conditions such as autoimmunity, allergy and hypersensitivity reactions. One of the cytokines is interferon-γ, which like interferon-α/β can inhibit virus replication. However, this is their only common attribute (see Section 14.5).

Delayed-type hypersensitivity or DTH reactions occur when T cells locate cognate antigen and secrete cytokines that cause macrophages and PMNLs to migrate to the site of antigenic stimulation. If the tissue site is immediately below the skin, it will be apparent as being swollen, red, hot and painful, and this is inflammation. Part of this reaction is the result of a transient increase in permeability of blood vessels in that

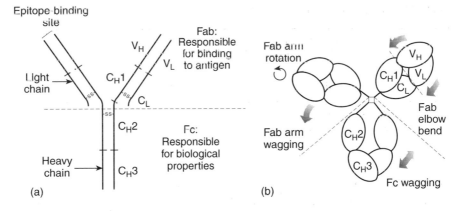

Fig. 14.4 Generalized immunoglobulin molecule, showing (a) the outline structure, consisting of two identical dimers formed of H–L (heavy and light) polypeptide chains. Note the sequence-variable (V) region of the H and the L chains. Each contains three hypervariable sequences each of 10–20 residues (not shown), that are folded together to form the unique epitope-binding site. The remaining H and L sequences are relatively constant. The constant region is subdivided into domains with sequence homology (C_H1–C_H3). IgM and IgE both have an additional domain (C_H4). The molecule is also divided into an N-terminal half that binds epitope (Fab), and a C-terminal half (Fc) that is reactive with various cell mediators. IgM and IgA form pentamers and dimers, respectively, linked by their C' ends with a joining (J) polypeptide. SS, disulphide bond. (b) The globular domains, with arrows indicating the flexibility, which allow the molecule to bind to two (identical) epitopes, can be different distances apart in three dimensions. (b, From Burton and Woof 1992.)

area, which results in an influx into the tissue of plasma that contains antibody and cells of the immune system. The same occurs in deep lying tissues, e.g. the lung. In moderation, this response helps to clear the infection, but in excess it can cause tissue damage—as with any immune response.

Humoral or antibody-mediated or B-cell-mediated immunity

(Fig. 14.3, upper centre)

Antibodies (immunoglobulins) are synthesized by cells variously called plasma cells, activated B cells or antibody-forming cells. Their immediate progenitors are B lymphocytes, which have been stimulated to divide and differentiate, by their contact with cognate epitope, to form a clone of plasma cells and memory cells. The plasma cells (now located in a lymph node) synthesize and secrete huge numbers of antibody molecules, which react specifically with the stimulating epitope. Immunoglobulins (Ig) have the basic structure seen in Fig. 14.4. The five types are distinguished by having different heavy chains called alpha (α), delta (δ), epsilon (ε), gamma (γ) and mu (μ), which characterize IgA, IgD, IgE, IgG and IgM, respectively. Light chains are either kappa (κ) or lambda (λ). There has evolved a division of labour among the antibodies, so that they have specialized functions or are found in certain locations in the body that make the immunological defence more efficient. The main immunoglobulins are as follows.

Immunoglobulin M

IgM is the first antibody synthesized when lymphocytes are activated. It is composed of five IgM monomers covalently linked at their C-termini. Its epitope binding site is of low affinity but, being pentameric, its overall avidity is high. It is found only in blood and along with IgG is the only antibody capable of activating complement (see below). The presence of virus-specific IgM in serum indicates that there has been a recent new infection—it contributes to neutralization.

Immunoglobulin G

IgG is the most abundant antibody and most important for neutralization. In humans there are in fact four IgG antibody classes—IgG1–IgG4—all with different properties. IgG is a monomer and is small enough to leave the bloodstream and bathe the tissues, and to cross the placenta where it helps to protect the developing foetus and later newborns from infection until its own immunity is fully developed. IgG activates complement and binds to Fc receptor proteins on the surface of phagocytic cells. Stimulation of mucosal surfaces gives rise to the synthesis of IgG at the mucosal surface and is known as 'local IgG'.

Immunoglobulin A

IgA is a key defender of the mucosae mentioned above, and exceeds IgG in total amount produced and in local concentration. It has the unique property of being secreted across the mucosal epithelium, so that it is found within the lumen of the gut, respiratory tract and urogenital tract. This main type of IgA is a dimer, but monomers also exist. It is particularly resistant to degradation by digestive enzymes. It also neutralizes viruses.

Neutralizing immunoglobulins bind to viruses and cause them to lose infectivity (see Section 14.4). The activation of complement by IgG or IgM bound to viruses can enhance neutralization or permit virus inactivation by non-neutralizing antibody. Complement is a nine-component system of soluble proteins found in blood. Activation of the first complement component activates the next by proteolytic cleavage and so on, with progressive amplification. It is a classic biochemical cascade system similar to blood clotting. Usually, each component is cleaved into a part that adheres to the antigen–antibody complex or close to where the antibody has bound, and a diffusible part, which forms a chemical gradient and attracts cells of the innate immune system into the vicinity of the antigen. The final stage is the insertion of pore structures (the membrane attack complex) into the cell membrane, which, in sufficient number, create an ionic imbalance and lyse that cell. Thus, enhanced neutralization is probably brought about by a build-up of complement proteins on the surface of the virus, so that attachment of virus to cell receptors is sterically prevented, although enveloped viruses may have pores inserted into their lipid bilayers that permit the entry of nucleases, etc.

However, there is no evidence that complement is an important part of antiviral defence *in vivo*, e.g. people who have a congenital complement deficiency do not have an increased number or severity of virus infections, although they are known to be at increased risk from certain bacterial infections. Other aspects of their immune response may, however, be increased to compensate for lack of complement activity.

By themselves, antibodies—which are the size of a single membrane protein—do no harm to an infected cell by binding to cognate virus antigen on its surface, and antibodies do not enter cells because they cannot cross lipid membranes. However, bound IgG or IgM can activate complement and, in sufficient amounts, this can lyse the cell as described above. Cells can repair minor damage and it is estimated that it takes about 10^5 complement pore structures inserted into the plasma membrane to kill a cell. Alternatively, bound IgG can act as a ligand for phagocytic cells (macrophages in the tissues) that have Fc receptor proteins on their cell surface. These receptors specifically recognize and bind to the Fc region of IgG, and lead to the phagocytosis and destruction of the infected cell. In these ways components of the innate immune system can be made to act specifically.

14.4 Understanding virus neutralization by antibody

Neutralization is the loss of infectivity that ensues when antibody binds to a virus particle. Viruses are unusual because neutralization is usually mediated by antibody alone, whereas larger organisms, such as bacteria, require the action of secondary effectors such as complement which recognize bound antibody. The antibody–antigen reaction is so specific that it is unaffected by the presence of other proteins. Hence, antibodies need not be extracted from crude serum, and impure virus preparations can be used to observe neutralization. An antibody molecule recognizes and combines with part of the antigen, called an *epitope* (see Fig. 14.1). A protein epitope is a planar surface of about 16 amino acids which interacts with a complementary surface that forms the binding site (or *paratope*) of the cognate antibody. Ten or so overlapping epitopes constitute an *antigenic site*, and several such antigenic sites are present on the surface of a virus particle. However, not all antibodies that bind to a virus particle are capable of neutralizing its infectivity. This is an epitope-specific phenomenon. The role of antibody in the recovery from viral diseases and prevention of reinfection is covered in Chapters 15 and 16, but it is appropriate here to discuss how antibody renders virus non-infectious (or neutralized).

After infection or immunization, an antiserum can contain antibodies to many different viral epitopes, as well as antibodies to unrelated antigens to which the animal has been exposed. Hence analysis of this complex mixture is difficult. An antibody-synthesizing B cell makes antibody of just one sequence (i.e. it is monoclonal), but such cells do not divide and are of little practical use. The problem was solved in 1975 by G. Köhler and C. Milstein who devised a way to immortalize antibody-synthesizing cells by fusing each one with a cell from a B-cell tumour that no longer made its own antibody. This *hybridoma* can then be cloned and grown in the laboratory. Each cloned cell line thus synthesizes a *monoclonal antibody* or MAb. The fusion is a random process so many clones must be generated, and the desired antibody is identified by its reaction with antigen.

Early work assumed that neutralizing antibody prevented virus from attaching to receptors on the cell surface. However, when this assumption was tested with influenza A virus and poliovirus, the majority of neutralizing MAbs did not block attachment, although attachment inhibition was the mechanism operating with most rhinovirus-specific MAbs. However, the mechanism of neutralization is antibody specific, not virus specific, and one virus can be neutralized in a variety of different ways, determined largely by the epitope to which the antibody binds. Rather surprisingly, no antibody is made to the attachment sites of influenza A virus, poliovirus or rhinovirus. These sites are contained in depressions on the surface of the virus, where they are hidden from the immune system. Rhinovirus-specific neutralizing antibody attaches to and bridges amino acids on either side of the rhinovirus attachment site and blocks it indirectly. Antibody attached to virions can be visualized by electron microscopy as a fuzzy outer layer. However, it can be diluted to an amount that is no longer detected by electron microscopy, but is still sufficient to neutralize infectivity thoroughly. Such a small number of antibody molecules per particle are unlikely to interfere with attachment. Interference with attachment of influenza virus would be particularly inefficient, because there is a high density (500–1000) of trimeric attachment proteins (the haemagglutinin) per virion, and an IgG molecule is slightly smaller than a haemagglutinin protein. These conclusions are also in accord with the biochemical evidence discussed below.

Not all antibodies that attach to virus particles neutralize infectivity. Such antibodies are *non-neutralizing antibodies* and they attach to *non-neutralizing epitopes*. This emphasizes just how specific the neutralization reaction is, and that antibodies bound to the surface of a virus do not necessarily interfere with the attachment process. There is a further dimension to this, because a non-neutralizing antibody bound close to a neutralizing epitope can sterically prevent the binding of neutralizing antibody, and thus allow the virus to evade the immune response.

So how do viruses lose infectivity? Recent studies have demonstrated that poliovirus neutralized by an MAb of one particular specificity attaches to cells and is taken up in an endocytic vesicle, but fails to uncoat and release the viral genome into the cell. In a second example, an influenza virus neutralized by one of its MAbs attaches and is endocytosed, but fusion of the viral envelope and cell membrane does not take place. Current thinking is that there are as many mechanisms of neutralization as there are postattachment processes that a virus undergoes before its genome can be expressed. Any mechanism of neutralization must be defined by all the relevant parameters, which includes the neutralization protein, the epitope, the isotype of antibody, the antibody:virus ratio and the cell receptor.

At first sight it is puzzling that the same MAb will neutralize influenza virus by a mechanism that does not prevent the attachment of neutralized antibody to cells in culture, and yet can cause haemagglutination inhibition (see Fig. 2.8), which does result from the antibody preventing attachment of virus to red blood cells. The explanation probably lies in the nature of the cell receptor unit, *N*-acetylneuraminic acid

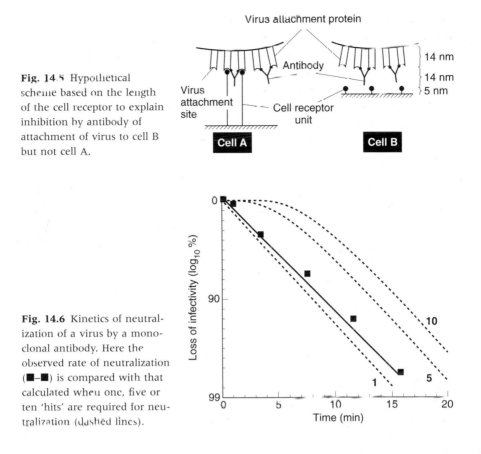

Fig. 14.5 Hypothetical scheme based on the length of the cell receptor to explain inhibition by antibody of attachment of virus to cell B but not cell A.

Fig. 14.6 Kinetics of neutralization of a virus by a monoclonal antibody. Here the observed rate of neutralization (■–■) is compared with that calculated when one, five or ten 'hits' are required for neutralization (dashed lines).

(NANA), or rather the molecule that carries the carbohydrate moiety of which NANA is a component. It is known that a protein carries most of the NANA on the surface of the red blood cell and that this is a small molecule which protrudes only a short way from the cell surface. Thus, antibody sterically prevents virus from attaching to the red blood cell. Presumably the (unknown) molecule that carries NANA on the cultured cell is longer and this permits the antibody to interdigitate with it, and enables the NANA to interact with the virus attachment site (Fig. 14.5).

Kinetics of neutralization

With the advent of MAbs, it became possible to tackle the problem of how many molecules of antibody per virion were needed to cause neutralization. This can be approached through a study of the kinetics of neutralization. These can be pseudo-first order, where there is a straight-line relationship between the $\log_{10}$ of the surviving fraction plotted against time, or second order, where the plot has a shoulder before there is any loss of infectivity (Fig. 14.6). First-order kinetics are usually interpreted as meaning that the infectivity of one virus particle is inactivated by the binding of one molecule of antibody, whereas second-order kinetics require that two or more molecules act cooperatively to bring about neutralization. Most neutralization reactions (Fig. 14.6) follow first-order kinetics. This has implications for the mechanism of neutralization because: (i) the number of attachment sites per virion ranges from 60 (picornaviruses) to around 3000 (influenza A virus); and (ii) a bound IgG molecule occupies a relatively small part of the surface area of a virion. Thus, it is inconceivable that one IgG molecule could block attachment. Rather, it appears that the IgG molecule either prevents uncoating of the virion from taking place or triggers the uncoating process prematurely; in either case, the virus particle is unable to initiate infection.

The concept of critical neutralization sites

Although kinetic data demonstrate that one molecule of antibody can neutralize a virus particle, the biochemical data for poliovirus and influenza virus show that, in fact, many more molecules have attached. If, indeed, one molecule of antibody per virion is neutralizing, the Poisson distribution predicts that, when there is an average of one molecule of antibody per virion, 37% of viruses have no attached antibody— and hence are not neutralized. Conversely, 63% of viruses will have bound one or more molecules of antibody and will be neutralized. However, assay of monoclonal antibody attached to the haemagglutinin (HA) of influenza virus that was 63% neutralized showed that there were in fact about 70 molecules of antibody per virion (Taylor *et al.* 1987). This paradox can be resolved by postulating that only one of the 70 antibody molecules neutralizes and is said to have bound to a *critical site*; the other 69 antibody molecules are bound to *non-critical sites*, which, by definition, are non-neutralizing. How do critical and non-critical sites differ? As a MAb was used, all epitopes are by definition identical with respect to binding antibody; the HA bearing the critical site is postulated to differ by its association with another component(s) of the virion (Fig. 14.7). The non-critical HA spikes are not so associated. These aspects of neutralization are still being elucidated.

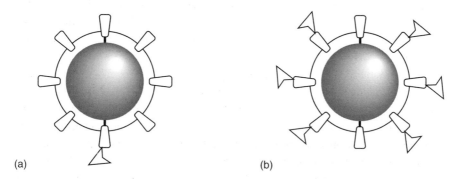

(a)   (b)

Fig. 14.7 The concept of critical and non-critical neutralization sites applied to influenza virus. All the haemagglutinin (HA) surface spikes are identical in structure but two are critical by virtue of their connection with internal virion structures. Binding of an antibody molecule to either of these (a) results in neutralization by an unknown mechanism. In (b) antibody has bound to all the non-critical sites and the particle is still infectious (see text).

There are conditions under which virions can bind neutralizing antibody and yet not be neutralized. This occurs when the concentration of antibody is too low, or the affinity of the antibody is so low that it dissociates, or because infectious virions are non-specifically aggregated with neutralized particles, protecting them from contact with antibody. Alternatively, there may be virions in a population that cannot bind a particular MAb and so are not neutralized. These are *neutralizing antibody escape mutants*, which occur naturally at a frequency of about 1 in 100 000. They usually have a point mutation, which results in the substitution of a crucial single amino acid within the epitope and abrogates binding of antibody. None of the progeny of an antibody escape mutant is neutralized either. Thus, wild-type virus is neutralized and the escape mutant replaces it in the population. Sequencing of the gene encoding the neutralization protein locates the mutation and in this way, with sufficient MAbs, the epitopes and hence the antigenic sites of a virus particle can be mapped.

14.5 Interferons

The interferons are host-cell proteins with antiviral and other activities, which were discovered in 1957 by Alick Isaacs and Jean Lindenmann. They incubated chorioallantoic membranes from embryonated chicken eggs with heat-inactivated, non-infectious, influenza virus in buffered saline, washed the membrane to remove any non-adsorbed virus and continued the incubation for 24 h. The membranes were then discarded and the buffer solution was tested for antiviral activity. This was done by placing a fresh chorioallantoic membrane in the buffer and inoculating with infectious influenza virus. Membranes so treated did not support the growth of active virus, in contrast to untreated membranes. It was concluded that an extracellular, virus-inhibitory product had been liberated in response to the heat-killed virus, and this substance was named interferon.

We now know that there are three types of interferon, which are classified as cytokines: interferon-α and interferon-β are similar in structure and function, and can potentially be synthesized by most cells. Expression of both is stimulated by replicating or abortively replicating virus. Interferon-γ is a completely different molecule, and is released by stimulation of activated T cells by their cognate antigen. The major problem of studying interferons was that they are, similar to many other chemical messengers of the body, present in very small amounts. Much effort was devoted to their purification and they were demonstrated to be glycoproteins, with carbohydrate groups that were essential for activity. Since the advent of genetic engineering, milligram concentrations of interferon can be produced by expressing cloned interferon genes in eukaryotic cells, usually yeast. The antiviral activity of interferons can be measured by inhibition of the incorporation of radioactive uridine into viral RNA in cells infected by an interferon-sensitive virus. Activity is expressed as the amount of interferon required to reduce the normal level of viral RNA by 50% and this is arbitrarily defined as one unit. Purified interferon has a very high specific activity of around 10^9 units/mg protein, which is of the same order as hormones.

Molecular cloning has allowed the determination of the sequences of interferon-α, -β and -γ. This has confirmed the amino acid sequence similarity (35% homology) of interferon-α and interferon-β and their difference from interferon-γ. Although interferon-β and interferon-γ are each represented by a single sequence in the human genome, there is a family of about 12 interferon-α genes which differ slightly in sequence, but apparently not in function. Now that pure interferon is available, it is clear that the antiviral activity is only one of several physiological effects so far discovered. Interferon enhances the activity of NK cells and macrophages (see above). In addition, interferons serve as both positive and negative regulatory controls in the expression of innate and adaptive immune responses. One very important effect is the upregulation of expression of MHC class I proteins by interferon-α and interferon-β, and of both MHC class I and class II proteins by interferon-γ. Interferon-γ also induces the *de novo* transient expression of MHC class II proteins on epithelial cells, fibroblastic cells, endothelial cells (which line blood vessels) and astrocytes (which provide nutrition for neurons in the central nervous system). Upregulation of MHC I proteins on cells increases the efficacy of their interaction with CD8$^+$ T cells, and the *de novo* expression of CD4 allows these cells to interact with CD4$^+$ T cells for the first time.

The ready availability of cloned interferons has permitted the investigation of their antiviral activity in humans in double-blind clinical trials. In such trials, half the participants are given a placebo, an innocuous substance without therapeutic effect, and the other half are given interferon. Neither the trial subjects nor the clinician who monitors the symptomatology of the disease knows who has received interferon or placebo, and thus there can be no subjective bias from either patient or doctor. Generally the results have been disappointing. However, treatment of persistent viral infections, such as those caused by hepatitis B virus and hepatitis C virus, diminishes virus replication, although these effects are often transitory and virus can reappear when treatment is withdrawn (see Section 16.8). Interferons also slow down the spread of some cancers, but they are less effective than conventional chemotherapy with antimitotic compounds. Future work entails testing the 12 cloned interferon-α, one interferon-β and one interferon-γ in various permutations, chemical modification to enhance their activity, and much more fundamental research directed towards increasing understanding of their properties.

In the last regard, use has been made of stem cell techniques specifically to disrupt (knock-out) genes in mice. Initially the animals are chimeras composed of knock-out and normal cells; a chimera is mated with a normal mouse to give rise to progeny that are uniformly heterologous for the knock-out, and then heterologous animals are mated to produce the homologous knock-outs. The resulting animals lack a specific gene function. Mice have been produced with no functional interferon-γ gene or no functional interferon-γ receptor gene. Such mice develop normally, but have impaired macrophage and NK cell function, and reduced amounts of macrophage MHC class II protein. CD8$^+$ cytotoxic T-cell activity, CD4$^+$ T-helper cell and antibody responses were normal, but the mice showed increased susceptibility to vaccinia virus, and to the bacteria *Mycobacterium bovis* and *Listeria monocytogenes*. Thus interferon-γ is important in the defence against these micro-organisms.

The direct antiviral action of interferon-α and interferon-β

This antiviral effect is called 'direct' to distinguish it from the indirect antiviral effect when interferon-α and interferon-β stimulate the expression of MHC I proteins, described above. Most viruses induce interferon and, once induced, this is active against the whole spectrum of viruses, not just the virus responsible for its induction. The action of interferon can be divided into two stages (Fig. 14.8). The first stage is *induction* in which the genes coding for interferon are derepressed and the interferon protein is released from the cell. Induction of interferon in human cells is controlled by chromosome 9. All multiplying viruses induce interferon and the specific inducer is double-stranded RNA (dsRNA). It is not necessary for the inducing virus to have a completely double-stranded genome, because single-stranded cRNA can at least partially base pair with its template or itself to form dsRNA. Some DNA viruses transcribe RNA from both genome strands, and these can then form dsRNA and act as an inducer of interferon (see Chapter 9).

The second stage is the creation of an *antiviral state* in neighbouring cells by the released interferon binding to interferon-specific receptors on their cell surface. In humans, the receptor is encoded by chromosome 21. However, the fully active antiviral state is not achieved until the cell is infected, and virally encoded dsRNA is again responsible. Induction of the antiviral state by interferon allows the cell to respond in two ways, both of which curtail virus replication. First, dsRNA stimulates the phosphorylation of certain cellular proteins, notably the eukaryotic initiation factor, EIF2α, which impairs their function in the initiation of protein synthesis. Second, dsRNA activates a ribonuclease (RNaseL) which degrades mRNA, and also stops protein synthesis. In fact, the mechanism is more complex (Fig. 14.9): binding of interferon to interferon receptors stimulates the synthesis of 2′,5′-A synthetase, but this is inac-

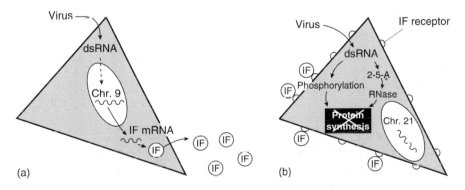

Fig. 14.8 (a) Induction of interferons-α/β (IF) by viral double-stranded (ds) RNA and its release from the cell. The dashed arrow indicates that we do not know whether dsRNA is required to enter the nucleus to initiate interferon mRNA synthesis or if it acts through an intermediate. (b) Binding of interferon by nearby cells creates in them an antiviral state through inhibition of protein synthesis by two different mechanisms. RNase, ribonuclease.

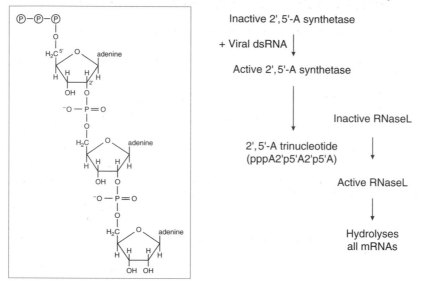

Binding of interferons to their receptors stimulates synthesis of:

Inactive 2',5'-A synthetase

+ Viral dsRNA ↓

Active 2',5'-A synthetase

Inactive RNaseL

2',5'-A trinucleotide
(pppA2'p5'A2'p5'A)

Active RNaseL

Hydrolyses
all mRNAs

Fig. 14.9 Action of interferons-α/β: activation of RNaseL. Inset: the formula of the 2',5'-A trinucleotide.

tive without the viral dsRNA, which is produced when the cells are infected. The dsRNA-activated 2',5'-adenine (A) synthetase then synthesizes the unusual triribonucleotide 2',5'-A (pppA2'p5'A2'p5'A), which has 2'p5' bonds instead of the normal 3'p5' linkage. This in turn activates RNaseL. During the interferon-induced antiviral state, all protein synthesis, both viral and cellular, is inhibited and if continued this would lead to death of the cell. However, when the virus infection is inhibited, synthesis of viral dsRNA ceases, the inhibitory mechanisms are no longer brought into play and the cell returns to its normal state.

Interferon-γ

When infected, cells of the immune system synthesize interferon-α/β, similar to any other cell. However, activated T cells release interferon-γ upon reaction of their TCRs with their cognate viral (or other foreign) epitopes. This can be demonstrated by infecting an animal with a particular virus and then, some days later, reacting its T cells *in vitro* with antigen from the same virus which has been purified free of nucleic acid. Interferon-γ has antiviral properties similar to those of interferons-α/β, but differs in sequence. It is unstable at pH 2, whereas interferons-α/β are acid stable. Interferon-γ may be of advantage to the host during infections by viruses that switch off host RNA and/or protein synthesis, and so prevent interferon-α and interferon-β being made. However, as interferon-γ is made only by activated T cells, it appears either late in infection or after a second infection or stimulation of the immune system, so that its use in preventing the first exposure to a virus is questionable.

14.6 Further reading and references

Ahmed, R. (ed.) (1996) Immunity to viruses. *Seminars in Virology* **7**, 93–155 (whole issue).

Alcamí, A. & Koszinowski, U. H. (2000) Viral evasion of immune responses. *Immunology Today* **21**, 447–455.

Balkwill, F. (2000). *Cytokine Molecular Cellular Biology: a practical approach*. Oxford: Oxford University Press.

Balkwill, F. (ed.) (2000). *Cytokine Cellular Biology: a practical approach*. Oxford: Oxford University Press.

Burton, D. R. & Woof, J. M. (1992) Human antibody effector function. *Advances in Immunology* **51**, 1–84.

Butcher, E. C., Williams, M., Youngman, K. & Briskin, M. (1998) Lymphocyte trafficking and regional immunity. *Advances in Immunology* **72**, 209–253.

Crowe, J. E. (1996) The role of antibodies in respiratory viral immunity. *Seminars in Virology* **7**, 273–283.

Dimmock, N. J. (1993) Neutralization of animal viruses. *Current Topics in Microbiology and Immunology* **183**, 1–149.

Dimmock, N. J. (1995) Update on the neutralisation of animal viruses. *Reviews in Medical Virology* **5**, 165–179.

Goldsby, R. A., Kindt, T. J. & Osbourne, B. (2000) *Kuby Immunology* (4th edn). New York: W. H. Freeman.

Griffin, D. E. (ed.) (1994) Cytokines in viral infections. *Seminars in Virology* **5**, 403–463 (whole issue).

Herbert, W., Wilkinson, P. & Stott, D. (eds) (1995) *The Dictionary of Immunology*. London: Academic Press.

Hussell, T. & Oppenshaw, P. J. M. (1996) Immunological determinants of disease caused by respiratory syncytial virus. *Trends in Microbiology* **4**, 299.

Kovarik, J. & Siegrist, C.-A. (1998) Immunity in early life. *Immunology Today* **19**, 150–152.

Krajcsi, P. & Wold, W. S. M. (1998) Viral proteins that regulate cellular signalling. *Journal of General Virology* **79**, 1323–1335.

Lehner, T., Bergmeier, L., Wang, Y., Tao, L. & Mitchell, E. (1999) A rational basis for mucosal vaccination against HIV infection. *Immunological Reviews* **170**, 183–196.

Mantovani, A. (1999) The chemokine system: redundancy for robust outputs. *Immunology Today* **20**, 254–257.

Marshall-Clarke, S., Reen, D., Tasker, L. & Hassan, J. (2000) Neonatal immunity: how well has it grown up? *Immunology Today* **21**, 35–41.

McMichael, A. J. (1998) T cell responses and viral escape. *Cell* **93**, 673–676.

Mims, C. A. & White, D. O. (1984). *Viral Pathogenesis and Immunology*. Oxford: Blackwell Scientific Publications.

Ogra, P. L., Mestecky, J., Lamm, M. E., Strober, W., McGhee, J. R. & Bienenstock, J. (eds) (1998). *Handbook of Mucosal Immunology* (2nd edn). London: Academic Press.

Pitha, P. M. (ed.) (1995) Interferon and interferon inducers. *Seminars in Virology* **6**, 141–213 (whole issue).

Powrie, F. & Coffman, R. L. (1993) Cytokine regulation of T-cell function: potential for therapeutic intervention. *Immunology Today* **14**, 270–274.

Ramsay, A. J., Ruby, J. & Ramshaw, I. A. (1993) A case for cytokines as effector molecules in the resolution of virus infection. *Immunology Today* **14**, 155–157.

Roitt, I. M., Brostoff, J. & Male, D. K. (eds) (1997) *Immunology* (5th edn). London: Mosby.

Sattentau, Q. J. (1996) Neutralization of HIV-1 by antibody. *Current Opinion in Immunology* **8**, 540–545.

Sattentau, Q. J., Moulard, M., Brivet, B., Botto, F., Guillemot, J.-C., Mondor, I., Poignard, P. & Ugolini, S. (1999) Antibody neutralization and the potential for vaccine design. *Immunology Letters* **16**, 143–149.

Sen, G. S. & Ransohoff, R. M. (1993) Interferon-induced antiviral actions and their regulation. *Advances in Virus Research* **42**, 57–102.

Smith, T. J. (ed.) (1995) Antibody recognition of viruses. *Seminars in Virology* **6**(5), (whole issue).

Smith, G. L. (ed.) (1998) Immunomodulation by viruses. *Seminars in Virology* **8**, 359–442 (several articles).

Staats, H. F., Jackson, R. J., Marinaro, M., Takahashi, I., Kiyono, H. & McGhee, J. R. (1994) Mucosal immunity to infection with implications for vaccine development. *Current Opinion in Immunology* **6**, 572–583.

Stark, G.R., Kerr, I. M., Williams, B. R., Silverman, R. H. & Schreiber, R. D. (1998) How cells respond to interferons. *Annual Review of Biochemistry* **67**, 227–264.

Taylor, H. P., Armstrong, S. J. & Dimmock, N. J. (1987) Quantitative relationships between an influenza virus and neutralizing antibody. *Virology* **159**, 288–298.

Also check Chapter 22 for references specific to each family of viruses.

Chapter 15
Animal virus–host interactions

It is useful to remember that viruses are parasites and that the biological success of a virus, like any other parasite, depends absolutely upon the success of the host species. Hence the evolutionary strategy of a virus in nature must take into account that it is disadvantageous to kill many hosts (although killing the odd one will not affect the survival of the species) or to impair their reproductive ability. In this chapter, we are concerned with viruses of eukaryotes and their host cells, drawing largely on the animal kingdom for examples. Particular emphasis will be placed on the complexity of these interactions and the multitude of factors that share responsibility for the final outcome of infection. These studies represent one of the most important frontiers of modern virology. Finally, to understand the topic fully, students who have not studied immunology should first read Chapter 14.

15.1 Cause and effect: Koch's postulates

Over a century ago, the bacteriologist Robert Koch enunciated criteria for deciding whether an infectious agent was responsible for causing a particular disease. With the continuing evolution of new diseases, these are equally relevant (with some modifications) today. In essence, Koch postulated that: the suspected agent must be present in particular tissues in every case of the disease; the agent must be isolated and grown in pure culture; and, lastly, pure preparations of the agent must cause the same disease when they are introduced into healthy subjects. As understanding of pathogenesis grew the following modifications were made.

Postulate 1. Koch originally thought that the agent should not be present in the body in the absence of the disease; this was abandoned when he realized that there were asymptomatic carriers of cholera and typhoid bacteria. As seen below, many viruses can cause such subclinical infections.

Postulate 2. Viruses were not known in Koch's time and some still cannot be grown in culture today, so this postulate was modified to say that it is sufficient to demonstrate that bacteria-free filtrates induce disease and/or stimulate the synthesis of agent-specific antibodies.

Postulate 3. Clearly, it is impossible to fulfil the third postulate when dealing with serious disease in humans, although accidental infection can sometimes provide the necessary evidence, e.g. unfortunate laboratory accidents have demonstrated that human immunodeficiency virus (HIV) is responsible for the acquired immune deficiency syndrome (AIDS).

15.2 A classification of virus–host interactions

A classification of the various types of virus–host interactions with some named examples is given in Table 15.1, and each category is dealt with in turn below. However, Table 15.1 is intended only as a guide, because there is a continuous spectrum of virus–host interactions, and divisions are imposed solely for the convenience of classification. The categories are distinguished on just four criteria: the production of infectious progeny, whether or not the virus kills its host cell, if there are observable clinical signs or symptoms, and the duration of infection. It can be seen that cell death does not necessarily correlate with disease, because the former can be compensated for without overt harm to the individual by normal cell replacement. It is also apparent that a single virus can appear in several of the categories, depending on the nature of its interaction with its host, and that the duration of infection appears to correlate inversely with the need for efficient transmission. It is also important to appreciate the three-way interaction of virus, host cell and immune system, which determines the outcome of all infections (Fig. 15.1). Thus, in acute and subclinical infections, the balance favours the host (i.e. the virus is cleared from the body), whereas in persistent and chronic infections it is tilted towards the virus (which is not cleared and does not have to face the hazards of finding a new susceptible host for a long time). Persistent, chronic and latent infections have in common the feature that the immune system cannot clear the viruses responsible from the body. However, it is likely that all infections have some immunomodulatory effects, and thus the immune system is never permitted to operate at its full potential. We shall discuss below some of the ways in which this may come about.

15.3 Acute infections

Acute infections are analogous to acutely cytopathogenic infections *in vitro*, except that the infecting dose of virus is always small and the virus goes through many rounds

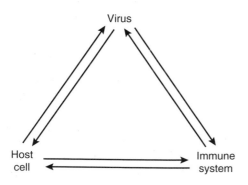

Fig. 15.1 Diagram to show the three-way interactions that decide the outcome of infection.

Table 15.1 A classification of virus infections at the level of the whole organism.

Type of infection	Production of infectious progeny	Cell death	Clinical signs of disease	Duration of infection*	Transmission	Examples†
Acute	+	+	+	Short	Must be efficient	Measles virus, poliovirus (1% of infections)
Subclinical	+	+	−	Short	Must be efficient	Poliovirus (99% of infections)
Persistent	+	− or +	−	Long (+ immune defect?)	Many opportunities	Rubellavirus
Chronic	+	+	+	Long (+ immune defect?)	Many opportunities	Hepatitis B virus
Latent	−	−	−	Long	Many opportunities	Herpesviruses
Slowly progressive disease						
(a)	+	+	Eventually	Long	Many opportunities	HIV-1, TSE agents
(b)	−	+	Eventually	Long	None	Measles virus (SSPE)
Tumorigenic						
Productive	+	−	+	Long	Many opportunities	Retroviruses, Hepatitis B virus, human papilloma- viruses, Epstein–Barr virus
Defective	−	−	+	Long	None	

*Short, approximately 3 weeks or less; long, up to a lifetime.

†Examples are given of viruses that have the given type of infection at some point of their life history, and are not intended to convey that they cannot also be classified elsewhere, e.g. herpesviruses switch between latency and acute infection (see text).

HIV-1, human immunodeficiency virus type 1; SSPE, subacute sclerosing panencephalitis; TSE, transmissible spongiform encephalitis (includes bovine spongiform encephalopathy and Creutzfeldt–Jakob disease); these are not viruses but agents of unknown provenance, and sometimes called prions (see Chapter 20).

of replication. Thus, the minimum time-scale is measured in days rather than hours. During an infection, viruses circulate around the body and come in contact with many different organs. However, most viruses do not cause generalized infections but attack particular organs or tissues, known as *target organs* or *target tissues*. Hepatitis viruses infect the liver and influenza viruses infect the respiratory tract, and the reverse never occurs. This specificity is achieved largely through the presence of specific cell receptors as discussed for *in vitro* infections in Chapter 13, but there may be intracellular restrictions on infection as well.

Infection of organisms can be described in terms of clinical *signs* (which are objectively assessed, such as an elevated temperature) and *symptoms* (which are subjective, such as a headache), and by a variety of laboratory tests. Without the last, no identification of the causative agent is complete. Laboratory tests (see Chapter 2) include isolation and titration of infectious virus, detection of viral antigens in blood and other tissues obtained by biopsy by a variety of immunological assays (especially ELISA), detection of viral nucleic acids by the polymerase chain reaction (PCR) for DNA or reverse transcriptase PCR (RT-PCR) for RNA—the direct identification of virus using the electron microscope—possibly in conjunction with antibody that will agglutinate cognate virus particles. Electron microscopy can be used for faecal specimens, nasal wash material or biopsy materials.

The course of acute infection and recovery from infection

An acute infection begins with infection of one or a few cells, infectious progeny are produced and the infected cells die. Further cycles of multiplication ensue with increasing numbers of infected cells, and eventually the first signs and symptoms of disease appear. Thus, the infection has been progressing for several days before we are even aware of it. Often virus is also being shed before we are aware that we have an infection, which can only enhance the spread of the virus to the susceptible people we meet. Fortunately, most people recover from acute virus infections within a few days or weeks. There are many different acute infections and the essentials are summarized in the key points. An excellent example is measles virus because almost all the infections that it causes are acute. By comparison, only 1% of poliovirus infections are acute and 99% are subclinical (see Section 15.4). The example shown in Fig. 15.2 is of that familiar infection, the common cold, caused by one of the 100 or so different serotypes of rhinovirus (a picornavirus). The damage that a rhinovirus does to its target tissue, the ciliated epithelium of the upper respiratory tract, is shown in Fig. 2.3.

Key points on acute infections

Infection → Incubation period → Signs or symptoms → Recovery/death

Infectious virus produced
(= transmission)

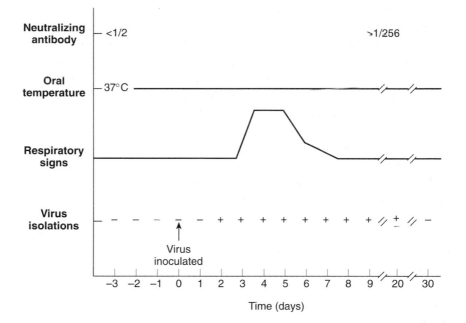

Fig. 15.2 Virus infections are measured by clinical signs and laboratory tests for the presence of virus. The figure illustrates an experimental 'common cold' in humans caused by deliberate intranasal inoculation of a rhinovirus. Note that there is no fever. Respiratory signs are quantified very simply by the number of paper tissues used per day.

At this point it is appropriate to consider the mechanisms that are responsible for allowing us to survive our first encounter with a particular virus, i.e. a primary infection. It is difficult to obtain unequivocal data from such a complex situation, far removed from the study of cells in culture. However, the innate immune system is the first line of defence. If it is successful, the infection progresses no further, but there is no immunological memory from the encounter. However, if the infection gets past innate immunity, adaptive immunity is activated (see Chapter 14). T cells, and in particular CD8+ cytotoxic T cells, and antibodies are the main defence against primary infections. Information on immune responses in humans is not readily available, but studies on people with natural immune deficiencies have been revealing. These have led to the surprising and important conclusion that a virus infection can be controlled only by one or other of the arms of the adaptive immune response, and that this antiviral effect is not interchangeable, e.g. people who have a congenital deficiency in antibody-mediated immunity, but normal T-cell-mediated immunity, are unable to combat primary infections with picornaviruses, orthomyxoviruses and paramyxoviruses (Table 15.2). The virus is not cleared from the body and may persist for years, and clinical disease may be exacerbated. Conversely, a congenital inability to mount T-cell-mediated immunity means that herpesvirus and poxvirus infections are not cleared, and can become generalized throughout the body and even life threatening. However, the conclusion that antibody has no role in recovery from these infections is only tentative, because the lack of CD4+ T-helper cells means that immunoglobulin M (IgM) is the only antibody present.

One important implication of these findings is for vaccines. In the past vaccines have been made empirically but, if this does not produce an agent to provide protection, there is a problem that cannot currently be solved (see Chapter 16). Immunization with a vaccine will always stimulate immune responses, but these may not be the one(s) that can deal with the infection. The problem today facing vaccinologists who are trying to make a vaccine against HIV is first to identify which immune response(s) is needed and then to determine how to stimulate it, and this is not possible at the present time.

B cells and T cells continue to be stimulated to divide and differentiate as long as cognate antigen is present, and adaptive immunity is said to be 'antigen driven'. Once virus antigen is eliminated, immune cell division stops, leaving a population of memory cells. However, antibody often persists long after the infection, and this is thought to reflect the presence of long-lived, non-infectious antigen. Memory cells are resting cells with no effector functions. Thus, after an infection for the rest of our

Table 15.2 Recovery from primary virus infections (when innate immunity has failed).

Immunity responsible for recovery	Virus family
Antibody from activated B cells (plasma cells)	Picornavirus Orthomyxovirus Paramyxovirus
Activated T cells	Herpesvirus Poxvirus

Key points on combating acute infections (in military terms)

1 Constant readiness to detect and repel attack: the sentries and non-specialized troops (the innate immune system). If this first defensive line is breached ...

2 Call up specialized reinforcements to carry out counter-attacks, including search-and-destroy missions to hit the factories where munitions are being manufactured (adaptive T-cell-mediated immunity and especially CD8+ cytotoxic T cells). To help in this effort ...

3 Call up more reinforcements with a different specialization to recognize and take out individual virus particles (adaptive antibody-mediated immunity).

4 With the enemy defeated, stand down the troops but maintain an increased defensive line to prevent a second attack from the same enemy (adaptive antibody-mediated immunity, memory T and B cells).

lifetime, we have a much larger population of cognate B and T cells than we had before the infection. In addition, memory cells have a lower threshold for activation by antigen than lymphocytes. Thus, the secondary immune response is mounted with less delay, reaches its maximum more rapidly and attains a higher level than the primary response. Antibodies are the main defence against reinfection by the same agent. If they are present in sufficient quantity, there will be no reinfection, but it is doubted in some quarters that such 'sterilizing immunity' actually exists and that, when we meet the virus a second time, we are actually infected, although this time the infection is low level and subclinical. Such reinfection may in fact be beneficial, because it means that immune responses are boosted from time to time.

15.4 Subclinical infections

Subclinical infections are also known as inapparent or silent infections. These are the most common infections and, as their name implies, there are no signs or symptoms of disease. In all other regards, these are the same as acute infections. Evidence of infection comes only from laboratory isolation of the virus or by a postinfection rise in virus-specific antibody. Any virus that causes a subclinical infection has evolved to a favourable equilibrium with its host. The enteroviruses (picornavirus family), which multiply in the gut, are one such group and the classic example is poliovirus, which causes no symptoms in over 99% of infections. A subclinical infection is arguably the expression of a highly evolved relationship between a virus and its *natural* host, because there are many examples of such a virus causing lethal disease when it infects a different host, e.g. yellow fever virus (a flavivirus) causes a subclinical infection in Old World monkeys but results in a severe infection in humans and is fatal in some New World monkeys. Subclinical infections have the same duration and are cleared by the same means as acute infections (see key points).

15.5 Persistent and chronic infections

Both persistent and chronic infections are acute or subclinical infections that are not terminated by the immune response (see Table 15.1). However, quite frequently persistent infections involve few cells and produce low levels of progeny virus, whereas chronic infections are more active. Hence, it is not surprising to find that persistent infections are largely subclinical and chronic infections are accompanied by clinical

---

Key points on subclinical infections

Infection → Incubation period → *No* signs or symptoms → Clearance of virus

Infectious virus produced
(= transmission)

Table 15.3 Some ways in which viruses evade immune responses and establish persistent and chronic infections.

Immunosuppression by infection and inactivation of functions of macrophages, B cells, T cells, etc.

Downregulation of expression of MHC proteins

Immune deficiency of the host, e.g. agammaglobulinaemia leading to persistent infection with live poliovirus vaccine viruses

Viruses that are poorly immunogenic, and/or fail to synthesize or display enough antigen on the cell surface

Viruses that synthesize excess soluble antigen that binds all available neutralizing antibody

Infection of fetus or neonate before it has a fully competent immune system, resulting in immunological tolerance, e.g. infection of the foetus *in utero* by rubella virus (a togavirus)

Viruses that stimulate non-neutralizing antibody that binds to virions and blocks neutralizing antibody

Viruses that stimulate insufficient interferon (and cells that synthesize low amounts of interferon)

Viruses that generate mutants through inaccurate transcription (includes synthesis of antigenic variants and defective–interfering (DI) genomes)

MHC, major histocompatibility complex.

disease. The pertinent question is why the immune response is ineffective. In general, it appears that viruses that cause persistent and chronic infections have inhibitory effects on some aspect of the immune response itself, or on the expression of major histocompatibility complex (MHC) proteins (see Chapter 14). Alternatively, such infections result when an individual with an abnormality of the immune system contracts a virus that would otherwise cause an acute or subclinical infection. Whatever the cause, the end result is that virus particles and infected cells are not cleared. There are many scenarios that have this effect and some are listed in Table 15.3. It is important to appreciate the three-way interaction of virus, host cell and immune system that determines the outcome of all infections (see Fig. 15.1). In persistent and chronic infections, this balance is tipped more to the virus, whereas in acute and subclinical infections it favours the host. At the moment, it is not known why interferon is ineffective in curtailing infection. However, viruses can inhibit host-cell macromolecular synthesis in general and even inhibit the functioning of individual cellular genes (see Section 21.2). Many viruses have specific functions that downregulate the expression of interferon-α/β, and also vary in their inherent sensitivity to the antiviral activity of interferon. Interferon must be present in a sufficiently high concentration to be effective.

A classic example of a chronic infection occurs when lymphocytic choriomeningitis virus (an arenavirus) is inoculated into newborn mice. The animals become

Table 15.4 Infection of *adults* by hepatitis B virus: chances of progression of liver infection and disease.

Approximate percentage of infected adults who develop the type of infection listed	Type of infection
100	Acute
↓	↓
10	Persistent
↓	↓
1.0	Chronic
↓	↓
0.1	Liver cancer (primary hepatocellular carcinoma)

immunologically tolerant to the viral antigens, and virus can be found in large quantities in the circulation and every tissue including the brain. One unexplained feature is that tolerance is rarely complete, but presumably the reduced immune response is unable to cope with the infection. Similarly, little interferon is made, although the viruses are sensitive to its action. Neonatally infected animals remain healthy, whereas animals infected as adults suffer an acute infection. They mount an extremely strong T-cell response to the virus which is lethal—a form of 'immunological suicide'.

Another example of a chronic infection occurs when humans are infected with hepatitis B virus (HBV, an hepadnavirus)—one of the causative agents of viral hepatitis. Infection is spread sexually, and from infected blood and saliva. After an initial acute infection of an adult, there is disease with liver damage, and most infections are completely cleared. However, in a small proportion of people (10%), virus persists in the liver for a lifetime, although most of these infections are subclinical (Table 15.4). This contrasts with the situation in the Far East where HBV is endemic (i.e. most people are infected), and is transmitted from a carrier mother to her children early in life, possibly via infected saliva. The younger the age at which infection takes place, the higher the chance of a persistent infection and, in very young infants, it rises to over 90% (Fig. 15.3). The number of other complications in the population remains proportional.

Hepatitis B virus persists by downregulating MHC class I proteins on the surface of infected liver cells (hepatocytes), with the result that $CD8^+$ cytotoxic T cells are unable to act. Upregulation of MHC class I proteins by interferon-α/β leads to a reduction of the number of infected cells through the action of $CD8^+$ cytotoxic T cells (see Section 14.6), but unless all infected cells are destroyed the relief is only transient. A small proportion of the persistent infections evolves to become chronic, with further liver pathology which can develop into total liver failure. This probably results from the balance tilting in favour of the $CD8^+$ cytotoxic T cells, which then destroy a large number of infected liver cells. In addition, during chronic infection, co-circulation of large amounts of virus antigens and virus-specific antibody causes the formation of

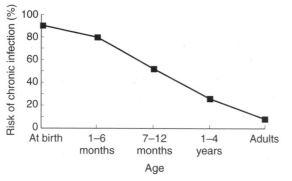

Fig. 15.3 Risk of becoming a carrier who is chronically infected with hepatitis B virus (HBV)—depends on the age at which a person is infected (horizontal axis).

Key points on persistent and chronic infections

• It is likely that these have resulted from the efficacy of the immune system being compromised in some way (e.g. immunosuppression, downregulation of MHC proteins, infection early in life).
• As chronic infections yield large numbers of infected cells and amounts of viral antigen for years, immune responses can have pathological instead of beneficial consequences.

---

antigen–antibody complexes. When deposited in the kidney, these circulating complexes can activate complement and result in the destruction of kidney tissue, leading to immune complex disease. Hence, the appearance of disease in this instance depends not on the cytopathic effects of virus but on the relative proportions of viral antigens and virus-specific antibody. Finally, a proportion of the chronic cases develops into liver cancer (see Section 17.7) with part of the viral genome integrated into the host DNA. The example of HBV shows how one virus can cause four different types of infection depending on host cell–virus–immune response balance.

15.6 Latent infections

By definition, all latent infections start and finish as acute infections. No infectious virus is present during a latent infection, and such infections are always subclinical. The herpesviruses are a diverse and ubiquitous family which all cause latent infections in many different animal species. Several cause common childhood infections. All become latent and such infections endure for life, making these viruses probably the best adapted of all to coexistence with their host. One of the most common and most successful is the human herpes simplex virus type 1 (HSV-1) because, by 2 years' of age, nearly everyone has been infected. This is caused by contact with infectious saliva (e.g. kissing the baby).

Latency and HSV-1

The acute infection is cleared by the immune system but, by this time, the virus has travelled up the sensory nerve supplying the infected area and become latent in the cell body of one or more of the sensory neurons that form the dorsal root ganglia supplying the original area of infection. For HSV-1 this is the trigeminal cranial nerve. All acute HSV-1 infections become latent.

The molecular biology and immunology of HSV-1 reactivation

Reactivation of HSV occurs regularly several times each year in about 10% of infected people (Table 15.5). It usually results in the familiar pathology of a cold sore, but up to 40% of infections are subclinical and thus can be passed on inadvertently. When latency is broken, the virus multiplies in one of the neurons in which it is latent, and infectious virus particles descend the sensory nerve; it then infects the epithelium that it supplies. Thus, because the virus is confined by the sensory nerve, the reactivated acute infection reoccurs each time in the *area of the skin supplied by this nerve*. In addition, infection is usually monolateral because the nerves are paired, with each supplying only the right or the left side of the body, and virus does not move from one to the other. HSV-1 has been isolated by co-cultivation of excised trigeminal ganglia from infected mice or humans with cultured cells, showing that virus is activated under the conditions of culture. The blisters that form the early cold sore contain infectious virus particles which can be passed on to any non-immune infant—time, after time, after time.

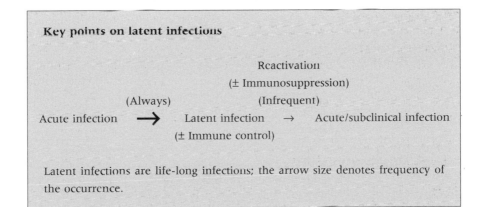

Key points on latent infections

Acute infection $\longrightarrow$ Latent infection $\rightarrow$ Acute/subclinical infection

(Always) Reactivation (± Immunosuppression) (Infrequent) (± Immune control)

Latent infections are life-long infections; the arrow size denotes frequency of the occurrence.

Molecular understanding of the distinction between acute and latent HSV infections is shown in Fig. 15.4. In an acute infection, activation of transcription depends on the interaction of a virion protein, VP16, with cellular activator protein(s). There is then progressive flow of transcription of the α, β and γ genes and the production of progeny virus. Latency arises when the activating cellular proteins are absent or unavailable, with the result that there is no transcription of α genes (see Section 9.6). During

Table 15.5 The establishment of latent infections in humans with herpesviruses and the breakdown of latency (reactivation).

Virus	Primary acute infection	Site of latency	Stimulus for reactivation	Reactivated acute infection
HSV-1	Stomatitis: infection of the inside of the mouth and tongue	Dorsal root ganglion of the trigeminal (cranial) nerve	For example, strong sunlight, menstruation, stress	Cold sore*
HSV-2	Genital lesions	Dorsal root ganglion of the sacral region of the spinal cord	Not known, though probably similar to HSV-1	Genital lesions (and infection of neonates)
VZV	Chicken pox: fluid-filled vesicles on the skin	Any dorsal root ganglion of the central nervous system	Release of immune control, e.g. in elderly people	Zoster (shingles)
Epstein–Barr virus	Child: subclinical Adult: glandular fever (IM acute)†	B cells or possibly epithelium of the throat	Not known; frequent	Subclinical (but can also cause cancer: nasopharyngeal carcinoma and Burkitt's lymphoma)
CMV	Prenatal:‡ Child: subclinical Adult: subclinical	Salivary glands and probably other sites	Frequent	In all body fluids, especially during the immunosuppression of pregnancy. A major cause of death in AIDS and transplantation surgery

*10–15% of reactivations are subclinical, but in some individuals this reaches 40%.
†Example (1) of more severe clinical disease when infection is delayed from childhood—a common microbiological problem of the (over-)sanitized world.
‡The foetus is only at risk when its mother acquires a primary infection. CMV is the main cause of brain damage (microcephaly) in babies; example (2) of more severe clinical disease is when childhood infections are delayed.
AIDS, acquired immune deficiency syndrome; CMV, cytomegalovirus; HSV, herpes simplex virus; IM, infectious mononucleosis; VZV, varicella–zoster virus.
'Cytomegalo' refers to the characteristic swollen cell cytopathology caused by CMV to cells of the kidney, lungs and liver, and 'mononucleosis' refers to the uncommonly large numbers of mononuclear (compare polynuclear) cells (mainly lymphocytes) that are found in the blood.

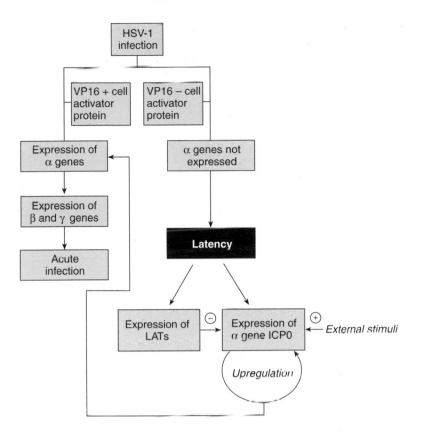

Fig. 15.4 Molecular control of latency in herpes simplex virus type 1 infection. Of 72 viral genes, only the latency-associated transcripts (LATs) are synthesized during latency, and external factors upregulate the synthesis of the viral non-structural protein, ICP0 (intracellular protein 0), which then enhances its own expression. This activates expression of the α genes and unlocks the pathway to acute infection (see text).

latency, only the latency-associated transcripts (LATs) are made. These are actually composed of spliced introns transcribed from the strand complementary to that encoding the α gene, ICP0 (intracellular protein 0), and do not encode any protein. They negatively regulate the transcription of ICP0 and so stabilize latency. In some unknown way, external stimuli (e.g. strong sunlight, local trauma, menstruation) upregulate the expression of ICP0 which then upregulates its own expression. This, in turn, enables the transcription of the α genes and reactivation of the acute infection.

Is latent HSV-1 DNA integrated into the host genome? This question was technically difficult to answer because there is only 0.15–0.015 HSV-1 genome per dorsal root ganglion neuron. This cell-associated DNA differed from the linear virion DNA because it has no free ends, meaning either that it was integrated or that it had circularized (Fig. 15.5). Careful work showed that it is not integrated and that latent HSV-1 DNA exists as a free, circular molecule in the cell nucleus, i.e. it is an episome.

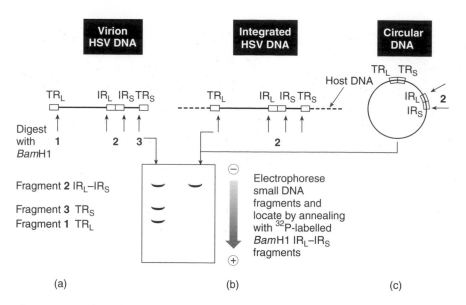

Fig. 15.5 Hypothetical scheme to illustrate the qualitative change in the structure of herpes simplex virus (HSV) DNA isolated from the central nervous system (CNS) of latently infected mice, which results from the loss of the free termini of the linear virion DNA. DNA from (a) virions or (b, c) CNS is digested with the restriction endonuclease *Bam*H1, which cuts at the position of the arrows. DNA fragments are electrophoresed in agar gel and annealed to a radiolabelled HSV DNA *Bam*H1 fragment from the IR_L–IR_S junction. As a result of the reiterated sequences at the ends of the long and short unique regions (see Fig. 4.4), the probe shows up the presence of both the IR_L–IR_S junction and the terminal fragments in the virion DNA. If the HSV DNA was integrated with host DNA (b), the IR_L–IR_S segment would appear as before but the terminal segments would now be attached to host DNA and be too large to enter the gel. Other data show that (c) is the form adopted during latent infection.

One of the difficulties in understanding HSV-1 reactivation is why some people with a normally functioning immune system suffer repeated acute infections which follow virus reactivation. About 10% of us experience the resulting cold sores. Although the immune system does not prevent the cold sores in the epithelium around the mouth, they are short-lived, indicating that the immune system does clear them. We also know that T cells control HSV-1, because the virus is life threatening in people who have a T-cell deficiency. What happens occurs in two stages (Fig. 15.6). First, there is a short-term, virus-mediated evasion of the immune response, which allows the reactivation to start and, second, there is immune activation which allows the infection to be cleared. Initially, in the early stages of infection, expression of MHC I proteins in infected cells is inhibited. Two mechanisms operate in tandem: the immediate early viral protein (ICP47) prevents the translocation of processed peptides into the endoplasmic reticulum, and hence the formation of MHC–peptide complexes and, second, the virion protein that shuts off host protein synthesis prevents the synthesis of MHC

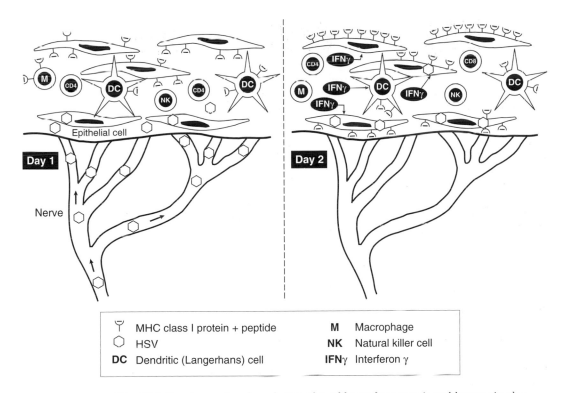

Day 1

Nerve

Epithelial cell

Day 2

	MHC class I protein + peptide	**M**	Macrophage
	HSV	**NK**	Natural killer cell
DC	Dendritic (Langerhans) cell	**IFNγ**	Interferon γ

Fig. 15.6 Development and resolution of a cold sore from reactivated herpes simplex virus type 1 (HSV 1). At time 0, reactivated virus is released from cells of the dorsal root ganglion and transported down the sensory neuron. It then infects adjacent epithelial cells. Expression of major histocompatibility complex (MHC) I and viral peptide antigen on the cell surface is inhibited (see text) and virus-specific memory CD8$^+$ T cells are not activated. Natural killer (NK) cells and virus-specific CD4$^+$ T cells move gradually into the lesion and release cytokines. By day 2, they are releasing enough interferon-γ to upregulate the expression of MHC class I protein. The CD8$^+$-memory T cells are then activated and virus-infected cells are cleared. (From Posavad *et al.* 1998.)

polypeptides. Thus virus peptides are not presented and virus-specific CD8$^+$ T-memory cells are not activated. In due course, CD4$^+$ cells and natural killer (NK) cells move into the cold sore lesion where lysis of infected cells is taking place. Viral antigen is processed and presented as MHC class II protein–virus peptide complexes on the surface of antigen-presenting cells (local dendritic (Langerhans) cells and macrophages). The CD4$^+$ T cells are activated and, in turn, through the cytokines that they liberate, NK cells are activated as well. By day 2, both cell types are producing interferon-γ. This upregulates the expression of MHC I proteins in infected epithelial cells. CD8$^+$-memory T cells are then activated, virus-infected cells are killed, and the infection resolves. It seems therefore that there is a very delicate symbiosis between the virus and the host that allows both to prosper.

Latency and some other human herpesviruses

Other important examples of human herpesviruses are shown in Table 15.5. Herpes simplex viruses can also be sexually transmitted—usually HSV-2; its cycle of acute–latent–acute infection is very similar to that of HSV-1, although the former centres on the genital area. In fact the location of infection is interchangeable and which virus is resident depends on the site of the initial acute infection. This, in turn, determines which dorsal root ganglion becomes the site of the latent virus. With a genital primary infection, the virus becomes latent in the sacral dorsal root ganglion which innervates that area.

Varicella–zoster virus (VZV) causes acute chicken pox of children and shingles (zoster) on reactivation. Apart from chicken pox being generalized over the entire surface of the body, the transition from acute to latent infection occurs exactly as with HSV-1. As a result of the generalized nature of the infection, almost any of the brain and spinal dorsal root ganglia can be latently infected. Reactivation is a rare event, and the lesions formed by the reactivated virus are restricted to the precise segment of the body that is innervated by the nerve in which virus was latent. Latent VZV is maintained in the presence of antibody and a T-cell response, both of which probably contribute to keeping the virus in its latent state because immuno-suppression (often the natural process that accompanies ageing) results in virus reactivation.

Epstein–Barr virus (EBV) normally causes a subclinical infection in infants and becomes latent in B cells (see Section 13.3) and possibly epithelial cells of the throat. However, if infection is delayed until adulthood, infectious mononucleosis (glandular fever) results. This unpleasant, drawn-out and debilitating disease is common in students, and probably contracted by kissing, because the reactivated virus is present in saliva. This suggests that either EBV has evolved to coexist peacefully with infants, and in adults this balance is no longer maintained, or the severity of the outcome is proportional to the size of the inoculum, with small children receiving a small dose. Putting off infections from childhood is a recent development of the over-sanitized age in which people in the developed world live. EBV also causes rare and geo-graphically limited forms of cancer (see below).

Another ubiquitous infection is cytomegalovirus (CMV). This is normally a sub-clinical childhood infection which becomes latent in the salivary glands. It reactivates frequently but the infection is controlled by the immune system and is completely subclinical. However, CMV infection can become generalized and life threatening when there is partial immunosuppression, such as is prescribed to avoid organ rejec-tion during transplantation surgery, and occurs naturally during HIV-1 infection. In addition, although adult infection is subclinical, the virus can pass through the placenta and infect the foetus. As a result, some foetuses suffer severe and per-manent brain damage (such as microcephaly). This is also thought to be the result of the immunosuppression that occurs naturally during pregnancy, which helps to prevent the foetus (a foreign graft) from being rejected by the maternal immune system.

Latency and HIV-1

During the subclinical phase of an HIV-1 infection which typically lasts for around 10 years, most of the virus is very actively multiplying and turning over, with approximately 10^{10} virions produced every day and dealt with by the immune response. However, there is an underlying minority of HIV-1 genomes that become latent in around 1000–10 000 memory CD4$^+$ T cells. This is important because most active virus multiplication, but not the latent virus infection, is eliminated by chemotherapy (see Section 19.8). Thus the latent infection is a potential reservoir for virus rebound if chemotherapy is stopped. This means that the current chemotherapy regimen may have to be maintained for life, with the attendant difficulties of expense, toxicity, non-compliance or the eventual breakthrough of resistant mutants.

15.7 Slowly progressive diseases

As their name implies, these are diseases that take years to manifest, although infection proceeds at the normal rate. There are two categories: slowly progressive diseases caused by viruses and the spongiform encephalopathies. The viruses are subdivided into those that make infectious progeny throughout (HIV-1) and those with genomes that become defective during the long incubation period and hence are non-infectious (measles virus—subacute sclerosing panencephalitis or SSPE). The transmissible spongiform encephalopathies are believed to be caused by a novel type of infectious agent, and are dealt with in detail in Chapter 20.

Slowly progressive virus diseases that are infectious

The classic example of such a virus disease is HIV-1, which predominantly infects CD4$^+$ T cells. From the detailed statistics that are available only for infections in the developed world, a typical adult infection starts with an acute influenza-like infection. The disease is then quiescent for an average of 10 years, although the virus continues to multiply to high titre (10^{10} virions produced per day), with high turnover. During this time there is a progressive loss of CD4$^+$ T cells, and these eventually reach such a low level that there is a virtual collapse of the immune response (which is AIDS (acquired immunodeficiency syndrome)). The patient is then overwhelmed by endogenous viral or bacterial infections which are normally kept in check by the immune system. In the developed world, these are most commonly *Pneumocystis carinii* (a yeast-like organism) or cytomegalovirus. We return to HIV-1 in Chapter 19.

Slowly progressive virus diseases that are non-infectious

Another example is a rare sequel of infection with measles virus (a paramyxovirus). Before a vaccine was available (see Fig. 16.3), most children contracted an acute measles virus infection, which was cleared in around 3 weeks. However, in a very small proportion, the virus established itself in the brain and, after a long incubation

Table 15.6 Comparison of the acute and slowly progressive measles virus infections.

Acute measles virus infection
Common acute childhood infection before immunization
Respiratory transmission
Systemic infection
100% disease (smooth skin rash)

Subacute sclerosing panencephalitis
Very rare, affecting about 6–22 per 10^6 cases of measles
A sequel to acute measles
Brain infection and disease
Incubation period of 2–6 years
Associated with measles infection early in life (<2 years old)
Invariably fatal

---

period, caused a pattern of degenerative changes in brain function, including loss of higher brain activity and inevitable progression to death (Table 15.6). The disease is called SSPE, and *post mortem* the areas of 'hardening' or 'sclerosing' of brain tissue can be seen that give the disease its name. Why only certain individuals contract the disease is not understood but a predisposing factor is infection early in life—usually before 2 years of age. Infectious viruses cannot be isolated, although cultured cells can be infected, still non-productively, by co-cultivation with brain extracts. Genomes of SSPE measles viruses obtained from these cells have now been sequenced, and these have accumulated many mutations and produce a defective internal virion (matrix) protein. The nature of the presumed defect in the immune response that fails to clear the initial acute infection is not known.

15.8 Virus-induced tumours

All viruses that cause tumours have DNA as their primary genetic material in the cell but some, members of the retrovirus family, have RNA in their virions, i.e. they belong to class 6. DNA 'tumour' viruses normally cause an acute infection and it is rare that a tumour results, e.g. nearly everyone is infected as a child or young adult by Epstein–Barr virus (EBV—a herpesvirus), which is subclinical or causes infectious mononucleosis (glandular fever), and yet occurrence of the tumours (Burkitt's lymphoma, nasopharyngeal carcinoma) with which the virus is associated is very rare. Demonstration of tumorigenicity in the laboratory is usually made under 'unnatural' circumstances that are known to be favourable to the development of the tumours. Important factors are the genetic attributes and age of the animal: certain inbred mouse lines develop tumours more readily than others, and young animals are more susceptible, because their immune system is not fully mature. If no tumours result, it may be necessary to transfer cells transformed by the virus *in vitro* into the animal to demonstrate viral tumorigenicity.

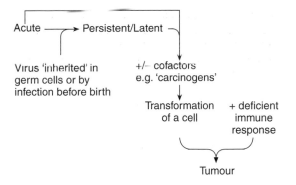

Fig. 15.7 The relationship between virus infection and tumorigenesis.

The diagnostic criterion for a tumour virus in the laboratory is the conversion of a normal cell into a transformed cell. This involves many complex changes in cellular properties, which are discussed more fully in Chapter 17. However, not all transformed cells are tumorigenic when transplanted into appropriate animals, and other steps in addition to transformation are needed to produce a cancer cell. Thus, transforming viruses that cause cells to divide more rapidly than normal are at one end of the spectrum and those causing cellular destruction are at the other. Tumours fall into two groups: those that produce infectious virus and those that do not. The latter are the more common and are caused by viruses from several different families. Such tumours contain viruses that are unable to multiply, viruses with their multiplication repressed in some way or those in which only a fragment of genome is present. The methods described above for detecting viral components in latently infected cells can be used successfully on tumour cells. However, some tumours are probably caused by a 'hit-and-run' infection which leaves no trace of virus behind.

As for other virus–host interactions that have been discussed above, the induction of tumours depends on the balance of a complex situation. This can be viewed as shown in Fig. 15.7, where, initially, infection may be acute or in a persistent or latent form. The transformation/tumorigenesis process initiated by infection requires additional factors (such as chemical carcinogens in Fig. 15.7). We do not know how often transformation of cells takes place in the animal, but it is likely that this is frequent and that, on most occasions, the immune system recognizes and destroys the transformed cell. Tumour formation probably requires that the immune system is in some way deficient, a state perhaps induced by infection or ageing. The apoptosis mechanisms present in all cells also act as a powerful protection against tumour development.

A number of viruses for which the principal interactions with their hosts can be classified into one or more of the categories described in Sections 15.3–15.7 can also, as a rare consequence of infection, trigger a sequence of events leading to the formation of a tumour. The in vitro parallel of this is the morphological transformation of cells in culture after virus infection. Examples of some common viruses that can cause cancer are shown in Table 15.7. These are chosen to emphasize that they all cause acute disease initially which is not cleared, and remain in the body in a variety of

Table 15.7 Some common human viruses that (rarely) cause cancer.

Virus	Initial infection	Subsequent infection	Cofactors	Cancer
HBV (hepatitis B virus: an hepadnavirus)	Acute (hepatitis)	Persistent or chronic (hepatitis)	Infection early in life, smoking, alcohol, fungal toxins	Liver (primary hepatocellular carcinoma*)
HPV (human papillomavirus: a papovavirus)	Subclinical	Persistent (anogenital warts)	Early and frequent sexual activity, smoking	Cervix of the uterus (carcinoma)
EBV (Epstein–Barr virus: a herpesvirus)	Child: subclinical Adult: acute (infectious mononucleosis)	Latent	Malaria, MHC genetics	Lymphoid tissue of the jaw (Burkitt's lymphoma) or nasopharyngeal carcinoma

*Carcinoma: a cancer of epithelial cells.
MHC, major histocompatibility complex.

ways. Then, depending on the circumstances of the infection and the cofactors experienced in life, a cancer may eventually be manifested.

15.9 Transmission of infection: interactions with viruses which take place outside the host

In this section, the discussion of virus–host interactions is continued by considering how viruses are transmitted from an infected to a susceptible host. As most viruses are inherently unstable and, when outside the host, lose infectivity rapidly, this is an important step in their life cycle. Knowledge of how viruses are transmitted may enable us to break the cycle at this stage and thus prevent further infections. We know approximately by what routes viruses are transmitted, but natural transmission in humans is difficult to investigate and the details are poorly understood. All transmission strategies adopted by viruses have, in common, the ability to circumvent the dead, impermeable, outer layer of the skin and to bring virus into contact with the naked cell surface. Normal person-to-person infections are classified as taking place by *horizontal* transmission, and take place by the following routes: respiratory, oral–faecal, sexual, via urine, conjunctival and mechanical, but those from mother to baby are put in a separate category and described as *vertical* transmission. Examples of viruses transmitted by these routes are shown in Table 15.8. It is no coincidence that most viruses are spread by the respiratory and oral–faecal routes, because, in order to carry out their normal physiological functions, the lungs and small intestine each have a

Table 15.8 Transmission of infection.

Type of transmission	Occurrence	Example
Horizontal		
Respiratory route	Common	Rhinoviruses, influenza viruses
Conjunctival route	Not known	Respiratory viruses
Oral–faecal route	Common	Polioviruses
Sexual	Used by specific viruses	HIV-1, HBV, papillomaviruses
Via urine	Used by specific viruses	Lassa fever virus, sin nombre virus, cytomegalovirus
Mechanical	Common in the tropics	Arboviruses, HIV 1, HBV
Vertical		
Mother to baby	Used by specific viruses	Rubella virus, HIV-1

HBV, hepatitis B virus; HIV-1, human immunodeficiency virus type 1.

surface of living cells with an area approximately equivalent to that of a tennis court (200–300 m^2). Finally, we discuss *zoonoses*, infections where the virus is transmitted from animals other than humans to humans.

15.10 Horizontal transmission

Transmission via the respiratory tract

Not only viruses that cause respiratory infections but also some viruses causing generalized infections, such as measles and smallpox, are contracted by this route. After virus has multiplied, it either reinfects host cells or escapes from the respiratory tract in the aerosol of liquid droplets that results from our normal activities, such as talking, coughing and sneezing (and particularly singing). These aerosols are inhaled and give rise to a 'droplet infection' of a susceptible individual. The size of droplets is important, because those of very large (> 10 μm) diameter rapidly fall to the ground and the very small (< 0.3 μm) dry very quickly, resulting in accelerated inactivation of virus contained within them. Thus, the middle-sized range of droplets consists of those that transmit infection more efficiently. The precise size of these will determine where the droplets are entrapped by the respiratory system of the recipient; because the 'baffles' lining the nasal cavities remove the larger airborne particles, the medium-sized ones get as far as the trachea (throat), whereas the smallest may penetrate deep into the lung (Fig. 15.8). However, particles have to escape being trapped on the flow of mucus that is driven upwards by the cilia lining the tubes of the respiratory tract, counter to the incoming air. This is the mucociliary flow. The mucus also serves as a physical barrier which prevents virus attaching to cell receptors.

The increase in nasal secretions that accompanies many respiratory infections favours the dispersal of the viruses responsible, and the increase in coughing and

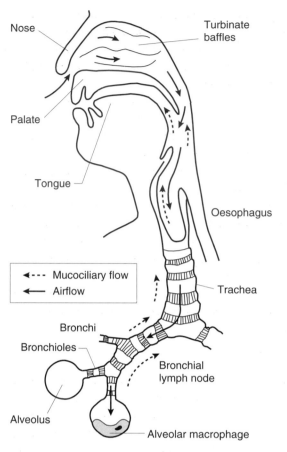

Nose

Turbinate
baffles

Palate

Tongue

Oesophagus

◄- - - Mucociliary flow
◄--- Airflow

Trachea

Bronchi

Bronchioles

Bronchial
lymph node

Alveolus

Alveolar macrophage

Fig. 15.8 Route taken by aerosol droplets containing virus particles that are breathed into the respiratory tract. Note the defences designed to trap and remove all sorts of small particles: turbinate baffles, the mucociliary flow that runs counter to inspired air and macrophages. The airways become progressively narrower (here much foreshortened) and end in the alveolus, which is formed by a single cell. An alveolar macrophage can be seen in one of the alveoli. (Adapted from Mims and White 1984.)

sneezing increases the production of infected aerosols. However, transmission experiments from people infected with a rhinovirus to susceptible individuals sitting opposite at a table proved singularly unsuccessful. Equally unsuccessful was the transmission of influenza from a naturally infected husband/wife to his or her non-immune spouse. This has led to the suggestions that only particular individuals may shed sufficient virus, produce excess nasal secretion and/or aerosols containing optimum-sized particles, and so act as efficient spreaders of infection or that, in order to be infected, the person has to be in a certain physiological state. Apart from the traditional stories of wet feet predisposing you to catch a cold (not true!), it has been shown that recently bereaved people are particularly susceptible to infectious diseases. Thus resistance is influenced by one's state of mind. Neuroimmunology is a developing field of study.

In addition to aerosol transmission via the respiratory tract, it is now believed that some respiratory viruses are spread by contact with the conjunctiva, the layer of cells covering the eye. Natural drainage to the throat would then carry progeny virus to the respiratory tract. Little information is available, but there is a report of an infection that resulted when a frozen chicken was being dismembered and a particle of ice flew into cook's eye and resulted in infection by a chicken virus. Another variation on standard respiratory transmission is the suspicion that large aerosol droplets that land on solid surfaces can then be transferred from fingers to the conjunctiva or direct to the nose. The importance of airborne aerosols relative to physically borne droplets for virus transmission remains to be determined.

Environmental factors that result in seasonal variation in the amount of illness or frequency of isolation of a virus are linked with transmission of respiratory infections, e.g. influenza is a winter disease, occurring around January in the Northern Hemisphere, June in the Southern Hemisphere, but nearly all the year round close to the Equator. One can point to variations both in the environment (e.g. temperature and humidity) and in social behaviour (crowding together in winter with poor ventilation) that could well affect virus stability and transmission, and hence the seasonal incidence of virus diseases, but a full explanation of these complex phenomena is not yet available. As viruses can survive for only a limited time outside the cell, it is fair to assume that infected individuals are present continuously somewhere in the population. Interestingly these seasonal patterns for influenza have not been altered by modern mass air travel, which transports thousands of people, many of whom will be incubating the virus, from winter to summer at the other end of the world.

Transmission via the oral–faecal route

Many viruses are ingested, and infect and multiply in the alimentary tract. The surface of the small intestine normally functions to adsorb nutrients and water, and can potentially be infected. One virus that invades by this route is poliovirus (a picornavirus), traditionally by ingestion of sewage-contaminated drinking water and more recently from the water of swimming pools that were inadequately disinfected. Another picornavirus commonly experienced by travellers is hepatitis A virus, often encountered when sewage-contaminated water has been used to wash fruit or salad vegetables, or used to prepare ice cubes for drinks. All these viruses are excreted in faeces, so their spread is favoured by poor sanitation and poor personal hygiene. Many of these viruses grow to very high titres (10^9 infectious units (IU)/ml), but, even if there were only 10^6 IU/ml of faeces, then just 1 µl of faeces contains 10^3 IU. (Try pipetting 1 µl next time you are in the laboratory—it's a very small volume!) Not surprisingly, when we remember how small children investigate strange objects by putting them in their mouth, many of these viruses cause infections in early childhood.

We would expect to see a reduction of disease caused by enteric viruses when sanitation is improved, and basically this has been the experience. However, there have been some unpleasant surprises. In conditions of poor sanitation, poliovirus infects young children as soon as they can crawl outside the house, and results usually in a subclinical gut infection. With improved sanitation, poliovirus is not contracted until adolescence, and associated with this shift in age distribution is an increase in the incidence of paralytic poliomyelitis. This is another example of increased severity of a disease seen when a virus infects an 'unnatural' host, and in this case age is the important difference.

Sexual transmission

Only a few viruses are spread in this way but, for them, it is a very successful route indeed. The main examples are HIV-1 and hepatitis B virus (HBV), herpes simplex viruses (HSV), type 2 predominantly but also type 1, and human papillomaviruses (a papovavirus). HIV-1 has been recognized as the causative agent of AIDS only since 1983 and is currently responsible for over 36 million infections worldwide. AIDS predisposes to both viral and non-viral infections of exaggerated severity, and has a mortality rate approaching 100% (see Chapter 19). HBV is responsible for primary hepatocellular carcinoma in 0.1% of infected individuals, and this makes it the most common human cancer because the virus is endemic in the very large population of China and the Far East (see Chapter 17). All three viruses are transmitted sexually, by both heterosexual (male to female and female to male) and homosexual intercourse. For HIV-1 the risk of infection is marginally less (by two- to threefold) if the female is the carrier. This may be because ejaculate contains not only virus but also around 10^6 lymphocytes, some of which may be infected and producing virus. Virus is also spread by homosexual behaviour and the risk for receptive anal intercourse (which is between 1 in 300 and 1 in 1000) is about the same as that in male–female intercourse when the male is infected. Sexual transmission of infection is affected by sociosexual behaviour and it is probably no coincidence that the spread of these viruses was increased by the promiscuity of the 1960s and 1970s. However, although rates of infection are stable in some countries, the worldwide pandemic that centres on Africa and now the Far East shows no signs of slowing down.

Transmission via urine

This transmission is rare because urine is usually sterile. However, a few viruses, such as those causing Lassa fever (an African arenavirus) and sin nombre (a US bunyavirus), are present in urine and are thought to be transmitted in this way. These viruses infect mice and, when these come into contact with people, food may become contaminated and eaten, or virus-contaminated dust particles may be breathed in. Both these viruses cause human fatalities. In addition, cytomegalovirus is excreted in urine in fairly large amounts and may be transmitted by this route among small children.

Transmission by mechanical means

Under this umbrella are included animal viruses that infect their hosts by means that directly puncture the normally impermeable outer skin layer. This route introduces *virus vectors* for the first time, i.e. intermediaries that transmit the virus to humans. They are usually biting arthropods (mosquitoes, which are insects, and ticks, which are arachnids) found in tropical parts, which feed on blood by piercing the skin with their mouth parts. Preparatory to the next meal, the arthropod injects saliva as an anticoagulant, and in so doing introduces the virus that was picked up previously. Viruses spread by arthropods are known collectively as *arboviruses*.

The main groups of animal viruses that are spread by mosquitoes belong to the alpha-, flavi-, reo-, rhabdo- and bunyavirus groups. Transmission of a virus is specific to particular mosquito species. Such viruses have a complex life cycle, and are adapted to multiply in the tissues of the invertebrate vector as well as those of the vertebrate host. Although arboviruses are found mostly in the tropics, examples from more temperate regions are the flaviviruses, tick-borne encephalitis virus, which is found in Europe from Austria eastwards, and louping-ill virus of grouse, which is spread by the tick *Ixodes* sp. and is found in the northern UK, Ireland and Norway. Sheep and humans in these regions can be infected with louping-ill virus through bites from infected ticks. Other arthropod-borne animal viruses do not multiply in their vector but are carried passively, e.g. myxomavirus (a poxvirus) that causes myxomatosis is spread between rabbit hosts by contaminated mouth parts of infected mosquitoes in Australia and rabbit fleas in the UK.

There is also transmission of serious blood-borne infections (HBV and HIV) between intravenous drug abusers. Virus can be carried from one person to another by sharing non-sterilized hypodermic syringes. The practice of drawing blood into the syringe increases the risk. For HIV the risk of infection is 1 in 150—higher than for sexual activity. However, there is also risk of transmission through any shared device that can cause minor cuts or abrasions and transfer blood, such as combs, razors, and needles used for acupuncture, tattooing or body piercing. Transmission has also occurred through poor hygiene in dental surgeries. In addition, HBV can be spread through infected saliva by biting. There is no evidence that these viruses can be carried passively between individuals by biting insects.

Most plant viruses are spread by mechanical means. They are transmitted into plant cells by animal vectors, such as plant-feeding aphids, leafhoppers, beetles and nematodes, by fungi, or by non-specific abrasions made to the plant tissue by the wind, etc. which expose cells sufficiently for them to be infected. Transmission by plant-feeding animals is a specific process and only certain species are implicated in the transmission of a particular virus (see Chapter 22). As with their animal virus cousins, some plant viruses multiply in the vector, whereas others are passively transmitted.

15.11 Vertical transmission

This is the transmission of virus from mother to baby, in contrast to the horizontal transmission between other individuals. Rubella virus (a togavirus) is the classic example of a vertically transmitted virus. In an adult, the infection (German measles) is manifest as a mild skin rash or is subclinical. However, the virus can cross the placenta and multiply in the foetus. As a result the foetus can die or be born with serious congenital malformations that affect the cardiovascular and central nervous systems, the eyes and hearing. The risk of malformations arising from infection decreases from levels of up to 80% as foetal development proceeds, to practically nil if the mother contracts rubella virus in the fifth month of pregnancy or later. Fortunately, an excel-

Table 15.9 Examples of vertical transmission of viruses from mother to baby and their consequences.

	Possible modes of infection			Possible adverse outcomes		
Virus	Transplacental	During birth*	After birth	Death of foetus	Clinical disease after birth	Persistent infection
Rubella	+	−	−	+	Congenital abnormality	+
CMV, primary	+	−	−	?	Congenital abnormality	+
CMV, reactivated	−	+ (2% of all babies)	+ (bm)	na	−	+
HIV-1	?	+	+ (bm)	−	Up to 15% of babies born to infected mothers: AIDS at age 2	+
HBV	+	+	+	−	−	+
HSV (genital)	+	+	+	+	Herpes lesions	+
HPV	−	+	−	−	−	+

*A caesarean delivery can help avoid infection.
CMV, cytomegalovirus; HBV, hepatitis B virus; HIV-1, human immunodeficiency virus type 1; HPV, human papillomaviruses; HSV, herpes simplex viruses.
bm, breast milk; na, not applicable; ?, not known.

lent single-dose vaccine is available which is given to children in their second year and protects them when they become mothers. A number of other viruses infecting the mother that can be passed vertically to her baby are listed in Table 15.9, together with the outcomes affecting the foetus or baby.

Analysis of the exact route of vertical transmission is not possible because the virus can be transmitted to the zygote from an infected oocyte or sperm, or the zygote can be infected from virus present in cells of the uterus, or via the bloodstream in placental mammals. Any virus that infects the mother while she is producing oocytes or offspring can in theory be vertically transmitted, whereas those with genomes that are integrated with the genomes of the reproductive cells of the host cannot help but be transmitted by this route. The genomes of mammals, including humans, contain large numbers of *endogenous retroviruses* in their germline cells. These remnants of infections that occurred long ago in evolutionary time do not produce an infectious virus. In addition, as Table 15.9 shows, infection can also result during birth. This risk can be minimized by caesarean section. Also included here, because they cannot be distinguished, is infection of the baby by the mother after birth as a result of virus-containing breast milk, or possibly via saliva. Accordingly, HIV-1-infected mothers are advised not to breast-feed their babies. The use of anti-HIV drugs is also beneficial, and the combined precautions reduce the risk of HIV-1 vertical transmission from around 15% to <2%.

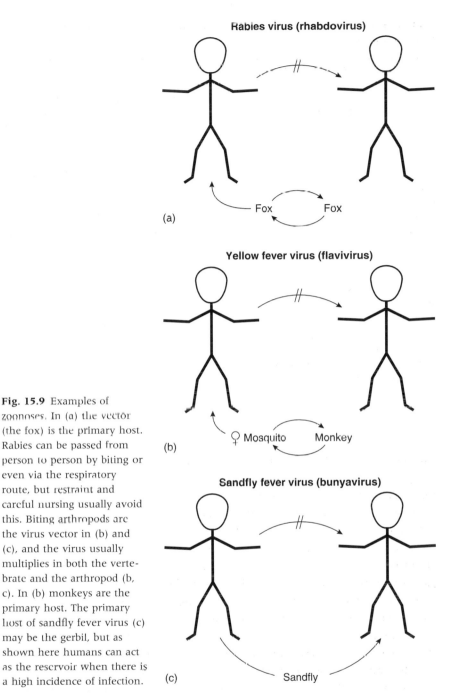

Fig. 15.9 Examples of zoonoses. In (a) the vector (the fox) is the primary host. Rabies can be passed from person to person by biting or even via the respiratory route, but restraint and careful nursing usually avoid this. Biting arthropods are the virus vector in (b) and (c), and the virus usually multiplies in both the vertebrate and the arthropod (b, c). In (b) monkeys are the primary host. The primary host of sandfly fever virus (c) may be the gerbil, but as shown here humans can act as the reservoir when there is a high incidence of infection.

15.12 Zoonoses

A zoonosis is an infection where the infectious agent is transmitted from an animal other than a human to a human. Usually the main host is another vertebrate and humans are only infected accidentally. Biting arthropods are often the vectors responsible for transmission, although rabies virus (a rhabdovirus) is contracted from the bite of an infected vertebrate. There are different animal reservoirs for rabies all over the world, e.g. the red fox in Europe, the skunk, racoon and other animals in North America, the vampire bat in South America, the yellow mongoose in Southern Africa and the jackal in India. This and other examples are shown in Fig. 15.9.

15.13 Recapitulation

• There are seven categories of *infection*.

• Any one *virus* may be found in more than one category, so the categorization of a virus and its infection must include the circumstances of the infection.

• Most infections result in a satisfactory end for both the host species (if not the individual) and the virus. In other words, infections are dealt with by the host's immune system so that the host recovers and can reproduce itself, but not before the virus has had time to multiply and perpetuate itself by being transmitted to other susceptible individuals.

Infections that kill (or reproductively incapacitate) large sections of the host species are rare, and these are likely to result from a relatively new virus–host interaction (such as HIV-1 in humans). Such infections evolve towards genetic compatibility and relatively peaceful coexistence (e.g. myxomavirus in the European rabbit—see Section 18.4).

15.14 Further reading and references

This is an enormous area to cover, but the following provide a way in and more references.

Alcamí, A. & Koszinowski, U. H. (2000) Viral evasion of immune responses. *Immunology Today* **21**, 447–455.

Beaty, B. J. & Marquardt, W. C. (eds) (1995) *The Biology of Disease Vectors*. Washington, DC: ASM Press.

Belshe, R. B. (1998) Influenza as a zoonosis: how likely is a pandemic? *Lancet* **351**, 460–461.

Braciale, T. J. (ed.) (1993) Viruses and the immune system. *Seminars in Virology* **4**(2), 81–122 (several articles).

Casadevall, A. & Pirofski, L.-A. (1999) Host–pathogen interactions: redefining the basic concepts of virulence and pathogenicity. *Infection and Immunity* **67**, 3703–3713.

Craighead, J. E. (2000). *Pathology and Pathogenesis of Human Viral Disease*. New York: Academic Press.

Dimmock, N. J., Griffiths, P. D. & Madeley, C. R. (eds) (1990) *Control of Virus Diseases (Society for General Microbiology Symposium 45)*. Cambridge: Cambridge University Press.

Fazakerley, J. K. & Buchmeier, M. J. (1993) Pathogenesis of virus-induced demyelination. *Advances in Virus Research* **42**, 249–324.

Fujinami, R. S. (ed.) (1990) Mechanisms of viral pathogenicity. *Seminars in Virology* **1**(4), (whole issue).

Gilden, D. H., Mahalingam, R., Dueland, A. N. & Cohrs, R. (1992) Herpes zoster: pathogenesis and latency. *Progress in Medical Virology* **39**, 19–75.

Glaser, R. & Kiecolt-Glaser, J., eds (1994) *Handbook of Human Stress and Immunity*. New York: Academic Press.

Glezen, W. P. (1999) Maternal immunization for the prevention of infections in late pregnancy and early infancy. *Infections in Medicine* **16**, 585–595.

Ho, D. Y. (1992) Herpes simplex virus latency: molecular aspects. *Progress in Medical Virology* **39**, 76–115.

Kaaden, O.-R., Czerny, C.-P. & Eichhorn, W. (eds) (1997) *Viral Zoonoses and Food of Animal Origin*. Vienna: Springer-Verlag.

Lederberg, J. (1999) Paradoxes of the host–parasite relationship. *American Society for Microbiology News* **65**, 811–816.

Liberski, P. P. (ed.) (1993) *The Enigma of Slow Viruses. Archives of Virology* (Suppl. 6). Vienna: Springer-Verlag.

Mims, C. A. & White, D. O. (1984) *Viral Pathogenesis and Immunology*. Oxford: Blackwell Scientific Publications.

Mims, C. A., Nash, A. & Stephen, J. (2000) *Mims' Pathogenesis of Infectious Disease*. London: Academic Press.

Morens, D. M. (1994) Antibody-dependent enhancement of infection and the pathogenesis of viral disease. *Journal of Infectious Diseases* **169**, 500–512.

Posavad, C. M., Koelle, D. M. & Corey, L. (1998) Tipping the scales of herpes simplex reactivation: the important responses are local. *Nature Medicine* **4**, 381–382.

Preston, C. M. (2000) Repression of viral transcription during herpes simplex virus latency. *Journal of General Virology* **81**, 1–19.

Rouse, B. T. (1992) Herpes simplex virus: pathogenesis, immunobiology and control. *Current Topics in Microbiology and Immunology* **179**, 1–179

Steven, N. M. (1997) Epstein–Barr virus latent infection *in vivo*. *Reviews in Medical Virology* **7**, 97–106.

Stevens, D. L. (1998) Immunity and the host response. *Current Opinion in Infectious Diseases* **11**, 269–270.

Timbury, M. C. (1997) *Notes on Medical Virology*. Edinburgh: Churchill Livingstone.

Tudor-Williams, G. & Lyall, E. G. H. (1999) Mother to infant transmission of HIV. *Current Opinion in Infectious Diseases* **12**, 27–52.

Wright, P. F. (ed.) (1996) Viral pathogenesis. *Seminars in Virology* **7**(4) (whole issue).

Zwilling, B. S. (1992) Stress affects disease outcomes. *American Society for Microbiology News* **58**, 23–25.

Also check Chapter 22 for references specific to each family of viruses.

Chapter 16
Vaccines and antivirals: the prevention and treatment of virus diseases

Humans are concerned to protect themselves, and their plants and animals, against death, disease and economic loss caused by virus infections. In animals (including humans), immunization with vaccines has been far more effective than antiviral chemotherapy, but vaccines against some viruses are less than ideal, and for other viruses (notably human immunodeficiency virus type 1 or HIV-1) there is as yet no vaccine that gives any worthwhile protection. These problems will be discussed below. The rationale of immunization is to raise virus-specific immunity without the individual having to experience the disease. As plants do not have an inducible immune system, prevention of plant virus diseases has relied on other means, such as selective breeding of plants that are genetically resistant to the virus or its vector, or by control of the vector. Control of animal virus diseases through improvements in nutrition, public health, personal hygiene and education is also a vitally important aspect of prevention, especially in the developing world.

16.1 Principal requirements of a vaccine

This section deals initially with conventional vaccines, and then takes a forward look at new approaches to immunization.

A vaccine should cause less disease than the natural infection

Conventional vaccines comprise either infectious ('live') or non-infectious ('killed') virus particles (Table 16.1). The process of producing a virus strain that causes a reduced amount of disease for use as a live vaccine is called *attenuation*. The disease-causing virus is referred to as the *virulent* strain and the attenuated strain as *avirulent*. Note that these are only relative terms—there are no absolutes. Avirulent vaccines have been obtained by selecting for naturally occurring avirulent variants. This is achieved empirically, and experience has shown that it can be helped by multiplication of the virus in cells unrelated to those of the normal host, by multiplication at a subphysiological temperature or by recombination with an avirulent laboratory strain. The alternative of using natural strains that are antigenically related to the virulent strain and that cause less disease in that host has been known since Edward Jenner, in 1798, used cowpox virus to vaccinate against the disease smallpox. Until recently,

Table 16.1 Some human and veterinary virus vaccines.

Virus	Family	Live or killed vaccine	Disease (if named differently from the virus)
Human viruses			
Hepatitis A	Picornavirus	Killed	Hepatitis
Hepatitis B	Hepadnavirus	Killed*	Hepatitis
Influenza A and B	Orthomyxovirus	Killed*	
Measles	Paramyxovirus	Live	
Mumps	Paramyxovirus	Live	
Polio	Picornavirus	Live, killed†	Poliomyelitis
Rabies	Rhabdovirus	Killed	
Rubella	Togavirus	Live	German measles
Vaccinia	Poxvirus	Live	Smallpox caused by variola virus
Yellow fever	Flavivirus	Live	
Veterinary viruses			
Canine distemper	Paramyxovirus	Live	
Equine influenza	Orthomyxovirus	Killed	
Foot-and-mouth	Picornavirus	Killed, live	Affects cattle and pigs
Newcastle disease	Paramyxovirus	Live	Fowl pest
Parvo	Parvovirus	Killed, live	Causes death in young dogs and cats
Pseudorabies	Herpesvirus	Killed, live	Aujeszky's disease of pigs
Turkey herpes	Herpesvirus	Live	Lymphoma in chickens‡

*A recombinant DNA subunit vaccine (the surface (S) antigen) is used for hepatitis B virus, and a natural subunit vaccine (haemagglutinin and neuraminidase antigens) for influenza.

†The killed vaccine is used successfully in Scandinavian countries, and is now used instead of the live attenuated vaccine in the USA.

‡Caused by Marek's disease virus, a herpesvirus. Thus, a virus with some related antigens can provide protection—compare vaccinia virus against smallpox. Marek's disease virus is also notable as the first vaccine against cancer. As hepatitis B virus can cause liver cancer, its vaccine doubles as the first vaccine against a human cancer.

a related virus, vaccinia virus (origin unknown), was used for the same purpose, and the term 'vaccine' has become synonymous with any immunogen used against an infectious disease.

Naturally enough, killed vaccines should cause no disease at all. However, as this type of vaccine is made from the virulent strain, it is essential to kill every infectious particle present (Fig. 16.1). An advantage of killed preparations is that any other unknown contaminating viruses will probably be killed in the same process. However, a major disadvantage is that the killed virus does not multiply, so that an immunizing dose has to contain far more virus than a dose of live vaccine, and repeated doses may be required to induce adequate levels of immunity. This increases both the cost

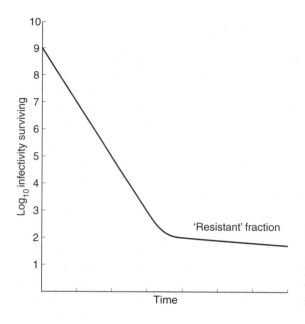

Fig. 16.1 Exponential kinetics of virus inactivation, for example by formaldehyde. Note the 'resistant' fraction which is inactivated more slowly.

and the amount of non-viral material injected; the latter may result in hypersensitivity reactions when these substances are experienced again.

The public perception of vaccines

Vaccines are beneficial only if the public is persuaded to take them. One problem is that by chance a person may become ill at about the time he or she has a vaccine, so the vaccine is blamed even though it was not at fault. The second problem is that no vaccine is perfect, and there will be a finite minor risk of rare side effects. The public seems to have little appreciation of risk, e.g. when poliovirus was endemic and every child was exposed there was a 1:100 chance of contracting paralytic poliomyelitis; the risk of poliomyelitis from the live vaccine as a result of reversion is around 1:1 000 000 per first dose of vaccine (see Section 21.1), and the other 999 999 are protected. This seems like a good deal, unless it is you or your child who develops the vaccine-associated disease. Parents in particular feel that they are to blame, and this dissuades other parents from allowing the vaccine to be given to their children. What is worse is that knowledge of an adverse event associated with one vaccine may dissuade them from taking up other vaccines as well. The problem is made worse by heavy media publicity. As a disease becomes rarer because of the success of the vaccine, the perception of the risk from vaccine use is heightened as the relative risk differential narrows. Vaccine manufacturers strive to make the vaccine safer but there is a risk in any medical intervention, and a vaccine will never be absolutely safe. As the risk cannot be completely eliminated, it is up to government to minimize bad publicity by ensuring that there is compensation for all those who appear to be damaged as the result of taking a vaccine.

Effective and long-lasting immunity

The requirement for these properties is both scientific and sociological. As explained in Chapter 15, different viruses are susceptible to different parts of the immune system; hence it is necessary for the vaccine to stimulate immunity of the correct type, in the correct location and in sufficient magnitude for it to be effective. This is a problem with the killed influenza virus vaccines currently in use, which are administered by injection and do not raise virus-specific IgA antibody at the mucosal (respiratory epithelial) surface where it is needed. However, sufficient of the immunoglobulin IgG to be protective is thought to diffuse from the bloodstream to the respiratory surface in about 70% of individuals immunized. If effective immunity is not achieved, the virus being immunized against may still be able to multiply. This can occur with foot-and-mouth disease virus of cattle (picornavirus). Worse still the partial immunity provides a selection pressure that favours the multiplication of new antigenic variants of the virus, and in time these replace the pre-existing strain. Even animals with effective immunity against the original strain are not protected from the new strain, and the long, expensive process of vaccine development and immunization has to begin again. Immunogenicity can be enhanced by mixing virus with an *adjuvant*. The virus either binds to the adjuvant or, with an oil based adjuvant, is emulsified with it. Adjuvants work by stimulating the immune system in ways that are poorly understood, and act as a depot and release the immunogen over a long period, which can result in a greater immune response. However, many adjuvants cause local irritation and only one, aluminium hydroxide gel, is licensed for use in humans.

Public faith in any preventive medicine is rapidly lost if that measure is not effective in most cases (see above). Consequently, to avoid bad publicity, a vaccine must protect most individuals who have received it. Immunization should result in an effective response (i.e. a sufficiently large immune response to protect against the disease) and this should also last for years, preferably a lifetime. However, as it is surprisingly difficult to persuade people to come forward to be immunized, a single-shot vaccine requiring only one visit to the clinic is ideal and has been realized only for measles, mumps, rubella and yellow fever viruses. All other vaccines require at least two doses. This is an even greater problem in the developing world, where there is little health-care infrastructure to implement a multidose immunization campaign. Some employers have circumvented this problem by offering immunization at the place of work. Vigorous advertising campaigns are widely used also (Fig. 16.2) to overcome public apathy. In tropical countries, where immunization is often most needed, there is the problem of ensuring that a vaccine has a good shelf-life (i.e. that its infectivity or immunogenicity is stable for periods when refrigeration is not available) or immunization will be ineffective.

Another hope for the improved efficacy of killed vaccines is their administration by mouth or intranasally, which has the added advantages of avoiding 'jabs' (see Section 16.2) and of stimulating mucosal immunity (see Fig. 14.2). Intranasal vaccines are looking more promising than oral vaccines as a result of the dual problems of digestive enzymes and local immunological tolerance of foreign antigens (food) in the gut.

**SALLY'S
MUM GOT
GERMAN
MEASLES.**

**LOOK WHAT
SALLY GOT**

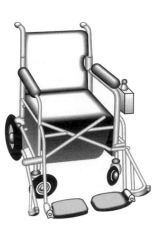

Fig. 16.2 Newspaper advertisement aimed at increasing public awareness and acceptance of immunization against rubella virus.

Genetic stability

An attenuated live virus vaccine must not revert to virulence when it multiplies in the immunized individual, but even the excellent poliovirus vaccine is not perfect in this regard (see Section 21.1). Vaccines that are formed by 'killing' virulent strains of viruses must not be restored to infectivity by genetic interactions with each other or with viruses occurring naturally in the recipient.

Individual vs herd immunity

It seems obvious that the entire susceptible population should be immunized and for every single individual to develop effective protective immunity. However, 100% immunity is not usually needed for the elimination of a virus from that population (which is as well, because this is never achieved). What is required is a sufficient percentage of immune individuals to break the chain of transmission of the virus. Thus, *herd immunity*, rather than individual immunity, is the key issue. A natural experiment that illustrates this point is measles virus in island populations. Natural measles infection leaves people immune to reinfection, and a population of at least 500 000 is needed to generate a continuous supply of susceptible children in sufficient number to prevent the virus from dying out (see Section 18.3). The exact percentage of immune individuals required for herd immunity is different for each infectious agent and can be calculated.

16.2 Advantages, disadvantages and difficulties associated with live and killed vaccines

Inactivation

A killed vaccine must have no infectivity and yet be sufficiently immunogenic to provoke protective immunity. Inactivating agents should inactivate viral nucleic acid function and not merely attack the outer virion proteins, because infectious nucleic acid could be released by the action of the cell on the coat. Second, inactivating agents must lend themselves to the industrial scale of manufacturing processes. Formaldehyde and β-propiolactone are two that have been used. The former reacts with the amino groups of nucleotides and cross-links proteins through ε-amino groups of lysine residues. β-Propiolactone inactivates viruses by alkylation of nucleic acids and proteins. The basic problem of preparing a killed vaccine is that every one of the 1 000 000 000 (10^9) or so infectious particles that are contained in an immunizing dose of virus must be rendered non-infectious, and this was the problem that faced Jonas Salk before the first poliovirus vaccine could be presented to the general public in 1953. Inactivation of the infectivity of polio- and other viruses is exponential and has the kinetics shown in Fig. 16.1. Frequently, it is found that a small fraction of the population is inactivated far more slowly than the majority, so that the whole virus population has to be kept in contact with the inactivating agent for much longer than is predicted by the initial rate of inactivation. This has the concomitant risk that immunogenicity will be destroyed.

Examination of the 'resistant' fraction shows that it results from inefficient inactivation of particles trapped in aggregates or clumps. Despite these difficulties, the inactivated Salk vaccine has been used very successfully, mainly by Scandinavian countries, to this day. Live vaccines present none of these problems, but the difficulties of producing a genetically stable attenuated strain are at least as great. For poliovirus, these were overcome by Albert Sabin, who attenuated all three serotypes to produce the successful live vaccine, which has been in use since 1957. None the less, the type 3 vaccine reverts to virulence at the rate of one case of poliomyelitis per 10^6 first doses of vaccine. This is an extremely low figure, and much lower than the risk of developing poliomyelitis in areas where poliovirus is endemic and infection is universal (1 in 100 children), but is very serious for the child involved. However, as the virus is eradicated from a country and the risk from natural infection is restricted to outbreaks resulting from virus imported by infected individuals, the goalposts shift, and the risk from the vaccine becomes unacceptable. For this reason, countries such as the USA have changed to the killed vaccine despite its greater expense (see below). The molecular explanation and ways of improving the vaccine are discussed later (see Section 21.1).

Routes of administration and a public relations problem

The unavoidable difficulty of killed vaccines is that they do not multiply. Hence they do not reach and stimulate mucosal immunity in those areas of the body where infectious virus normally enters (usually the intestinal and respiratory tracts—see Fig. 14.2). However, provided that the killed vaccine is a potent immunogen, the high levels of serum IgG that result from injection of the vaccine can often serve in place of local immunity, presumably because there is a sufficient concentration to diffuse to the extremities.

The means of administration is an important factor in persuading people to accept a vaccine. Injections via the traditional 'jab' (hypodermic syringe and needle) are painful and unpopular, particularly with small children and their mothers.

A compressed-air device that also impels vaccine through the skin has been used for mass immunization, mainly of military personnel in the USA, and is being developed for general use. A disadvantage of killed viruses is that in theory they require two or more injections to build up a sufficient secondary immune response, whereas live vaccines, similar to the natural infection, should take only one. However, in practice most vaccines require more than one dose.

Cost

The production and use of vaccines in humans are subject to legislation by the Food and Drug Administration in the USA and by the Department of Health in the UK. The precautions laid down for the production of a safe vaccine are necessarily stringent. However, this means that the cost of the final product will be correspondingly high. Precautions increase as knowledge is obtained about hitherto undetected viruses in cells used in vaccine manufacture, or other potential hazards, and as they do so the costs increase yet again. Yet another problem to manufacturers, particularly in the USA, is the cost of defending lawsuits where vaccine-induced damage is alleged; this is a trend now spreading to other countries. Regulations governing the use of veterinary vaccines are not so demanding and the products are correspondingly cheaper. The other major aspect that directly affects the cost is the quantity of vaccine that has to be produced. Live vaccines are inexpensive compared with killed vaccines, because only a small infectious dose is needed to initiate an infection, whereas an immunizing dose of killed vaccine costs many times more, e.g. live and killed poliovirus vaccines cost around 7 cents and $US3 per dose, respectively. Finally, the costs involved with packaging, distribution, administration, and the syringe and needle if injection is needed are considerable. These all add up to make immunization a luxury that poor countries, usually those in the greatest need, cannot afford. Provision of vaccines is one of the ways in which the richer countries can, and often do, provide valuable aid.

Multivalent vaccines

The immune system can respond to more than one antigen at a time—hence it is possible to immunize with a vaccine 'cocktail'. Such a multivalent vaccine has advantages in minimizing the number of injections and the inconvenience. In practice, satisfactory immunity results from a cocktail of killed viruses, but problems with multiple live vaccines may arise as a result of mutual interference in multiplication, possibly caused by the induction of interferon (see below). However, the live poliovirus vaccine contains the three serotypes, and a triple live vaccine of measles, mumps and rubella viruses (MMR) is in routine use.

16.3 Infectious diseases worldwide

Table 16.2 puts in context the impact of major virus diseases (poliomyelitis, measles, rotavirus-induced diarrhoea, some respiratory viruses, HIV and hepatitis B virus

Table 16.2 The world's leading infectious diseases.

Disease	Comments
Vaccine-preventable diseases (polio, tetanus, measles, diphtheria, pertussis, tuberculosis)	Each year 46 million infants are not fully immunized; 2.8 million children die and 3 million are disabled
Chronic hepatitis (hepatitis B virus)	300 million infectious carriers; these have a 200-fold greater risk of developing liver cancer than the population as a whole
Diarrhoeal disease	750 million children are infected annually and 4 million die; 25% of these are caused by viruses (mainly rotaviruses)
Acute respiratory infections	4 million children die annually
Sexually transmitted diseases including HIV-1	1 of 20 teenagers and young adults contract such diseases each year, and 36 million are infected with HIV-1
Tuberculosis	1600 million people carry the bacterium; annually, there are 10 million new cases and 3 million deaths
Malaria	100 million cases annually; almost half the world's population lives in malarial areas
Schistosomiasis	200 million cases annually

HIV-1, human immunodeficiency virus type 1.

(HBV)) relative to bacterial diseases (tetanus, diphtheria, pertussis, tuberculosis and some sexually transmitted organisms), protozoal diseases (malaria) and the minute snail-borne worm, which causes schistosomiasis.

16.4 Elimination of virus diseases by immunization

The World Health Organization (WHO) is responsible for the worldwide control of virus diseases. Table 16.3 shows some of the achievements and current goals, made possible through the development of suitable vaccines.

Eradication of smallpox

Smallpox has been totally eradicated by vaccination. The live vaccine comprising vaccinia virus was administered intradermally by scarifying the skin. Virus multiplies at that site, causing a reaction about the size of a boil and stimulating both antibody-mediated and T-cell-mediated protective immunity. The last naturally occurring case was recorded in Ethiopia in October 1977. Yet less than 200 years ago smallpox was endemic throughout the Old World. The child mortality rate from smallpox in England

Table 16.3 Goals, past and future, for the control of some virus diseases.

Virus	Commencement of control programme	Achievement/goal
Variola (smallpox)	1966	Eradicated worldwide in 1977
Polio types 1, 2 and 3	1988	Eradication worldwide by 2002
Measles	1991	Eradication from Europe by 2000
Hepatitis B	1992	Universal immunization of infants in Europe by 1997

Table 16.4 Criteria for the worldwide eradication of human viruses.

Criterion	Smallpox virus	Measles virus	Poliovirus types 1–3
There must be overt disease in every instance so that infection can be recognized	+	+	−
The causal virus must not persist in the body after the initial infection	+	+	+
There must be no animal reservoir from which reinfection of humans can occur	+	+	+
Immunization must provide effective and long-lasting immunity	+	+	+
Viral antigens must not change	+	+	+

was up to 25%, and in India alone in 1950 over 41 000 people died. The decline of smallpox in those countries where it was recently endemic has been a triumph for the vaccination programme administered by the WHO. Smallpox is the only infectious disease that has ever been deliberately eliminated and this experience has served to emphasize that successful elimination is possible only if a disease fulfils certain criteria (Table 16.4).

Evidently, smallpox virus fulfilled all of these criteria. The elimination of smallpox took several decades as gradually countries could be declared smallpox free. The final assault on the virus in the Indian subcontinent and Africa was not done by mass vaccination but by 'ring' vaccination. When a new case of smallpox was identified, all people in that location were then vaccinated to form a ring of immunity. The virus could not break out to infect new susceptible hosts and so died out. Fortunately, other poxviruses (e.g. whitepox of monkeys, which is antigenically similar to smallpox virus) are apparently unable to step into the ecological niche vacated by smallpox virus and replace it as a human pathogen.

A problem arises with laboratory stocks of smallpox virus. As vaccination is no longer required, the population is becoming increasingly susceptible to the danger of an escaped laboratory virus. If we destroy the virus we lose, for ever, approximately 10^5 base-pairs (bp) of irreplaceable genetic information. We have no way of knowing the value of the smallpox virus genome, but it encodes proteins that are similar to those found in eukaryotic cells and could conceivably be of use in the future. The problem has been partly resolved by cloning the entire viral genome in fragments, and virus proteins can now be expressed in the absence of infectious virus. However, the genome can always be stitched back together. Discussions about destroying the remaining stocks of infectious virus in the USA and the Soviet Union/Russia have been in progress for many years and the matter is still unresolved.

Control of measles

Before the development of a vaccine, measles was an inevitable childhood infection, with 130 million cases annually worldwide and a mortality rate of 2.3% (3 million). However, in the developing world, the child mortality rate was, and still is, as high as 80%. The difference was not the virus but the result of other factors, in particular malnutrition, making the infection more virulent. Measles immunization reached nearly all countries by 1982, but the coverage is still very patchy and in half the African countries less than 60% of children are immunized. However, since immunization commenced, the occurrence of measles has fallen steadily. Nevertheless, there are still 29 million cases per year in children worldwide, mostly in developing countries, with 1 million deaths. In developed countries, death from measles results from respiratory and neurological causes in about 1 in every 1000 cases. Encephalitis occurs at the same rate and survivors are often permanently brain damaged. Rare cases of subacute sclerosing panencephalitis (SSPE) are also caused by measles virus (see Section 15.7).

Some way ahead of the world scene, the Department of Health, Education and Welfare announced, in 1978, their intention of eliminating measles virus from continental USA. The criteria for successful elimination were met by measles virus (Table 16.4): It causes an acute infection nearly every time; in SSPE, persistent virus is not infectious; there is no animal reservoir; there is a good live vaccine (95% effective); and there is only one serological type. The measles eradication programme in the USA started from a recorded number of approximately 500 000 cases annually, although the actual number was estimated at 3–4 million. This was so successful that, by 1983, only 1500 cases of measles were reported—and measles was now a rare disease in the USA (Fig. 16.3). However, a high immunization rate was necessary, and to achieve this legislation was passed prohibiting children from attending school unless they had been immunized or had had a natural infection. (The ethics of this decision make for an interesting debate.) Despite these measures, there was a disturbing increase in the annual number of measles cases in 1989–91 to nearly 28 000 (Fig. 16.3), which is thought partly to reflect the problems of maintaining a high level of immunization, particularly in socially deprived inner-city areas (where immunization falls to around 50%), but also a failure of some children aged less than 1 year to be protected by maternal antibody. This was caused by the mother's immunity resulting from immunization rather than infection, with a lower titre of antibody and thus a smaller

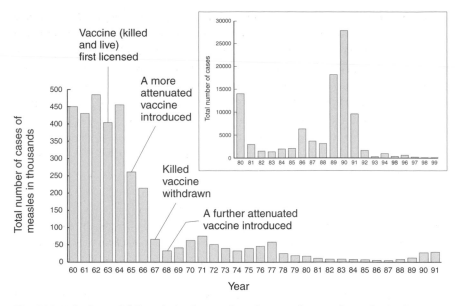

Fig. 16.3 Fall, rise and fall again in the number of reported cases of measles in the USA after immunization 1960–99. Note change of scale in the inset. (http://www.cdc.gov/nip/publications/pink/meas.pdf)

amount of antibody being transferred across the placenta to the foetus. Thus, immunity waned more rapidly than before and left the infants susceptible to infection at a younger age than in the past. However, more thorough immunization since 1989 with two doses of live vaccine has resulted in just hundreds of cases per year since 1995 and just 86 cases in 1999. At the present time, the immunization programme must be continued to prevent the epidemic spread of the virus following its inevitable introduction by infected visitors to an otherwise susceptible population. Measles in non-immune adults is more severe than in children, so the consequences of ceasing measles immunization would be far worse than if it had never been started. The WHO had planned to eradicate measles from Europe by 2000, but this has not been achieved. A target date for worldwide eradication of the disease has not yet been set.

The planned eradication of poliovirus

In 1985, the Pan-American health authorities announced a plan to eradicate poliovirus from North, Central and South America. However, Table 16.4 shows that poliovirus falls down on one of the key criteria for eradication because 99% of infections are subclinical. In addition, although there is no natural animal reservoir, Old World primates are susceptible to infection, and chimpanzees in the wild have been infected. In 1988, the policy was extended worldwide by WHO. The live oral vaccine that comprises the three attenuated serotypes was used. Laboratories have been set up in all countries, good public relations encouraged, and intensive immunization mounted whenever there is an outbreak. The campaign has been very successful, and the last case of natural disease (compare vaccine-caused disease—Section 16.2) in the Americas was in Peru in 1991. In addition, the following areas are poliovirus free: Europe including the former Soviet Union (last case in Turkey in 1998), the western Pacific region including the People's Republic of China (last case in Cambodia in 1997), and most of the Middle East and large parts of southern and northern Africa. There are currently three major foci of infection left in the world: southern Asia (Afghanistan, Pakistan and India), West Africa (mainly Nigeria), and central and other parts of Africa (Angola, the Democratic Republic of Congo, Ethiopia, Somalia and Sudan). Eradication measures include 'National Immunization Days' with house-to-house immunization to help build up adequate levels of protection. In countries where poliomyelitis has been abolished, the main danger is complacency, either because the virus has not been completely eliminated or because of the problem of its introduction from elsewhere by infected travellers. Immunity still needs to be maintained.

As total elimination of poliovirus approaches, attention is focused on the day that immunization will cease. Before that can happen, all other sources of virus also have to be eliminated. These include the frozen stocks of poliovirus in many laboratories where poliovirus was (and still is) studied for research purposes, and frozen poliovirus-containing faecal samples in hospital pathology laboratory freezers all over the world. All these have to be destroyed or made secure because the aim is to have a world without poliovirus, where people have no need of poliovirus immunity. A snapshot of the progress resulting from immunization is summarized in Fig. 16.4. Finally, when poliovirus immunization is no longer needed, the world will save $US1.5 billion, money that can be directed towards other health priorities.

Control of other virus diseases

Worldwide eradication of other virus diseases is unlikely to be easy, but the WHO is studying the feasibility of eliminating rabies virus. This presents a new level of difficulty because it is widely endemic in many common wild animals: foxes in Europe, racoons in the USA, vampire bats in South America, wolves in Iran, and jackals and mongooses in India. These transmit virus directly or indirectly to humans via domestic animals, notably dogs, but such infections are incidental to the normal life cycle of rabies virus. The WHO has experimented with immunizing wild animals in inaccessible mountainous terrain by dropping pieces of meat doctored with live vaccine from aircraft. The immunization of foxes in Belgium, using vaccinia virus genetically engineered to express the rabies virus envelope protein, has been notably successful (see Section 16.7 for an account of this type of vaccine). The eradication of hepatitis B virus is being considered by the WHO as part of its cancer prevention programme (see Section 16.6).

Fig. 16.4 Progress of poliovirus eradication worldwide, showing the countries in which virus was endemic (dark shading) in 1988 (top) and in 1999. (Courtesy of WHO, Expanded Programme on Immunization.)

16.5 Clinical complications with vaccines

Listed below are some of the circumstances in which the normal immunization procedure is ineffective, inadvisable or even dangerous. In doubtful circumstances a killed vaccine is considered safer than a live vaccine.

1 Very young children sometimes develop a more severe infection with the vaccine than occurs in an older child.

2 Maternal antibody is transmitted to offspring via the placenta and can prevent stimulation of the immune response, so immunization is carried out after this has waned, at about 2–6 months of age.

3 Certain clinical conditions can make an infection more severe than normal, e.g. people with eczema are prone to a generalized infection with vaccinia virus instead of a local infection at the site of inoculation of the vaccine.

4 As foetal development can be deranged by virus infection, e.g. by rubella virus, immunization should be avoided during pregnancy.

5 An existing virus infection can sometimes interfere with the multiplication of a live vaccine, with the result that effective immunity is not established.

Potentiation of disease

Under certain circumstances, it has been found that the immunity acquired from administration of a killed vaccine potentiated the disease when the wild virus was contracted. This occurred in a clinical trial of a potential human respiratory syncytial virus (HRSV) vaccine (a paramyxovirus), which causes a lower respiratory tract infection in very young children. There are two main theories to account for this situation. The first involves the ratio of the two viral surface proteins, F and G. Both stimulate antibodies, but antibodies to F are the more protective. Here it is believed that inactivation of HRSV infectivity also destroyed the immunogenicity of the F protein, and that the resulting antibodies to the G protein formed immune complexes in the respiratory tract when natural HRSV was encountered. These, in turn, stimulated a hypersensitivity response, and an influx of fluid and immune cells to that site, thereby impeding breathing. The second theory revolves around the stimulation of CD4$^+$ T-helper cells (see Section 14.3). These cells are in fact classified on a functional basis into T-helper type 1 (Th1) cells and Th2 cells. Any immunogen stimulates a balance of these two cell types. Th1 cells stimulate mainly cell-mediated immunity (the response that has evolved to deal with intracellular infections), which includes CD8$^+$ cytotoxic T cells, and Th2 cells stimulate hypersensitivity (the response that has evolved to deal with infections by parasitic worms). It is thought that the batch of killed vaccine used in the trial was altered by the inactivation process in such a way that it produced a response biased to Th2 cells, rather than the Th1 cells seen in a natural infection. Thus, when the immunized children contracted a natural RSV infection, only the Th2 arm had immunological memory; this response exacerbated the infection and some of the children died.

In this example, disease was potentiated only in infants, the group that was in greatest need of protection. This demonstrates the insoluble difficulty that possible untoward effects of a vaccine will not be revealed until it is administered to people, and the need to test vaccines in all age groups.

Postexposure vaccines

Once signs and symptoms of an infection are apparent, it is too late to immunize. It might appear that the very effective postexposure immunization against rabies is an exception, but this is done at the time of the suspected *inoculation* of virus by the bite of a rabid animal. It works because transport of virus along peripheral nerves to the spinal cord and brain takes several weeks, and the disease does not start until it gets there. (Incidentally, the modern rabies vaccine is produced in tissue culture and has none of the problems of the infamous infected rabbit brain suspension devised by Louis Pasteur, which required multiple injections into the abdomen and was prone to cause life-threatening hypersensitivity reactions.)

Passive immunization

One way to treat an already established disease is to administer immunoglobulin with activity against the infecting virus. This is really chemotherapy with antibody, because no immune response is stimulated and such passively acquired or *passive immunity* lasts only as long as the immunoglobulin survives in the body. A further complication is that one must be prepared to treat a possible anaphylactic response if successive doses of the same 'foreign' immunoglobulin are given. Better alternatives are to use pools of immunoglobulin from humans who are immune to the particular virus, or mono-clonal immunoglobulins from human hybridomas or cloned human antibody (see Section 21.8). The down-side of any such therapy is ensuring that these biologically derived products are free of infectious agents, and particularly as yet unknown agents for which there is no test, which may be present. Other emergency measures not involving the immune system also involve chemotherapy, including treatment with interferon (see Section 16.8).

16.6 New approaches to vaccines

From the foregoing pages, the reader will appreciate that the manufacture of vaccines is a difficult process. The central problem is that of safety—making sure that a killed vaccine contains no residual virulent particles or that a live attenuated vaccine does not revert to virulence. Neither can be guaranteed absolutely. Safety and the attendant problems of expense and efficacy have encouraged research into alternative ways of producing vaccines.

Understanding antigenic determinants

Most killed vaccines (see Table 16.1) consist of whole virus particles that in the main stimulate an antibody-based protective immunity, which is directed to only a minor part of the surface structure of the virus. The rest is either non-immunogenic or stimulates an immune response that is not protective. (This observation is not intended to discount the importance of cell-mediated immunity to other parts of the virion or non-virion proteins, which is actively being sought for future vaccines.) However, as a result of the limited immunogenicity of the virion surface, research has pursued a reductive approach to determine whether a vaccine can be made that consists only of those vital immunogenic regions of the virus particle.

X-ray crystallographic analysis of the influenza virus haemagglutinin (HA) protein, and of natural or laboratory-derived antigenic mutants, has shown that only five

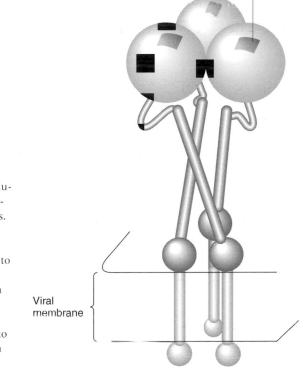

Virus attachment site

Fig. 16.5 Diagram of an influenza A virus haemagglutinin spike deduced from X-ray crystallographic analysis. The spike is composed of three identical monomers. Only when antibody binds to the solid areas is the virus neutralized. These occur on all monomers but are indicated only on one. The attachment site that binds to *N*-acetylneuraminic acid on the host cell is marked.

Viral membrane

regions of the molecule (antigenic sites) bind neutralizing antibody (Fig. 16.5). Antibodies that bind to other regions of the HA do not neutralize influenza virus infectivity so, in theory, immunization with one or more peptides corresponding to epitopes within the antigenic sites would be sufficient to induce protecting antibody. However, life is not so simple and the theory would be borne out only if we were dealing with a continuous epitope (Fig. 16.6a). In fact most naturally occurring B-cell epitopes have an exact conformation, which is unlikely to be achieved by an isolated peptide (Fig. 16.6b). Even more complex is the type of epitope that exists only when parts of two different polypeptides, or two parts of the same polypeptide, come together (Fig. 16.6c). In practice, peptides are generally poor at stimulating neutralizing antibodies, and this is particularly so for influenza virus.

From the discussion so far, the reader might conclude that a vaccine could, in theory, consist solely of peptides in the correct conformation. However, there is more to the story and we should now take a step back to consider how the immune system is stimulated to synthesize antibodies. Analysis of the immune system (see Chapter 14) has shown that CD4 T-helper cells are needed to assist B lymphocytes to become plasma cells, and that T-cell epitopes usually differ from those that bind neutralizing antibodies. Thus, a minimal peptide vaccine has to have two epitopes, one to stimulate the helper T cell and the other to stimulate the B lymphocyte.

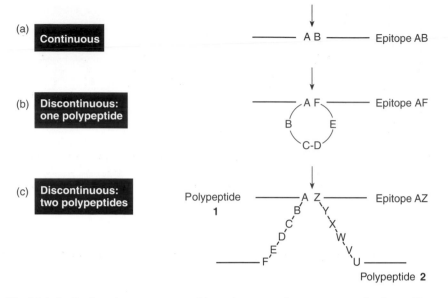

Fig. 16.6 Antibody epitopes constructed in various ways from segments of polypeptide sequence represented by an arbitrary two-letter code and viewed by the immune system in the direction arrowed.

Prediction of epitopes

Although epitopes have been located for influenza virus HA, as described above, such work is laborious and expensive. One short cut is based on the understanding that epitopes are likely to be parts of the molecule that are rich in hydrophilic amino acids. These will extend into the aqueous environment and be easily 'seen' by cells of the immune system. Regions of the protein that are rich in hydrophobic amino acids are generally folded into the internal parts of the molecule, where they are hidden from sight. All one needs to have to determine likely antigenic sites is the amino acid sequence and a computer program to search for hydrophilic regions, such as that devised in 1982 by J. Kyte and R. F. Doolittle (Fig. 16.7). Hydrophilic peptides are then synthesized chemically, usually covalently linked to a protein carrier molecule to provide the necessary CD4+ T-helper cell epitope, and used to raise antibody. The resulting antiserum reacts with the stimulating peptide, but the key tests to determine whether the computer has guessed right are reactivity with the *native protein*, the ability to neutralize virus *in vitro*, and above all the ability to protect against infection *in vivo*. The procedure has been used successfully with hepatitis B surface antigen (HBsAg). Recent refinements have suggested that antigenic determinants are likely not only to be hydrophilic but also to be those with the highest mobility. Apparently, lymphocytes are like fish and are attracted by the most lively bait. Even though epitopes can be predicted, it is still not possible to make a peptide with a desired conformation (e.g. as in Fig. 16.6). A free peptide will take up all possible conformations at random, and antibodies will be made to all of these. It is a matter of luck that there

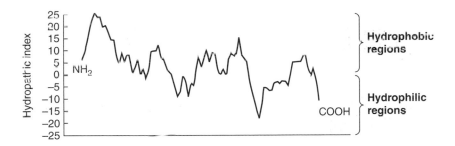

Fig. 16.7 Identification of hydrophilic regions (putative epitopes) of a protein by calculating the hydropathic value of each segment of amino acid sequence across a moving window of seven residues, i.e. the amino acid in question and the three residues on either side.

will be enough antibody present to react with the conformation that the peptide takes up in the native protein.

Hepatitis B virus vaccine

There are estimated to be more than 300 million carriers of hepatitis B virus (HBV— a hepadnavirus) worldwide. The virus infects people of all ethnic groups, but is endemic in the People's Republic of China and the Far East as a result of perinatal infection. In the developed world, HBV is transmitted mechanically or sexually through contact with infected body fluids. As its name indicates, HBV causes acute liver disease, an extremely debilitating infection that can progress to a chronic life-long carrier state which may prove fatal (see Section 15.5). In addition, the chronic infection may lead to one of the most common types of human cancer— primary hepatocellular carcinoma (see Section 17.7).

The problem with making an HBV vaccine was that the virus could not be grown in cell culture (and today still grows only poorly), and infects only humans and higher primates, such as chimpanzees. This was solved by cloning and expressing HBsAg, which elicits protective immunity, in yeast. This was the first *recombinant vaccine*, and it has proved to be very effective. Importantly, children born to infected mothers can be immunized at birth and protected. The immunology is complicated and not well understood, but it is self-evident that the infant's immune system is sufficiently developed and has not been made tolerant to HBsAg, and there is no interfering maternal antibody. Today, this is the standard, commercially available vaccine. Three injections are required and, at a total cost of around £56, it is one of the more expensive vaccines.

16.7 Genetically engineered vaccines

Killed vaccines required a considerable amount of virus, and it is a major problem to produce this cost-effectively and ensure that no infectious virus survives the inacti-

Table 16.5 Genetically engineered vaccines.

Advantages	Problems
Non-infectious	Identification of neutralization antigen and/or epitope
Large-scale production methods available	Need for proper co- and post-translational modifications of viral polypeptide
Cost-effective to produce	Need for the expressed protein to achieve proper conformation to avoid poor immunogenicity
Can use genes from non-cultivable viruses	Purification of viral protein from cell constituents

vation procedure. There are three solutions, now that DNA technology can be used to identify the part of a viral genome that encodes the particular virus protein against which protective immunity (usually antibody) is directed. These are: (i) expression of the entire protein; (ii) expression of a fragment of the protein containing the antigenic site, alone or fused with a carrier protein gene; or (iii) chemical synthesis of a peptide that contains the antigenic site. For (i) and (ii), viral DNA, or a DNA copy if the virus has an RNA genome, can be excised and inserted into an appropriate eukaryotic expression vector, together with control (promoter, stop and polyadenylation) signals. Thus, we have a small part of the viral genome, by definition non-infectious, which by insertion into a cell growing on an industrial scale will produce very large amounts of protein very cheaply. The ultimate is chemical synthesis (iii above), which needs no living cell. Lastly, with any of these approaches, the virus does not have to grow in culture for a vaccine to be developed, a great advantage for viruses such as HBV.

Does it work? Table 16.5 summarizes the state of what is still very much an experimental development. Genetic engineers would very much like to use bacterial expression systems that are well understood from the viewpoints of both their genetics and their large-scale industrial production. However, human and veterinary diseases are caused by viruses, the newly synthesized polypeptides of which undergo eukaryotic-type co-translational and post-translational modifications, such as glycosylation and proteolytic cleavage, which prokaryotic cells cannot accomplish. Thus eukaryotic cells are essential. However, in addition to cultured cells from higher animals (for which technology has now been intensively developed), there is several thousand years' experience in the bulk culture of yeasts, also eukaryotes, for brewing and baking. Yeast has well-understood genetics and offers an expression system with many of the advantages of a prokaryote in its handling, but with the essential eukaryotic features. Its genome has now been completely sequenced.

The method works well if a linear epitope gives protection immunity, but antibody is usually directed to conformational determinants (see Fig. 16.6). At the moment, we do know how to instruct a peptide to fold in a particular way. Thus, success is achieved only by chance. Nevertheless, as mentioned above, the first recombinant vaccine, comprising the HBsAg, has proved itself in clinical trials and, in 1986, came on to the commercial market.

A chemically synthesized, experimental, peptide vaccine against foot-and-mouth disease was also successful; it both raised neutralizing antibody and protected cattle, the main species at risk. Good luck and good science came together here, because it was only realized after the event that the peptide used, residues 146–161 of the coat protein VP1, was a linear B-cell epitope that also contained a helper T-cell epitope.

There are many problems still to be overcome before the conventional vaccines could be replaced by recombinant ones. Some have already been attended to (HBV— see above). Others require a more profound understanding of the immune response that is effective in a particular disease situation and how to stimulate it. One intrinsic problem is that subviral entities are generally poorly immunogenic.

Yet another recent development is the *DNA vaccine*, where DNA encoding an immunogen is injected rather than the immunogen itself. DNA vaccines are still experimental. The use of cytokines (soluble protein mediators of the immune system) as adjuvants is another area of research (see below).

Genetically engineering a virus as a vaccine

Live vaccines evoke the most effective immunity and are the cheapest to produce but in practice are very difficult to make. Hence, the idea of inserting the gene for the desired neutralization antigen into a pre-existing live vaccine, so that it is expressed naturally as the virus multiplies, is particularly attractive. This has already been achieved experimentally for antigens of numerous viruses, including influenza, rabies, herpes simplex type 1 and hepatitis B viruses, using vaccinia virus as the live vaccine.

Vaccinia virus, once used universally as a vaccine, is no longer employed since smallpox was eradicated in 1977. Thus, there is little pre-existing immunity to compromise its use in this way. Animals infected with recombinant vaccinia carrying genes for HBsAg or influenza virus HA respond with excellent antibody-mediated and T-cell-mediated immune responses, so vaccines of this type are an exciting prospect. The original vaccinia virus vaccine, once obtained from fluid that exuded from localized intradermal infections of sheep, is now produced in tissue culture and would be very cost-effective to produce—about 7 US cents per dose, although recouping large development costs would inflate this price.

The mechanics of producing a recombinant vaccinia virus vaccine, outlined in Fig. 16.8, start by inserting a cloned gene into a plasmid under the control of the promoter of the vaccinia thymidine kinase (TK) gene. Upon transfection into vaccinia virus-infected cells, recombination takes place, because of the homologous TK sequences that both possess. The TK gene is inactivated by insertion of the foreign sequence, which means that recombinants can be selected, because they are unable to use (and hence be inhibited by) the DNA synthesis inhibitor bromodeoxyuridine. Consequently, only recombinant virus produces plaques.

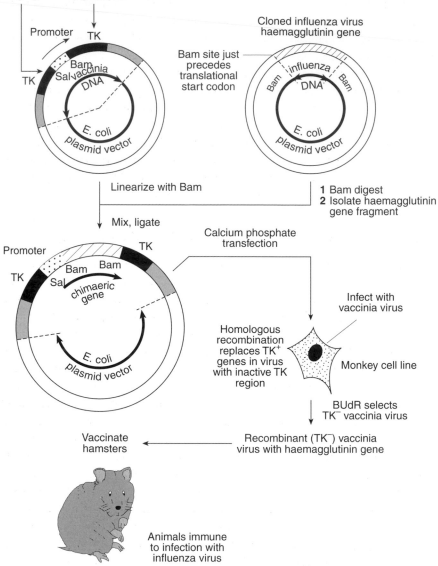

Fig. 16.8 Construction of an infectious vaccinia virus recombinant expressing the influenza virus haemagglutinin. Bromodeoxyuridine (BUdR) inhibits DNA synthesis. Bam and Sal, the restriction enzymes *Bam*H1 and *Sal*1; TK, thymidine kinase.

One problem with vaccinia virus-based vaccines is the rare life-threatening situation of generalized vaccinia infection, in which vaccinia virus infection spreads to the entire body; this occurs at an incidence of about 1 in 25 000 doses. Although this was an acceptable risk against the far higher likelihood of death from smallpox, it could not be tolerated, for example, in a vaccine against influenza. However, insertion of foreign genes reduces the virulence of vaccinia by an order of magnitude. In addition, co-expression of some inserted cytokine genes by vaccinia virus (Interleukin-2 and interferon-γ) completely abolishes its lethality for even the highly susceptible athymic (nude) mice that lack T-cell-mediated immunity and all antibody-mediated immunity except IgM. Another problem is that immunity directed against the vaccinia virus antigens themselves may preclude the vaccinia vector from being used on a subsequent occasion. However, the antivaccinial immunity is relatively short-lived and, in the days of smallpox, vaccination was repeated every 3 years. One possible way to circumvent the difficulty lies in the fact that the vaccinia genome can accommodate around 25 kbp of DNA without losing infectivity. This means that the recombinant virus could express 10–20 foreign antigens and serve as a one-off polyvalent vaccine, which would immunize against a broad spectrum of virus diseases. In addition, other pox virus vectors, such as canarypox virus, are being investigated.

16.8 Prophylaxis and therapy with antiviral drugs

Ever since the successful introduction of antibiotics to control bacterial infections, there has been the hope that similar treatments for virus infections (*antivirals*) could be found but only now, after some 50 years of work by the pharmaceutical industry, is this hope being realized. There are two main difficulties. First, there is the problem that, by the time clinical signs and symptoms appear, virus replication has reached such a peak that the antiviral has little *therapeutic* effect. In some circumstances, such as in the face of an approaching epidemic, it may be best to treat people *before* they are infected, i.e. to use the antiviral as a *prophylactic* or preventive measure. Prophylactics are not widely used but amantadine and rimantadine, a methylated derivative (Table 16.6), are employed in this way in communal homes for elderly people to combat the spread of influenza. The other problem is that virus multiplication is tied so intimately to certain cellular processes that most antivirals cannot discriminate between them. However, viruses do have unique features so, in theory, specific antivirals should be able to serve as effective chemotherapeutic agents. An antiviral would be effective if it inhibited any stage of virus multiplication, i.e. attachment, replication, transcription, translation, assembly or release of progeny virus particles. The current major antivirals act in one of these ways and examples are shown in Table 16.6. Another problem is the selection of resistant mutants and this is discussed below. Finally, nearly all current antivirals act only against replicating virus, and are ineffective, for example, against latent infections. Exceptions are experimental drugs such as Pleconaril and specific immunoglobulins that bind virus particles and prevent the initiation of infection.

Antivirals are found by empirical screening but rational design of drugs has arrived

The search for new antivirals has traditionally been by high-throughput empirical screening (trial and error), but this is moving successfully to tailor-made drugs as

Table 16.6 Antivirals in clinical use or at an advanced stage of development.

Antiviral	Mode of action	Usage
Inhibitors of virus replication: nucleoside analogues	**Incorporated into nascent DNA**	
Zovirax (aciclovir)*	DNA chain termination as the deoxyribose has no 3'-hydroxyl	Herpes simplex viruses 1 and 2; also varicella–zoster virus but less sensitive
Ganciclovir*	As above	Human cytomegalovirus infection of the eye during AIDS
Zidovudine (AZT)*	As above	HIV-1; retards virus replication and progress to AIDS (see Chapter 19); now used more effectively in combination with two protease inhibitors
Lamivudine*	As above	Hepatitis B virus
Ribavirin* + interferon-α* in combination	Ribavirin: as above; interferon-α—see below	Hepatitis C virus
Ion channel blockers Amantadine* Rimantadine*	Blocks the M2 proton channel	Type A influenza viruses (see Section 5.1)
Interferons Interferons*-α/β	Upregulate MHC class I	Although generally antiviral, are effective *in vivo* only against selected infections—chronic hepatitis B and C
Interferons*-α/β	Create the antiviral state (see Section 14.6)	Warts caused by human papillomaviruses
Inhibitors of attachment/ entry of virus into target cells Fleconaril	Binds to virus and causes a conformational change that prevents attachment/uncoating	Several different picornaviruses
Antibody*	Neutralization	Although antibodies specific for all viruses exist, this therapy is only used for life-threatening infections, e.g. Ebola virus

Continued

Table 16.6 *(Continued)*

Antiviral	Mode of action	Usage
Inhibitors of protease activation of viral proteins		
Includes saquinavir, indivir, ritonivir, nefinavir	Prevent post-translational cleavage	HIV-1; used in 'triple therapy' in combination with AZT; retard virus replication and progress to AIDS (see Chapter 19)
Inhibitors of virus release		
Zanamivir* (inhaled); Oseltamavir* (oral)	Analogues of the viral neuraminidase substrate	Influenza A and B viruses

*Compounds in clinical use.
AIDS, acquired immunodeficiency syndrome; AZT, 3'-azido-2',3'-dideoxythymidine;
HIV-1, human immunodeficiency virus type 1; MHC, major histocompatability complex.

understanding of viral processes and properties increases. The new anti-influenza drug, zanamivir (Relenza), was designed *de novo* as a competitive inhibitor for the viral NA substrate, *N*-acetylneuraminic acid. After budding, progeny virus attaches to the cell in which it was made through its HA proteins, and subsequently releases itself by the action of its NA proteins. Relenza binds to, and inhibits, NA activity with the result that virus remains sequestered at the cell surface and is unable to infect new cells. However, rational design and empiricism will continue side by side for the foreseeable future, with a recent upsurge of interest in natural plant products as potential antivirals.

Inhibition of viral DNA synthesis by nucleoside analogues

An important class of antivirals is the nucleoside analogues that are incorporated into DNA. The first versions were used to treat DNA virus infections of the conjunctiva and cornea of the eye, because these cause permanent scarring and hence loss of vision. These were incorporated into nascent DNA and being unable to base pair properly, prevented its replication and transcription. However, this first generation of nucleoside analogues was incorporated non-specifically into any DNA, and killed all dividing cells—hence use was effectively restricted to the poorly vascularized surface of the eye. In practice, 0.1% 5'-iodo-2'-deoxyuridine (IUdR), an analogue of thymidine, improved healing in 72% of infections with herpes simplex type 1, vaccinia and adenoviruses. In extreme circumstances, such as life-threatening cases of encephalitis (infection of the brain) caused by herpes virus type 1 or vaccinia viruses, analogues of cytidine (cytosine arabinoside) or adenosine (adenine arabinoside) were used systemically despite their toxicity. However, these have been superseded by the next

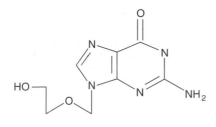

Fig. 16.9 The formula of the chain-terminating nucleoside, aciclovir. Note that most of the cyclic sugar ring is missing—hence the name. The hydroxyl group shown has to be phosphorylated before aciclovir can be incorporated into nascent DNA.

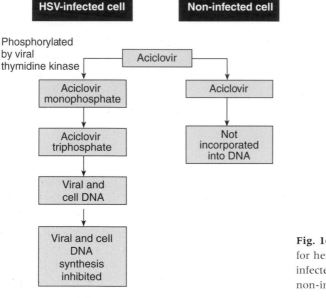

Fig. 16.10 Aciclovir is toxic for herpes simplex virus-infected cells but not for non-infected cells.

generation of nucleoside analogues such as aciclovir (see below) and ganciclovir. Another important nucleoside analogue is AZT (3′-azido-2′,3′-dideoxythymidine or zidovudine), which was the first anti-HIV drug, but the rate of viral mutation and generation of resistant mutants rendered it ineffective. However, it is now used to very good effect in combination with another nucleoside analogue and a drug that inhibits the viral protease (see Section 19.8).

Aciclovir: a selective nucleoside analogue

Properly known as 9-(2-hydroxyethoxymethyl)guanine (Fig. 16.9), aciclovir (Zovirax) was the first really effective antiviral to be discovered and, as with other pyrimidine analogues, it is particularly effective against acute infection with herpes simplex viruses. It cleverly overcomes the danger of being incorporated into cellular DNA in uninfected cells, because only the viral thymidine kinase can phosphorylate the hydroxyl group on the sugar ring, a necessary step before aciclovir can be used in DNA synthesis (Fig. 16.10). Thus, in aciclovir, the ultimate aim of creating a compound that is selectively toxic in infected cells has been achieved. Aciclovir has no

Table 16.7 Treatment of common colds caused by a rhinovirus infection with interferon-α/β.

	Number of people with colds/ number of people inoculated	Number of people virus positive/number of people inoculated
Interferon treated	0/16	3/16
Mock (placebo) treated	5/16	13/16

activity against latent infections because there is no ongoing viral DNA synthesis, but when used early on reactivated virus it is effective against cold sores, corneal eye infections and genital infections. It is also invaluable in treating life-threatening generalized infections that can occur in immunocompromised individuals, particularly those treated with immunosuppressants in the course of transplantation surgery or cancer chemotherapy. Aciclovir-resistant mutants do arise but do not seem to be a significant problem.

Interferon therapy

Their universal antiviral activity and high specific activity make interferon-α, interferon-β and interferon-γ ideal therapeutic agents in theory, but clinical trials have not been encouraging. In early experiments at the Common Cold Research Unit in Salisbury, UK, it was necessary to give volunteers 14×10^6 units of interferon-α,β as a nasal spray four times per day, including the day before infection, in order to combat a rhinovirus, one of the causative agents of the common cold (Table 16.7). The problem was then thought to be the difficulty of obtaining interferons in sufficient concentration but, today, cloned interferons can be expressed in high yield and purified with ease. Unfortunately, even with high doses, therapy is effective only if started before the virus is inoculated, and is thus not applicable to natural rhinovirus infections. In addition, interferons also have side effects, causing fever, local inflammation, muscular pain, fatigue and malaise—effects that result from their action as cytokines and regulators of the immune system.

Experimental treatment of a variety of human virus infections with interferons has been generally disappointing and it is still not clear why this should be. However, interferon has been of particular use in the treatment of persistent infections caused by human papillomaviruses (HPV, papovavirus), hepatitis B virus (hepadnavirus) and hepatitis C virus (flavivirus). The HPVs cause benign tumours (warts) on the surface of the body, the larynx and the anogenital regions. Genital warts are a particular concern, because they may develop into cancer of the cervix (see Section 17.4). Over half the warts injected with interferon regress, but they tend to recur when interferon is stopped. Warts also recur after their surgical removal, but combined surgery and interferon treatment is the best option.

After an acute HBV infection, a proportion of patient develop a chronic infection (see Table 15.4), which can lead to immune complex disease and destruction of the

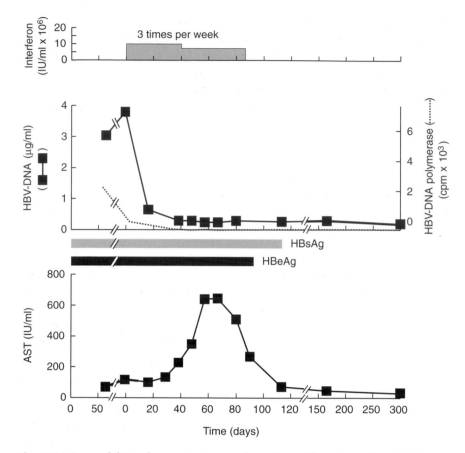

Fig. 16.11 Successful interferon-α treatment of a patient with a chronic hepatitis B virus (HBV) infection. Note the fall in viral DNA, viral DNA polymerase and the viral antigens HBeAg and HBsAg. These all became undetectable, showing that the infection had been cleared. The transient rise in aspartate aminotransferase (AST), a marker of liver damage, signifies the destruction of HBV-infected liver cells by virus-specific CD8+ cytotoxic T lymphocytes (CTLs) after the upregulation of major histocompatibility complex (MHC) I antigens (see text). (From Thomas 1990.)

liver (cirrhosis), and there is a 10% chance of the chronic infection developing into liver cancer. Virus-specific, activated, CD8+, cytotoxic T lymphocytes (CTLs) are present, but these cannot clear the infection, because the virus depresses the expression of major histocompatibility complex (MHC) class I proteins on infected liver cells. Treatment with interferon-α is beneficial, but it acts primarily through enhancing the expression of MHC I proteins (see Section 14.5) and not through its antiviral activity. HBV-specific peptides are then presented in sufficient amount by MHC I proteins to allow the CD8+ CTLs to lyse the HBV-infected liver cells. The treatment is successful in about 50% of patients treated and, under optimum circumstances, can clear the infection completely (Fig. 16.11). Lamivudine therapy is also used (see Table 16.6). Hepatitis C virus, although a member of a different family, causes a progression of infections and liver disease which, if anything, is more severe than that caused by HBV (see Section 21.6). About 70–90% of acute infections become persistent and 20–30% of these become chronic with a proportion progressing to liver cancer. Treatment with interferon-α is expensive ($US12 000–15 000), takes 6–18 months and has only a 25% success rate in clearing virus. Virus levels are always reduced but, if not completely cleared, there is a *rebound* and virus returns rapidly to its original concentration.

Selection of drug-resistant mutants and the advantage of multiple drug regimens

The problem of acquired drug-resistance is common in all rapidly multiplying biological entities and is well known in bacteria. It is acute in all viruses and particularly those with RNA genomes, because they have no proofreading mechanism for checking the fidelity of the replication of their genomes and so their populations contain large numbers of mutants. A demonstration of this—and a remedy—has been found during the treatment of HIV-1 infections with the nucleoside analogue, AZT. Although a very effective inhibitor *in vitro*, resistance to AZT appeared so rapidly in infected people that the drug was ineffective. The problem is a high mutation rate of approximately 10^{-5} mutations/base-pair per replicated genome. These are mostly $G \rightarrow A$ transitions. The result is that any mutant that happens to have a selective advantage will rapidly outgrow wild-type virus and become dominant. The answer is to use two or more drugs simultaneously, because the chance of two mutations occurring in the same genome is very low—the product of the mutation rate, i.e. 10^{-10}. However, as the virus population is intrinsically heterogeneous, one of the mutations may already be present, and the double mutant could be created by a single mutational event. Thus, the best option, as now used with HIV-1 (see Table 16.6), is a combination of three drugs ('triple therapy'—usually two nucleoside analogues and one protease inhibitor). This is covered in more detail in Chapter 19. Drug-resistant herpesvirus mutants also arise when aciclovir is used, although these do not seem to be replacing wild-type virus. It would be a tragedy if its largely cosmetic use in preventing cold sores resulted in it being no longer of any use when there were life-threatening herpesvirus infections, and for this reason there was concern when aciclovir became a non-prescription drug. Influenza A virus mutants are also seen when amantadine is used.

16.9 Further reading and references

Ahmed, R. (eds) (1996) Immunity to viruses. *Seminars in Virology* **7**, 93–155.
Almeida, A. J. & Alpar, H. O. (1996) Nasal delivery of vaccines. *Journal of Drug Targeting* **3**, 455–467.
Anonymous (1998) Vaccine. *Nature Medicine Vaccine Supplement* **4**.
Arvin, A. M. (2000) Measles vaccines—a positive step towards eradicating a negative strand. *Nature Medicine* **6**, 744–745.

Behbehani, A. M. (1991) The smallpox story: historical perspective. *American Society for Microbiology News* **57**, 571–576.

Ben-Yedidia, T. & Arnon, R. (1997) Design of peptide and polypeptide vaccines. *Current Opinions in Biotechnology* **8**, 442–448.

Beverley, P. C. L. (1997) Vaccine immunity. *Immunology Today* **18**, 413–415.

Birmingham, K. (1998) Polio eradication enters home stretch. *Nature Medicine* **4**, 873.

Birmingham, K. (1999) Report calculates value for money of US vaccine R & D. *Nature Medicine* **5**, 469.

Blair, E., Darby, G., Gough, G., Littler, E., Rowlands, D. & Tisdale, M. (1997) *Antiviral Therapy*. Oxford: BIOS Scientific Publishers.

Cox, J. C. & Coulter, A. R. (1997) Adjuvants—a classification and review of their modes of action. *Vaccine* **15**, 248–256.

Donnelly, J. J., Ulmer, J. B., Shiver, J. W. & Liu, M. A. (1997) DNA vaccines. *Annual Review of Immunology* **15**, 617–648.

Englund, J., Glezen, W. P. & Piedra, P. A. (1998) Maternal immunization against viral disease. *Vaccine* **16**, 1456–1463.

Fiddian, A. P. (2000) Antiviral developments: progress with neuraminidase inhibitors. *Reviews in Medical Virology* **10**, 135–137.

Fox, J. L. (1997) Education seen as key in bringing vaccines to adults. *American Society for Microbiology News* **63**, 29–33.

Fox, J. L. (2000) Rotavirus vaccine withdrawn from use. *American Society for Microbiology News* **66**, 5–6.

van Ginkel, F. W., Nguyen, H. H. & McGhee, J. R. (2000) Vaccines for mucosal immunity to combat emerging infectious diseases. *Emerging Infectious Diseases* **6**, 123–132.

Glezen, W. P. & Alpers, M. (1999) Maternal immunization. *Clinical Infectious Diseases* **28**, 219–224.

Griffiths, P. D. (1999) Destruction of smallpox stocks. *Reviews in Medical Virology* **9**, 217–218.

Hilleman, M. R. (1995) Viral vaccines in historical perspective. *Developments in Biological Standardization* **84**, 107–116.

Keating, M. R. (1999) Antiviral agents for non-human immunodeficiency virus infections. *Mayo Clinic Proceedings* **74**, 1266–1283.

Kinchington, D. & Schinazi, R. F. (eds) (1999) *Antiviral Methods and Protocols*. Totowa: Humana Press.

LeClerc, C. & Ronco, J. (1998) Introduction: new trends in vaccine research and development: adjuvants, delivery systems, and antigen formulations. *Research in Immunology* **149**, 9.

McMichael, A. (1998) Preparing for HIV vaccines that induce cytotoxic T lymphocytes. *Current Opinion in Immunology* **10**, 379–381.

Marques, A. R. & Straus, S. E. (1998) Advances in the treatment of chronic hepatitis B virus infection. *Reviews in Medical Virology* **8**, 223–234.

Marsden, H. S. (ed.) (1992) Antiviral therapies. *Seminars in Virology* **3**(1), 1–75.

Marwick, C. (2000) Merits, flaws of live influenza virus flu vaccine debated. *JAMA* **283**, 1814–1815.

Nichol, K. L. (1999) Complications of influenza and benefits of vaccination. *Vaccine* **17** (Supplement 1), S47–S52.

Pawelec, G., Adibzadeh, M., Pohla, H. & Shaudt, K. (1995) Immunosenescence: ageing of the immune system. *Immunology Today* **16**, 420–422.

Spier, R. A. (1998) Ethical aspects of vaccines and vaccination. *Vaccine* **16**, 1788–1794.

Sturmhoefel, K., Lee, K., Toole, M. O., Swiniarski, H. M. & Wolf, S. F. (1998) Interleukin 12 as vaccine adjuvant. *Research in Immunology* **149**, 37–39.

Thomas, H. C. (1990) Management of chronic hepatitis virus infection. In: *Control of Virus Diseases. Society for General Microbiology Symposium 45*, pp. 243–259. Dimmock, N. J., Griffiths, P. D. & Madeley C. R. (eds). Cambridge: Cambridge University Press.

Tisdale, M. (2000) Monitoring of viral susceptibility: new challenges with the development of influenza NA inhibitors. *Reviews in Medical Virology* **10**, 45–55.

Wild, T. F. (1999) Measles vaccines, new developments and immunization strategies. *Vaccine* **17**, 1726–1729.

Zambon, M. (1999) Active and passive immunization against respiratory syncytial virus. *Reviews in Medical Virology* **9**, 227–236.

Also check Chapter 22 for references specific to each family of viruses.

Chapter 17
Carcinogenesis and tumour viruses

The great majority of viruses of vertebrates are not oncogenic*—that is to say, they do not have the ability to initiate a cancer. However, for many years it has been recognized that certain viruses can induce tumours in appropriate experimental animals. More recently, good evidence has implicated some viruses in the development of specific human cancers or naturally occurring cancers in animals. Indeed, it is estimated that viruses are a contributory cause of 20% of all human cancers. As a result of the importance of understanding human cancer in order to find effective treatments, those viruses that caused such disease experimentally were subjected to intensive study from the 1960s onwards. Our detailed understanding of these viruses today owes much to this driving motivation for research.

There is no single mechanism by which viruses induce tumours. Moreover, for all the viruses associated with human malignancies, infection does not lead inevitably to the disease. In each case, infection is just one factor that contributes to disease development, i.e. carcinogenesis is multifactorial and hence a rare event even though the associated virus infection may be common. Thus, to call a virus a 'tumour' virus is rather misleading, because the name refers to one relatively infrequent aspect of its normal life cycle. Various factors, such as host genotype, diet, environmental carcinogens (other than viruses) and other invading organisms, all contribute to the process of virus-associated oncogenesis.

17.1 The link between tumorigenesis and transformation

Members of the papovavirus, adenovirus, herpesvirus and retrovirus families are also able to immortalize and transform primary mammalian cells in culture and/or to transform already immortalized cells. Immortalized cells are defined by their ability to be passaged indefinitely through cell culture, unlike normal cells which senesce and die after a fixed number of divisions (about 50). Transformed cells are recognized by further changes in phenotype (Table 17.1) which usually result in visible changes in appearance (morphological transformation). For both immortalization and transformation, specific genetic alterations in the cells, which can include the input

*A glossary of terms is provided at the end of the chapter.

274

Table 17.1 Some altered
properties of transformed
cells.

Multiply for ever (immortalization)
Grow to higher saturation density
Have reduced requirement for serum growth factors
Grow in suspension in soft agar
Form different cell colony patterns
Grow on top of normal cell monolayers
Readily agglutinated by lectins

of certain viral genes, underlie these changes. Several observations link transformation in cell culture with tumorigenesis *in vivo*. First, transformed cells, but not normal or simply immortalized cells, can often form tumours in suitable animal hosts. Second, the same genetic changes that have been found to underlie transformation of cells in culture are also found frequently in naturally occurring tumours. Third, several of the defining properties of transformed cells reflect decreased cell adhesion and increased mobility, which are essential features of invasive malignant cells *in vivo*.

As well as the acquisition by cells of an appropriate array of genetic alterations, tumour formation involves complex interactions with the immune system of the host animal. This system is important because it provides a check on the spread of cells bearing viral or non-self tumour antigens (even tumour cells arising without virus involvement often display novel antigens on their surface). Hence transformation in culture is not an infallible indicator of malignancy *in vivo*. However, the study of transformation has been very revealing of the events required for the development of malignant potential by cells.

17.2 The genetics of transformation and oncogenesis

There is now overwhelming evidence that the mechanisms underlying oncogenesis and transformation involve alteration to a cell's DNA, i.e. mutation. In principle, mutation could lead to disease through either the loss of functions that are required for normal cell behaviour or the acquisition of functions that disrupt normal behaviour, mediated by either changes to existing genes or the arrival of new genes. Genes, the loss of which is tumorigenic, are known as tumour-suppressor genes whereas those genes that mutation can activate to promote disease are known as oncogenes. These genes are involved in controlling cell signalling and regulating the cell division cycle. Applying the term 'oncogene' to a normal gene that has such activity only when mutated is rather confusing; therefore they are sometimes called proto-oncogenes, to distinguish them from their mutated, oncogenic forms. The involvement of viruses in oncogenesis is through altering the expression of these and other host-cell genes and/or the provision of new virus-encoded functions. However, transformation and tumorigenesis are multistage processes, in which the input of viral genes is just one possible event (Fig. 17.1).

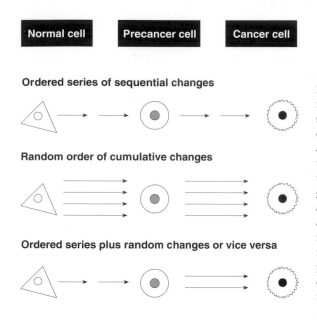

Fig. 17.1 Alternative scenarios for multistep progression from a normal to a cancer cell occurring by the acquisition of genetic alterations, sometimes including the effect of specific virus genes. The number of arrows does not indicate the actual number of events involved. The specific nature of early events in the pathways probably predisposes cells to suffer further mutations.

17.3 Oncogenic viruses

All the viruses listed in Table 17.2 are known or suspected to be oncogenic in humans or animals. Particularly good evidence links hepatitis B and C viruses (HBV, HCV) with primary hepatocellular carcinoma (PHC), Epstein–Barr virus (EBV) with nasopharyngeal carcinoma and Burkitt's lymphoma, and human papillomaviruses (HPVs) with cervical cancer. Complete observation of Koch's postulates (see Section 15.1) is out of the question so other criteria, such as statistical data, are also used to make these associations.

Representatives of most major families of DNA viruses are associated with cancer, with the exception of the poxviruses. As these viruses replicate in the cytoplasm, they do not have the same opportunity to affect the function of the nucleus as do the other DNA viruses; they are also acutely cytopathic. It is important to realize that the genes of a virus can initiate or contribute to a tumorigenesis process only if the virus infection does not kill the cell. For almost all the viruses identified as tumorigenic in Table 17.2, a natural mechanism exists whereby cells can potentially survive an infection, giving the chance that they will develop an altered phenotype subsequently. The herpes- and retroviruses routinely establish latent and non-cytopathic infections, respectively, in their natural hosts. The papillomaviruses, with their bipartite life cycle (see Section 9.4), maintain their genomes in cells that do not support virus production. Adeno- and polyomaviruses normally establish chronic productive infections in their natural hosts and so cause tumours only in species that are non-permissive for virus production. The exception is polyomavirus in laboratory mouse strains; in this case, tumorigenicity is linked to a defective immune response to the virus. Finally, it is important to consider the genetic status of the virus. Within a population of virus particles of any type, there will typically be many defective particles which, through

Table 17.2 Viruses known or suspected to have oncogenic properties in humans, and some selected non-human tumour viruses.

Virus family	Type	Tumour	Animal*	Cofactor
Adenovirus	9	Benign fibroadenoma, sarcoma of breast tissue	Female rat	Newborn only, not natural host
	12	Sarcoma	Hamster	Newborn only, not natural host
Flavivirus	Hepatitis C	Primary hepatocellular carcinoma	Humans	? Alcohol, smoking
Hepadnavirus	Hepatitis B	Primary hepatocellular carcinoma	Humans	Age at infection, aflatoxin, alcohol, smoking
Herpesvirus	Epstein–Barr	Burkitt's lymphoma;	Humans	Malaria, genome rearrangements;
		Nasopharyngeal carcinoma,	Humans	Dietary nitrosamines, HLA genotype
		Hodgkin's disease	Humans	?
		Immunoblastic lymphoma	Humans	Immunodeficiency
	HHV-8	Kaposi's sarcoma Primary effusion lymphoma	Humans	Immunodeficiency (e.g. HIV-1) infection), HLA genotype
	Herpesvirus saimiri	Experimentally induced lymphomas and leukaemias	Owl monkey	Species not the natural host
	Marek's disease virus	T-cell lymphoma	Chicken	Species
Papova	Human papilloma type 16, 18, 31	Cervical neoplasia	Humans	Smoking
	Bovine papilloma type 4	Warts, fibroepithelioma	Cattle, hamster, rabbit	Age (newborn), consumption of bracken fern
	SV40	Possibly neural tumours, mesothelioma	Humans	Not the natural host
		Gliomas, fibrosarcomas	Hamsters	Newborn, not the natural host
	Polyoma	Sarcoma, carcinoma, multiple tissues	Mouse, other rodents	Newborn, mouse strain, MHC genotype
Retro	Human T-cell lymphotropic virus type 1	Adult T-cell leukaemia, lymphoma	Humans	?
	Rous sarcoma	Sarcoma	Chicken	
	Mouse mammary tumour	Adenocarcinoma	Mouse	Strain, sex, hormone status
	Feline leukaemia	T-cell lymphosarcoma, leukaemia, fibrosarcoma	Cat	?

*Tumours are restricted to the species mentioned; this is the natural virus host unless indicated otherwise.
HHV-8, human herpesvirus type 8; HIV-1, human immunodeficiency virus type 1; HLA, human leukocyte antigen (i.e. human form of major histocompatibility complex, or MHC antigen).

random mutation, lack some functions essential for productive infection. If the genes that are important for transformation or tumour formation remain intact, such a particle should be capable of transforming what is normally a permissive cell.

Survival of the infected cell is only part of the story. If viral genes are to contribute actively to an altered cell phenotype, which includes uncontrolled cell growth, they must be successfully inherited by all the daughter cells of the originally infected cell. This can be achieved most easily by integration of the viral DNA into the host genome. Of the transforming viruses, only the retroviruses have an integration function as part of their normal life cycle. Alternatively, the viral DNA may carry an origin of replication which allows it to be copied during S phase (the DNA synthesis phase) of the cell cycle and partitioned between daughter cells at subsequent mitosis. Both Epstein–Barr virus (and possibly other oncogenic herpesviruses) and the papillomaviruses can do this. Failing this, the relevant viral genes must be inserted into the host genome by random non-homologous recombination, which is a rare event, occurring experimentally in roughly 1 in 10^5 infected non-permissive cells. This is how adenovirus, SV40 or polyomavirus DNA is retained in transformed and tumour cells.

17.4 Papovaviruses and adenoviruses

As discussed in earlier chapters, the two genera of the papovaviruses—polyomavirus and papillomavirus—and the adenoviruses are completely distinct in the molecular details of their gene expression. They also have completely different disease profiles, adenoviruses causing a variety of respiratory, gastrointestinal tract or eye infections, polyomaviruses causing urinary tract infections and papillomaviruses causing warts on various epithelial surfaces. It is particularly interesting, therefore, that the mechanisms by which these viruses can immortalize and transform appropriate cells in culture should be so similar.

Events leading to transformation by papovaviruses and adenoviruses

Integration of viral DNA

Formally, a virus that 'causes' transformation or tumour formation could do so either through provision of a gene function which is continuously required for the altered phenotype to be observed, or through a 'hit-and-run' mechanism where, for example, the virus causes a mutation in a host-cell gene and its presence is not subsequently needed. In fact, all these viruses operate through the first of these mechanisms. Viral DNA is present in transformed cells and can be shown, by Southern blotting with probes for viral DNA, to be integrated into the host genome without any apparent preference as to the integration site. Analysis through 200 cell generations shows that viral DNA retains the same relationship with flanking host DNA sequences and is therefore stably integrated.

Typically, a stably transformed cell line will not contain the entire genome of the transforming virus. Characterization of the viral genes that are found consistently in

Table 17.3 The oncogene products of the small DNA tumour viruses.

Virus	Gene(s)/protein(s) implicated in cell transformation
SV40	Early region, large T antigen
Polyoma	Early region, large T antigen plus middle T antigen
Papilloma	E6 and E7 proteins (in epithelial cells)
Bovine papilloma type 1	E5 (in fibroblasts)
Human adenovirus	E1A and E1B

many independently transformed lines gave the first indication of which genes from each of these viruses are involved in virus-mediated transformation. These results were then confirmed by isolating the relevant regions of the viral DNA and using these segments alone to demonstrate transformation. The viral genes involved in transformation in cell culture are summarized in Table 17.3.

Genetics of viral transformation

All the genes implicated in transformation by SV40, polyoma-, papilloma- and adenoviruses are essential for the normal replication of their respective viruses. Thus, deletion mutants lacking these genes are essentially non-viable. How do the roles that these proteins play in the normal growth cycles of these viruses relate to the functions that they provide in cell transformation? An early genetic study with SV40 showed that these two functions were not identical. A collection of virus mutants was isolated, each of which was temperature sensitive (ts) for virus growth because of a mutation in the large T antigen gene. Non-permissive rodent cells were then transformed by each of these viruses at the permissive temperature and the effect of subsequently shifting the cells to a higher temperature, at which the large T antigen would be defective, was determined. Some of the cell lines reverted back to a normal growth phenotype whereas others remained transformed. The interpretation of this result is that some of the normal replicative functions of large T antigen are needed for a cell to maintain its transformed phenotype, whereas others are not.

Studies of the adenovirus transforming genes were important in developing the idea that multiple genetic changes are needed to observe transformation or a tumour. Using the E1A gene alone, or in combination with the E1B gene or other potential oncogenes, it was found that at least two separate functions are needed to transform baby rat kidney (BRK) cells fully (Fig. 17.2). Neither E1A nor E1B alone was sufficient to produce permanently transformed cell lines, but both genes together could efficiently transform these cells. Only these fully transformed cells were tumorigenic in congenitally T-cell-immunodeficient mice (known as athymic nude mice). E1A alone produced abortive transformants, cells that appeared transformed but that could not be stably maintained. Similarly, polyomavirus large T and middle T together were needed for full transformation.

In contrast, SV40 large T alone was sufficient to transform BRK cells fully. Further experiments showed that it contains three separate transformation functions, one or

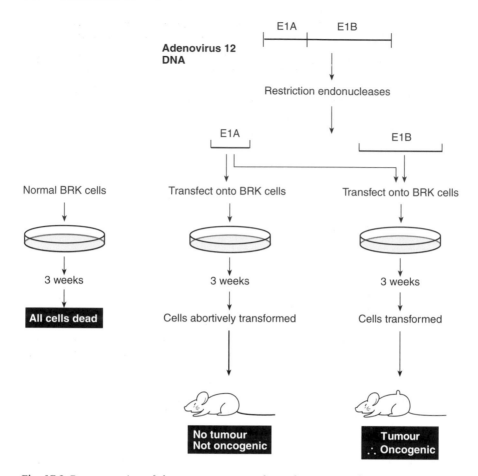

Fig. 17.2 Demonstration of the separate nature of transformation and oncogenesis: while the E1A region alone of adenovirus type 12 is transforming, only cells also carrying adenovirus type 12 E1B are tumorigenic in syngeneic rats. E1B alone gives no transformation. BRK, baby rat kidney.

more of which are needed to achieve cell transformation depending on which cells are used for the experiment. Thus, some T-antigen mutants are defective for transformation of one rodent cell line but not another. This result is taken as a reflection of the different level of mutation already present in the cell genome before introducing the T antigen. Even the most 'normal' cell line in culture has undergone some genetic changes during its establishment, and so the further changes that are needed before a transformed phenotype is observed will vary.

The roles of virus-encoded proteins

Through the last few years, details of the mechanisms by which SV40 T antigen, adenovirus E1A, E1B, etc. contribute to transformation have become much clearer. In each case, the protein achieves its effect(s) through one or more specific interactions

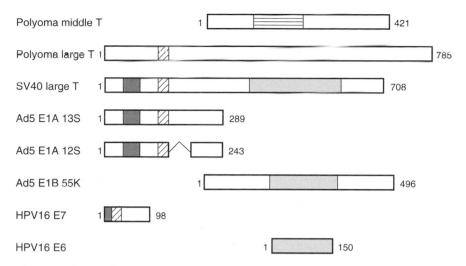

Fig. 17.3 The transforming proteins of SV40, polyoma-, papilloma- and adenovirus. Horizontal bars represent the proteins indicated, not to scale. The E1A 13S and 12S polypeptides are highly related, differing only in the removal by mRNA splicing of 46 residues from the shorter protein, and have equivalent functions in transformation. The other proteins are essentially unrelated, except in function as indicated. For expression and sequence relatedness of the SV40 and polyoma proteins, see Section 9.3. The various shaded blocks indicate functional rather than sequence relatedness, although there is sequence similarity between the indicated Rb-binding sequences. Diagonal hatching, Rb-binding; light tint, p53 binding; dark tint, accessory region required to release transcription factors from bound Rb; horizontal hatching, Src protein binding.

with host-cell proteins involved in regulating the mammalian cell cycle. Remarkably, these interactions are very similar for each virus, revealing how fundamental the disruption of the cell cycle is to their normal life cycles. The domain structures of these proteins is summarized in Fig. 17.3.

One of the targets of the viral transforming proteins is a key cell cycle regulator in mammalian cells known as Rb or p105. This protein, a tumour-suppressor gene product, binds to a transcription factor that is needed to drive expression of various genes, the products of which are needed for a successful S phase of the cell cycle. At the point in the cycle where a cell 'decides' to go through a round of division, Rb is specifically phosphorylated, so inactivating it and releasing its bound transcription factor, which allows S phase to proceed. SV40 and polyomavirus large T, the adenovirus E1A proteins and papillomavirus E7 protein all inactivate Rb as well. Each one binds to Rb via a short sequence motif that they have in common: Leu–X–Cys–X–Glu. Binding releases the transcription factor and so allows S-phase progression. As discussed in Chapter 6, each of these small DNA viruses needs the cell it infects to enter S phase to support its own replication, and so these viral functions are crucial to infection *in vivo*, where newly infected cells typically will not be dividing rapidly or even at all. However, when allowed to act outside the normal context of infection, these activities promote transformation because they remove a key negative regulation on continuous cell division. The more frequently and rapidly a cell divides, the more likely it is to acquire further mutations in the process. The fact that the Rb gene is a tumour-suppressor means that its inactivation by direct mutation has been observed to lead to tumours, again demonstrating the mechanistic link between transformation and tumorigenesis.

Another mechanism common to these viruses involves the very important cell protein p53. This protein acts as a sensor of the replicative health of the cell. Whenever the cell's DNA is damaged, p53 activity is induced and this leads first to the cell cycle being halted, to allow DNA-repair systems a chance to fix the damage and, second, if repair fails, to the induction of apoptosis. When the restraint on cell proliferation normally imposed by Rb is inactivated, p53 activity is induced. As S-phase activity is essential for these viruses to replicate, they need a mechanism for blocking the p53 response. SV40 large T (but not polyomavirus large T), papillomavirus E6 and adenovirus E1B M_r 55 (55K) proteins all bind to p53, inactivating and/or targeting it for proteolytic degradation. Also, a second E1B protein (M_r 19) acts independently to block apoptosis. The cooperation between adenovirus E1A and E1B to achieve full transformation can therefore be understood: E1A overcomes cell cycle control and E1B is needed to block the resulting apoptosis response. Once again, outside the context of a productive infection, inactivation of the p53 response powerfully predisposes a cell to accumulate further mutations and so the p53-inactivation functions of these viruses are transforming functions. The relevance of p53 inactivation to tumour formation is emphasized by the very high frequency with which naturally occurring human tumours contain a mutated p53 gene.

Unlike the other viruses considered in this section, polyomavirus does not produce a protein that binds to p53. It has long been a mystery how this virus can survive without a function that is apparently essential in the other viruses. In transformation assays, polyomavirus middle T and large T together are required for transformation to occur. Middle T binds to the product of the cellular proto-oncogene, *src* (the gene acquired as cDNA by Rous sarcoma virus—see Section 17.6), and upregulates its intrinsic tyrosine protein kinase activity. This phosphorylates middle T, allowing it to bind to further host-cell proteins. As a result, a complex set of effects on the signalling cascades within the cell ensues, which has been understood ultimately to affect the cell's growth, although the details are incomplete. Recently, a mechanism has been proposed whereby this signalling induces expression of a natural negative regulator of p53, leading effectively to inactivation of the p53 response pathway by an indirect mechanism. Thus, it appears that all the small DNA viruses target both Rb-mediated and p53-mediated regulation of the cell cycle, and these functions represent important parts of the transforming activities of these viruses.

When SV40 is asked to transform growth-arrested rodent cells, the process requires small t as well as large T antigen. This discovery led to the definition of another interaction with host-cell proteins which has counterparts in other viruses. SV40 small t and polyomavirus small and middle T antigens bind to and inactivate protein phosphatase 2A (PP2A), an enzyme that is important in cell cycle regulation. PP2A is a heterotrimeric protein, with active A and C subunits regulated by a variety of B subunits. The small t antigens act as alternative B subunits, displacing existing B subunits to alter enzyme activity. It is now known that adenoviruses also make a PP2A-modulating protein, in this case from the E4 gene.

Factors affecting adenovirus oncogenicity

Although the E1A and E1B genes of a range of different human adenovirus serotypes, e.g. Ad5 and Ad12, can transform rodent cells in culture, only a few of these, e.g. Ad12, cause tumours in normal immunocompetent animals. The same distinction is seen when the tumorigenicity of Ad5- and Ad12-transformed BRK cells is compared in syngeneic immunocompetent rats. (Syngeneic means genetically identical; such animals are used to avoid rejection of the tumour cells through immunological recognition of foreign rat antigens.) BRK cells were transformed with Ad5 E1A and E1B and Ad12 E1A and E1B in different combinations, and the resulting cells assessed for tumorigenicity. The experiment showed clearly that tumorigenicity of the cells was determined by the type of virus from which the E1A gene came (Fig. 17.4).

What additional or different function is carried by Ad12 but not Ad5 E1A? The main clue comes from data that show that, although both Ad5 and Ad12 express viral antigens in transformed cells, only the Ad5-transformed cells are lysed by activated cytotoxic T lymphocytes (CTL). The Ad12-transformed cells are not killed because their expression of major histocompatibility complex (MHC) class I antigen, which is essential for antigen presentation to CTL, is turned off by Ad12 E1A. It seems that Ad12 tumorigenicity resides in the ability of the virus to 'hide' the cells that it has transformed from the immune system—a form of immunosuppression. This result shows

Transfected DNA	Transformation	Oncogenesis
$E1A_{12}$ — $E1B_{12}$	+	+
$E1A_{12}$	+*	−
$E1A_{12}$ — $E1B_5$	+	+
$E1A_5$ — $E1B_5$	+	−
$E1A_5$	+*	−
$E1A_5$ — $E1B_{12}$	+	−

Fig. 17.4 Transfection into baby rat kidney (BRK) cells of E1A and E1B DNA from oncogenic adenovirus type 12 and non-oncogenic adenovirus type 5 in permutation. The type origin of the E1A DNA determines whether or not transformed cells can form a tumour in syngeneic immunocompetent rats. Cells transformed by E1A alone showed an abortively transformed phenotype (*), indicating that E1B plays a contributory role in transformation.

how important the immune system is in preventing the growth of tumours that would otherwise arise. Again it is important to remember that this Ad12 E1A function has the primary purpose of optimizing the survival of the virus during natural infections rather than enhancing its tumorigenicity.

Human papillomaviruses and carcinogenesis

There is a growing body of evidence that HPVs are implicated in cervical cancer and, more rarely, other cancers of the genitalia in men and women. There are over 80 HPV types (defined by a minimum level of sequence difference because no system for serotyping exists) and none can be grown in cell culture. These viruses cause warts (which are benign tumours) and can be subdivided into those infecting the skin and those infecting internal epithelial cell layers. Among the latter, the most prevalent are types 6, 11, 16 and 18, which are associated mainly with genital warts, the occurrence of which correlates with sexual activity from an early age and a high number of sexual partners. The incidence of cervical cancer also correlates with these indicators and analysis of tumour biopsies routinely reveals HPV DNA of one of a small group of types, most commonly HPV16, -18 and -31. These types seem selectively to be involved with this serious disease. It is thought that the specific molecular characteristics of these 'high-risk' HPVs promote progression from a benign wart to an invasive carcinoma. This progression involves successive changes in the properties of the cells and is invariably associated with a change from maintenance of the HPV DNA as a free episome to its integration into the host genome and loss of expression of all but the E6 and E7 genes. However, infection by a high-risk HPV type is not sufficient on its own for tumour formation, indicating that one or more cofactors affect disease development. As these viruses are now believed to be the trigger for cervical carcinoma in almost all cases, there is considerable interest in developing a vaccine that would protect against infection.

SV40 as a possible human tumour virus

SV40 is not naturally an infectious agent for humans. However, in cell culture, human cells are semipermissive for this virus, indicating a low level of replication. During the 1950s, when the mass immunization campaign for polio was beginning, the live Sabin vaccine was grown in cultures of African green monkey kidney cells which, it was subsequently realized, contained SV40 as an inapparent infection. Thus, many millions of people were potentially infected with SV40 during this period. (Once SV40 was discovered, it was eliminated from the vaccine production process, so modern vaccines are clean.) Clearly, since that date there has not been a massive epidemic of SV40-induced human tumours. However, there have been sporadic reports of SV40 sequences being isolated from choroid plexus tumours and ependymomas (brain), and more recently mesothelioma (a tumour of the body cavity especially associated with asbestos exposure). A true causative role for SV40 T antigen in any of these cases has yet to be proved.

17.5 Herpesviruses

Epstein–Barr virus

Epstein–Barr virus infects everyone. Childhood infection is subclinical, but infection of young adults can cause prolonged debilitation (glandular fever or infectious mononucleosis, named after the extensive proliferation of host B and T lymphocytes in the blood). Whether or not the primary infection is symptomatic, EBV establishes a life-long latent infection of circulating B cells. It has been suggested that primary infection occurs in epithelial cells, but this is now doubted. However, in both these cell populations, EBV infection is associated with specific tumours (see Table 17.2).

The stages by which EBV causes cancer are only now beginning to be unravelled. Natural infection results in a permanent pool of infected B cells, proliferation of which is held in check only through the action of the T cells of the immune system. This explains the incidence of EBV-associated B-cell-proliferative disease after organ transplantation, because such organ recipients have to receive some immunosuppressive therapy to prevent rejection of their grafts. Inadvertent immunosuppression may also play a part in the development of Burkitt's lymphoma, another B-cell tumour. This disease shows a geographical distribution markedly coincident with the distribution of the malaria parasite, an agent that also causes immunosuppression (see Section 21.3). Another feature of Burkitt's lymphoma cells is that they characteristically contain one of three chromosome translocations, linking a specific region of chromosome 8, usually to a point in chromosome 14 or more rarely to sites in chromosome 2 or 22. These translocations place a cellular proto-oncogene, *myc* (see Section 17.6), under the transcriptional control of the regions encoding the immunoglobulin heavy and light chains, respectively. These, of course, are highly active in B cells, so the effect of the translocations is to deregulate *myc* expression and cause inappropriate production of its protein product, a DNA-binding protein responsible for aspects of transcriptional control in the cell. This then impairs the cell's ability to control its division. Nasopharyngeal carcinoma involves epithelial cells and again shows a marked prevalence in a restricted geographical area, in this case south-east Asia. Here, consumption of salted fish from an early age, which is thought to expose the infected cells to various carcinogens, is a probable cofactor in disease development. Possession of certain HLA haplotypes also acts as a risk factor.

Human herpesvirus 8

Human herpesvirus 8 (HHV-8) is the most recently recognized human herpesvirus. It was identified through the cloning of sequences found consistently in a tumour, Kaposi's sarcoma, which has become a defining feature of acquired immune deficiency syndrome (AIDS), particularly in homosexual men (see Chapter 19). HHV-8 is now thought to be significantly involved in the development of this tumour, which appears only very rarely in the population in the absence of HIV-1-induced immunodeficiency. Another rare tumour associated with HHV-8 is primary effusion lymphoma, a body cavity tumour of B-cell origin. The virus is now known to be present at varying levels of prevalence in different populations, but always at a frequency greater than the malignancies believed to be associated with infection. Thus, as with the other examples already described, HHV-8, although suggested to be a cause of certain human cancers, is not sufficient of itself to elicit a tumour.

17.6 Retroviruses

The first retroviruses to be discovered were oncogenic (isolation of Rous sarcoma virus was reported in 1910) and they were much studied for this reason. However, it is now clear that not all retroviruses are oncogenic, e.g. the lentivirus genus, which includes human immunodeficiency virus (HIV) (see Chapter 19), is primarily associated with immunodeficiency diseases. Remarkably, the first human retrovirus, T-cell lymphotropic virus type 1 (HTLV-1), was not recognized until 1980, followed soon after by HIV.

Highly oncogenic retroviruses that also transform cells in cultures

Rous sarcoma virus (RSV) is an example of a subset of the oncogenic retroviruses that transforms immortalized cells in culture and causes tumours in appropriate animals with high efficiency and rapid onset. Each of these viruses has, in its genome, sequences that have been acquired from the host, and genetic analysis reveals that these are responsible for the cell transformation and tumorigenic phenotypes. The acquired genes represent cDNA copies (reverse-transcribed mRNA) from cellular proto-oncogenes (see Section 17.2); indeed, analysis of such viruses was the way in which these cellular counterparts were first defined as proto-oncogenes. Deregulated expression from the powerful viral long terminal repeat (LTR) promoter is at least partly responsible for their acquiring oncogenic properties (Fig. 17.5a). Another factor is that, when compared with their cellular counterparts, the acquired oncogenes often have mutations that affect protein function; classic examples are the altered *ras* genes of Kirsten and Harvey murine sarcoma viruses. The same *ras* mutations have also been found in the DNA of human bladder carcinomas which have no viral involvement in their development.

   With the exception of RSV, which acquired the *src* gene 3′ to its *env* gene, all highly oncogenic retroviruses have lost essential viral genes (*gag*, *pol* or *env*—see Section 8.1) during the acquisition of their oncogenes. They are therefore defective and can grow only with support from a standard retrovirus, such as the avian (ALV) or murine (MLV) leukaemia viruses, from which they derived. Although the study of the oncogenes from these defective viruses has been hugely informative about the mechanisms that control the growth of mammalian cells, the oncogenes play no part in the virus life cycle. The detailed mechanism by which these viruses arose is unclear. Models proposed involve integration of an intact provirus adjacent to a cellular gene, subsequent production of a hybrid mRNA containing viral and cell sequences, and then

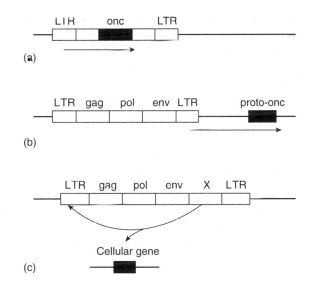

Fig. 17.5 Models of retrovirus oncogenesis involving integration of viral genomes into cellular DNA. (a) Integration of a defective viral genome carrying an oncogene (onc), the transcription of which is under the control of the strong promoter in the viral long terminal repeat (LTR). (b) Integration of a virus without an oncogene, so that its right-hand LTR drives expression of a cellular proto-oncogene (proto-onc). (c) Expression of a transcription-enhancing product from viral gene X which affects both viral and cellular transcriptional control.

recombination during reverse transcription to create a provirus with its essential flanking LTRs (see Chapter 8). There is no obvious evolutionary advantage to the virus in acquiring cellular sequences at the expense of its own essential genes, so these events may best be viewed as rare accidents of the retroviral life cycle.

Other oncogenic retroviruses

If immortalized fibroblastic cells in culture are infected with standard retrovirus, such as ALV or MLV, very little morphological change occurs. The cells are not transformed nor do they show cytopathology. This type of retrovirus does produce malignant disease, in this case lymphoma, when introduced into a suitable host but, unlike the highly oncogenic retroviruses, the efficiency is low (only a few animals show disease) and the time taken to observe disease is much greater. Genetic analysis does not reveal any viral oncogene to which the oncogenic properties of these viruses can be allocated. Thus, the ability of the virus to go through a productive cycle or to cause lymphoma cannot be separated by mutation.

The clue to understanding how these viruses cause disease came from an analysis of DNA from the tumours that they induced. Although the sites of provirus insertion were not identical, in each case insertion had occurred in the same region of the host genome. Characterization of the surrounding DNA led to the discovery of the *myc*

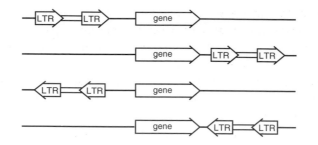

Fig. 17.6 Four variations on the retroviral insertional mutagenesis model shown in Fig. 17.5b. Transcriptional regulation of a cellular gene will be altered by the provirus in all four scenarios, because the enhancers in each long terminal repeat (LTR) operate in a position and orientation-independent manner to upregulate transcription from the gene's endogenous promoter.

gene, already known from its presence in cDNA form in a highly oncogenic retrovirus. This result led to the concept of retroviral insertional mutagenesis (Fig. 17.5b). By inserting upstream of the *myc* gene, the right viral LTR, which contains a powerful promoter identical to that in the left LTR (see Sections 8.1 and 9.9), drives its high level expression. This alters the developmental timing and level of expression of *myc*, so causing aberrant cell behaviour (activation of *myc* expression is also involved in Burkitt's lymphoma—see Section 17.5). In fact, insertional mutagenesis can work even when the insertion is downstream of the target gene, or in the wrong orientation to drive transcription directly via the LTR promoter. This is because the LTRs also contain enhancer elements which will upregulate expression from a gene's endogenous promoter whenever they are placed in reasonable proximity to it (Fig. 17.6). Why does disease caused by these viruses have a long latency? This is because retroviral integration is a random process. Many rounds of infection will typically be needed to give any probability of a provirus inserting in the correct region of the genome in order to have an oncogenic effect.

Retroviruses and naturally occurring tumours

Retroviruses do not have precise integration sites. The insertion of retroviral DNA into the host genome may therefore have a variety of effects on cellular gene expression. Thus, natural retroviral infections might be expected to confer a risk of mutational activation of cellular proto-oncogenes, as is seen experimentally with the avian and murine leukaemia viruses. One example of this is the disease observed with mouse mammary tumour virus (MMTV), which is endemic in certain mouse strains. When mammary tumours are examined for MMTV DNA, the proviral integration sites are in reasonable proximity to one of a small number of genes, known as the *int* genes. As a result, the expression of the *int* gene will come under the influence of the viral LTR (see Figs 17.5b and 17.6). The MMTV LTR is sensitive to steroid hormone regulation, perhaps explaining why the onset of disease is linked to pregnancy in infected female animals. The fact that specific targeting of these regions by provirus is seen in tumour cells is evidence that the effect on neighbouring genes is relevant to the disease process, although the ways in which altered expression of these otherwise unrelated *int* genes affects cell behaviour is not known.

HTLV-1 is associated with adult T-cell leukaemia, a disease with a geographical distribution concentrated in Japan and the Caribbean islands which closely matches that of HTLV-1 infection although, as with all the other examples of human viruses associated with cancer, only a minority (about 1%) of those infected show the disease. The virus can also immortalize T cells in culture. The probable mechanism is that shown in Fig. 17.5c, and is somewhat similar to that of the small DNA tumour viruses. The gene responsible for T-cell immortalization in culture is essential for natural infection and encodes a transcriptional activator. This acts on the viral LTR, but also on host-cell genes, where it is believed to alter the cell's growth properties so as to predispose it to accumulate further mutations as cell division continues.

17.7 Hepatitis viruses and liver cancer

Infection by HBV is strongly associated with liver cancer (primary hepatocellular carcinoma) in later life. People who become chronically infected have a 200-fold increased risk of this cancer over the rest of the population; chronic infection is particularly likely when the infection is acquired at an early age. With 300 million carriers, this makes HBV probably the most potent human carcinogen after smoking. In most cases of HBV-positive liver cancer, the tumour cells contain integrated HBV sequences. The integration site is the same in all cells of a tumour, indicating that integration occurred in a single cell which subsequently divided to form the tumour. There is, however, no consistent integration site among different patients. These observations are suggestive of a direct role of virus function(s) in the development of disease. As it can both transactivate transcription and disrupt p53 function (see Section 17.4), the X protein has been suggested as the likely culprit. However, its expression is usually undetectable in tumour cells and the integrated sequences are usually 'scrambled' by rearrangements so as to prevent most viral gene expression. Also, any proposed mechanism must take account of the long time lag (perhaps 40 years) between primary infection and the appearance of cancer. Therefore, an alternative theory is that the virus exerts an indirect effect, through the long-term damage to the liver that it inflicts. In a chronic infection, cells are lost through immune-mediated destruction of infected cells. The liver responds by regeneration and this persistently increased rate of division among hepatocytes may then predispose the cells to accumulate mutations, ultimately leading by chance to the emergence of a cell with a set of mutations that gives it cancerous properties.

A persuasive piece of evidence for the theory that HBV causes liver cancer through the indirect effects of chronic liver damage has come from the more recent discovery and characterization of another hepatitis virus—hepatitis C virus (HCV)—infection by which also predisposes to liver cancer after a long delay. HCV is the only tumour virus listed in Table 17.2 that does not have any DNA involvement in its replication cycle.

As its genome is always in the form of RNA, none of the mechanisms by which genetic material might persist in a dividing population of precancerous cells (see Section 17.3) is available to it. Therefore 'hit-and-run' or indirect mechanisms such as that proposed for HBV are the only options. Given the long delay between infection and disease, the latter is the more likely model. This conclusion is also supported by the fact that other, non-virological, agents that cause chronic liver damage, such as alcohol, also give rise to an increased risk of hepatocellular carcinoma. The great extent of HCV infection in developed countries is only now becoming apparent. With no vaccine or reliably effective treatment for HCV yet available, this virus promises to be a major public health problem.

17.8 Summary and prospects for the control of cancer

Vaccines that prevent infection with tumour viruses are thereby vaccines against cancer. This has been amply demonstrated with Marek's disease virus of chickens, a herpesvirus, which causes a cancer of cells of the feather follicles, and with successful experimental vaccines against herpesvirus saimiri and EBV. Immunization against tumour viruses is as easy (or as difficult) as immunization against any non-tumour virus. Any problems are simply those of producing a successful vaccine. An excellent vaccine for humans against HBV is already on the market, but to combat other known human cancer-causing viruses, several new vaccines would be needed (to EBV, HCV, and HPV-16 and -18). One problem is that each tumorigenic virus only rarely causes cancer. Even with a disease as emotive as cancer, would there be sufficient interest to persuade people to avail themselves of immunization? Would potential recipients understand that, at best, such vaccines would protect only against virally induced tumours? If not, to the common eye, the vaccines would be deemed a failure and acceptance by future generations would be compromised. Thinking about vaccines (see Chapter 16) must therefore proceed in tandem with the more traditional anticancer approaches involving the elimination of cofactors, education about the risk of cofactors to inform people's lifestyle choices, improved early diagnosis and better treatment.

17.9 Glossary

Adenocarcinoma: a carcinoma developing from cells of a gland.
Benign: an adjective used to describe growths that do not infiltrate into surrounding tissues. Opposite of malignant.
Cancer: malignant tumour—a growth that is not encapsulated and that infiltrates into surrounding tissues, the cells of which it replaces with its own. Its cells are spread by the lymphatic vessels to other parts of the body (metastasis). Death is caused by destruction of organs to a degree incompatible with life, extreme debility and anaemia or by haemorrhage.
Carcinogenesis: complex multistage process by which a cancer is formed.
Carcinoma: a cancer of epithelial tissue.
Fibroadenoma: tumours of mixed cell type.

Fibroblast: a cell derived from connective tissue.

Fibroepithelioma: tumours of mixed cell type.

Leukaemia: a cancer of white blood cells.

Lymphoma: a cancer of lymphoid tissue.

Malignant: a term applied to any disease of a progressive and fatal nature. In the context of carcinogenesis, an adjective describing a tumour that grows progressively and invades other tissues. Opposite of benign.

Mesothelioma: a tumour of the body cavity, surrounding the lungs.

Neoplasm: an abnormal new growth, i.e. a cancer.

Oncogenic: tumour causing.

Sarcoma: a cancer developing from fibroblasts.

Transformation: a constellation of phenotypic changes of cells in culture (see Section 17.1).

Tumour: a swelling, caused by abnormal growth of tissue, not resulting from inflammation. May be benign or malignant.

17.10 Further reading and references

Bishop, J. M. (1983) Cellular oncogenes and retroviruses. *Annual Review of Biochemistry* **52**, 301–354.

Brehm, A. & Kouzarides, T. (1999) Retinoblastoma protein meets chromatin. *Trends in Biochemical Sciences* **24**, 142–145.

Campo, M. S. (1992) Cell transformation by animal papillomaviruses. *Journal of General Virology* **73**, 217–222.

Dalgleish, A. G. (1991) Viruses and cancer. *British Medical Bulletin* **47**, 21–46.

Fanning, E. (ed.) (1994) Transforming proteins of DNA viruses. *Seminars in Virology* **5**(5), (whole issue).

Fanning, E. & Knippers, R. (1992) Structure and function of simian virus 40 large tumor antigen. *Annual Review of Biochemistry* **61**, 55–85.

Farrell, P. J. (1995) Epstein–Barr virus immortalizing genes. *Trends in Microbiology* **3**, 105–109.

Hesketh, R. (1995). *The Oncogene Facts Book*. London: Academic Press.

Kim, C.-M., Koike, K., Saito, I., Miyamura, T. & Jay, G. (1991) HBx gene of hepatitis B virus induces liver cancer in transgenic mice. *Nature (London)* **351**, 317–320.

Ko, L. J. & Prives, C. (1996) p53: puzzle and paradigm. *Genes and Development* **10**, 1054–1072.

Kubbutat, M. & Vousden, K. H. (1998) New HPV E6 binding proteins: dangerous liaisons? *Trends in Microbiology* **6**, 173–177.

Lane, D. P. (1992) p53, guardian of the genome. *Nature (London)* **358**, 15–16.

Levine, A. J., Momand, J. & Finlay, C. A. (1991) The p53 tumour suppressor protein. *Nature (London)* **351**, 453–456.

McCance, D. J. (ed.) (1998) *Human Tumor Viruses*. Washington, DC: ASM Press.

Messerschmitt, A. S., Dunant, N. & Ballmer-Hofer, K. (1997) DNA tumor viruses and Src family tyrosine kinases, an intimate relationship. *Virology* **227**, 271–280.

Minson, A., Neil, J. & McCrae, M. (eds) (1994) Viruses and cancer. *Society for General Microbiology Symposium* **51**. Cambridge: Cambridge University Press.

Moran, E. (1994) Mammalian cell growth controls reflected through protein interactions with the adenovirus E1A gene products. *Seminars in Virology* **5**, 327–340.

Pardoll, D. M. (1993) Cancer vaccines. *Immunology Today* **14**, 310–316.

Raab-Traub, N. (1996) Pathogenesis of Epstein–Barr virus and its associated malignancies. *Seminars in Virology* **7**, 315–323.

Schwartz, D. & Rotter, V. (1998) p53-dependent cell cycle control: response to genotoxic stress. *Seminars in Cancer Biology* **8**, 325–336.

Teich, N. (ed.) (1991) Viral oncogenes. Part I. *Seminars in Virology* **2**(5), (whole issue).

Teich, N. (ed.) (1991) Viral oncogenes. Part II. *Seminars in Virology* **2**(6), (whole issue).

de Villiers, E.-M. (ed.) (1999) Human papillomaviruses. *Seminars in Cancer Biology* **9**(6).

Weiss, R. A. & Boshoff, C. (eds) (1999) Human herpesvirus 8/Kaposi's sarcoma herpesvirus. *Seminars in Cancer Biology* **9**(3).

White, E. (1998) Regulation of apoptosis by adenovirus E1A and E1B oncogenes. *Seminars in Virology* **8**, 505–513.

Also check Chapter 22 for references specific to each family of viruses.

Chapter 18

The evolution of viruses

Like all other living organisms, viruses undergo evolutionary change. Their genomes are subject to mutations at the same rate as all other nucleic acids and, where conditions enable a mutant to multiply at a rate faster than its fellows, that mutant virus will be selected and will succeed the parental type. Where viruses have an evolutionary advantage is that many polymerases (notably in the RNA viruses) lack proofreading mechanisms, so that mistakes in replication are not corrected. Thus mutants accumulate more rapidly. In any discussion of evolution, one naturally starts at the earliest time possible, and it is pertinent to ask where viruses first came from (see Section 1.6). The absence of any fossil records and scarcity of other evidence have not, of course, prevented scientists from speculating about their origins! The two prevailing opinions are that viruses have either arisen (i) from degenerate cells that have lost the wherewithal for independent life, or (ii) from escaped fragments of cellular nucleic acid.

The molecular biology of viruses of eukaryotes and their host cells differs considerably from that of bacteriophages and their prokaryotic host cells, to the extent that it is not possible to grow bacteriophages in eukaryotic cells or eukaryotic viruses in bacteria. Thus, it appears that phages and viruses of eukaryotes have arisen independently or diverged at a very early stage.

Whatever their origins, viruses have been a great biological success, because no group of organisms has escaped their attentions. Viruses are largely species specific and most do not cross species boundaries; it seems likely that every animal species has the same range of viruses as occurs in humans. In Chapters 12–15, we discussed the various ways in which viruses interact with their hosts, and we saw how viruses cause a variety of changes, ranging from the imperceptible to death. Evolution of any successful parasite has to ensure that the host also survives. The various virus–host interactions alluded to earlier can be thought of as ways in which they have solved this problem. In this chapter, we discuss, first, the evolutionary implications of the distribution of morphologically similar viruses throughout a range of different hosts and, second, examples of virus evolution that have occurred in relatively recent times.

Although there may be no fossil record, the history of viruses extends as far back as human historical written or pictorial records. Among the earliest references is a 3500-year-old bas-relief sculpture from Egypt depicting a man supporting himself with a crutch. Careful examination shows that the calf muscle of his right leg is withered in the characteristic aftermath of paralytic poliomyelitis. In addition, slightly later excavations of the same civilization have found the mummified body of the pharaoh, Rameses V, whose face bears the characteristic pockmarks of smallpox. Such evidence tells us of the long association of these viruses with humankind.

18.1 The potential for rapid evolution in RNA viruses: quasispecies and rapid evolution

The lack of a proofreading function in the polymerase of RNA viruses means that base substitutions (mutations) occur at the rate of between 10^{-3} and 10^{-5} per base per genome replication. Not all of the resulting mutants will be viable but many are, resulting in an extremely heterogeneous population of viruses. This is the quasispecies. There are a number of implications, e.g. it is not possible to define the genome sequence of that virus population precisely, and any sequence in a database will represent only one of the members of the quasispecies. The quasispecies phenomenon also endows RNA viruses with the ability rapidly to adapt to and exploit any environment that they occupy. This is seen on both a micro and a macro level, and arises because one or more members of the quasispecies will inevitably have a selective advantage over others. The micro level is infection of the individual, and virus evolution during the life time of the individual is seen particularly in life-long infections with human immunodeficiency virus type 1 (HIV-1—see Chapter 19) and hepatitis C virus (see Section 21.6). It is likely that, in these infections, the immune system is a major selective force, and that the quasispecies phenomenon allows these viruses to persist in the face of that immune response. The macro level of evolution is seen in viruses with a worldwide distribution that infect people who have different genetic backgrounds and especially the major histocompatibility complex (MHC) haplotypes that control T-cell immunity. This leads to the establishment of different genotypes or clades in different parts of the world, and a classification that is based on sequence rather than serology (e.g. HIV-1—see Chapter 19).

Influenza A virus does not cause persistent infections but virus variants evolve under the selection pressure of antibody during acute infection, which are then passed on to susceptible individuals. Thus the virus evolves so that, in approximately 4 years, it is not recognized by the immune response made by the original host, and he or she can be reinfected (see Section 18.5). This is called *antigenic drift*, and essentially it results in the formation of new serotypes. However, not all RNA viruses exhibit such obvious variation, even though they have the same potential for change as HIV-1 and influenza virus, e.g. the overall antigenicity of polioviruses types 1, 2 and 3, and measles virus has not changed over their known history of approximately 50 years, and the original vaccines still provide the same level of protection. Thus, by comparison with HIV-1 and influenza virus, the polioviruses and measles virus are completely stable. However, base changes do occur and minor changes in antigenicity of the virion can be detected by probing individual epitopes with monoclonal antibodies. In addition, viruses from different global areas can be distinguished by their genotype. It is not understood why, for example, measles virus is antigenically stable and influenza virus is not, because in many ways these two infections are similar (both viruses are highly infectious, spread by the respiratory route and cause acute infections). Presumably measles virus must be under a less effective evolutionary selection pressure than influenza virus, but exactly what these pressures are is not known.

18.2 Rapid evolution: recombination

Recombination is the other major force in virus evolution and it takes place in a cell that is infected by two viruses. Usually the two genomes are highly related, with regions of homology between their genomes that permit the replicating enzyme to move from one strand to another. Thus, at a stroke, the daughter molecule has some of the properties of both parents. However, both parts of the resulting genome have to be sufficiently compatible to be functional. Both DNA and RNA genomes undergo recombination. Detailed monitoring of the HIV-1 pandemic shows new recombinants arising between different clades, and clade E is itself a recombinant with *gag* from clade A and the other parent unknown. Such plasticity makes the prospect of control by vaccine ever more problematic (see Chapter 19). Viruses with segmented genomes can also undergo recombination by replacing an entire genome segment with that from another virus, and this occurs with a much greater frequency. This is also known as *reassortment*. The effect is enormous because a virus can acquire an entirely new coat protein at a stroke. The prime example is influenza A virus where this process, called *antigenic shift*, is responsible for pandemic influenza in humans (see Section 18.5). However, viable recombinants are known to require compatibility among all eight genome segments, and this thankfully limits the creation of new viruses.

18.3 Evolution of measles virus

Measles virus (a paramyxovirus) infects only humans, and the infection results in life-long immunity from the disease. F. L. Black studied the occurrence of the disease in island populations (Table 18.1), and found a good correlation between the size of the

Table 18.1 Correlation of the occurrence of measles on islands with the size of the island population.

Island group	Population $(\times10^{-3})$	New births per year $(\times10^{-3})$	Months with measles (%)
Hawaii	550	16.7	100
Fiji	346	13.4	64
Solomon	110	4.1	32
Tonga	57	2.0	12
Cook	16	0.7	6
Nauru	3.5	0.17	5
Falkland	2.5	0.04	0

population and the number of cases of measles recorded on the island throughout the year. A population of at least 500 000 is required to provide sufficient susceptible individuals (i.e. new births) to maintain the virus in the population. Below that level, the virus eventually dies out, until it is reintroduced from an outside source.

On the geological time scale, humans have evolved only recently and have existed only in populations of over 500 000 for a few thousand years. In answer to the question about where measles virus was in the days of very small population groups, we can conclude that it could not have existed in its present form. It may have had another strategy of infection, such as persistence, which would allow it to infect the occasional susceptible passer-by, but we have no evidence of this. However, F. L. Black has speculated upon the antigenic similarity of measles, canine distemper and rinderpest viruses. The last two infect dogs and cattle, respectively, animals that have been commensal with humans since their nomadic days. Black suggests that these three viruses have a common ancestor which infected prehistoric dogs or cattle. The ancestral virus evolved to the modern measles virus when changes in the social behaviour of humans gave rise to populations large enough to maintain the infection. The first such population occurred 6000 years ago when the river valley civilizations of the Tigris and Euphrates were established in what is now Iraq.

18.4 Evolution of myxomavirus

Myxomavirus (a poxvirus) causes a benign infection in its natural host, the South American rabbit, producing wart-like outgrowths (benign tumours) as the only visible evidence of virus multiplication. However, in the European rabbit, myxomavirus causes an infection (myxomatosis) that is 99% lethal. This is a generalized infection, with lesions over the head and body surface. In nature, the disease is spread by arthropod vectors (mosquitoes in Australia and fleas in the UK) which carry virus on their mouth parts. However, this virus does not multiply in the vector.

Myxomavirus was released in England and Australia upon wholly susceptible host populations of the European rabbit in an attempt to eradicate this serious agricultural pest. This experiment in nature was carefully studied with respect to the changes occurring in the virus and the host populations. As we shall see, it provides an object lesson in the problems of biological control.

In the first attempts to spread the disease in Australia, myxomavirus-infected rabbits were released in the wild but, despite the virulence of the virus and the presence of susceptible hosts, the virus died out. It was then realized that this was the result of the scarcity of mosquito vectors, whose incidence is seasonal and depends on rainfall for breeding. When infected rabbits were released at the peak of the mosquito season, an epidemic of myxomatosis followed. Over the next 2 years, the virus spread 3000 miles across Australia and across the sea to Tasmania. However, during this time it became apparent that fewer rabbits were dying from the disease than at the start of the epidemic. The investigators, led by Frank Fenner, then compared the virulence of the original virus with virus newly isolated from wild rabbits by inoculating standard laboratory rabbits. Two significant facts emerged: (i) rabbits infected with new virus isolates took longer to die, and (ii) a greater number of rabbits recovered from infec-

Table 18.2 Evolution of myxomavirus to avirulence in the European rabbit after introduction of virulent virus into Australia in 1950*.

Mean rabbit survival time (days)	Mortality rate (%)	Year of isolation			
		1950–51	1952–53	1955–56	1963–64
<13	>99	100†	4	0	0
14–16	95–99		13	3	0
17–28	70–95		74	55	59
29–50	50–70		9	25	31
>50	<50		0	17	9

*Viruses are isolated in the wild and their virulence is tested by infecting laboratory rabbits.
†Percentage of viruses tested.

tion. From this Fenner inferred that the virus had evolved to a less virulent form (Table 18.2). The explanation was simple: mutation produced virus variants that did not kill the rabbit as quickly as the parental virus. This meant that the rabbits infected with the mutant virus were available to be bitten by mosquitoes for a longer period of time than rabbits infected with the original virulent strain. Hence the mutant would be transmitted to a greater number of rabbits. In other words, there was a strong selection pressure in favour of less virulent mutants which survived in the host in a transmissible form for as long as possible.

The second finding concerned the rabbits themselves, and the question was raised whether rabbits were evolving that were genetically resistant to myxomatosis. To test this hypothesis, a breeding programme was set up in the laboratory. Rabbits were infected and survivors were mated and bred. Offspring were then infected and the survivors mated, and so on. Part of each litter was tested for its ability to resist infection with a standard strain of myxomavirus. The result confirmed that the survivors of each generation progressively increased in resistance. The genetic basis for this is not understood.

This work shows how a virus that is avirulent and well adapted to peaceful coexistence with its host can cause lethal infection in a new, although closely related, host, and how evolutionary pressures rapidly set up a balance between the virus and its new host which ensures that both continue to flourish. This fact remains a stumbling block to the advocates of biological control of pests that attack animals or plants. Today, in the UK and Australia, rabbits remain a serious agriculture problem.

18.5 Evolution of influenza virus

Background

There are three genera of influenza virus (an orthomyxovirus) called types A, B and C, which are distinguished by the antigenicity of their internal virion nucleoproteins.

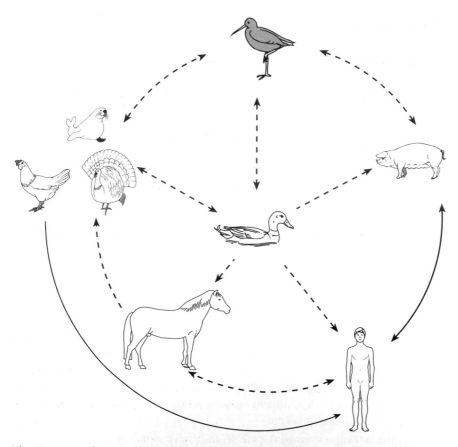

Fig. 18.1 Animal species that are naturally infected with influenza A viruses. Wild birds of the sea and shore form the natural reservoir (top, shaded). Known routes of transmission are indicated by continuous arrows and probable routes of transmission by dashed arrows.

Type A viruses cause the worldwide epidemics (pandemics) of the respiratory disease influenza, and both type A and B viruses cause epidemics. Type C viruses cause only minor upper respiratory illness and are not discussed further. In terms of natural history, the primary hosts of influenza A viruses are wild aquatic birds (ducks, terns and shore birds). In wild birds, viruses usually multiply in the gut and infections are subclinical. Influenza B viruses infect humans, with infections of non-human animals being reported only rarely. Both A and B viruses cause disease in humans only in the winter, usually January and February in the Northern Hemisphere, and June and July in the Southern Hemisphere. At the equator, virus is present at a low level throughout the year. The cause of this periodicity is not known. In addition, a limited number of other species are naturally infected with type A influenza viruses. These and the directions of transmission are shown in Fig. 18.1.

Typical influenza is a lower respiratory infection with fever and muscular aches and pains, but can range from the subclinical to pneumonia (where the lungs fill with

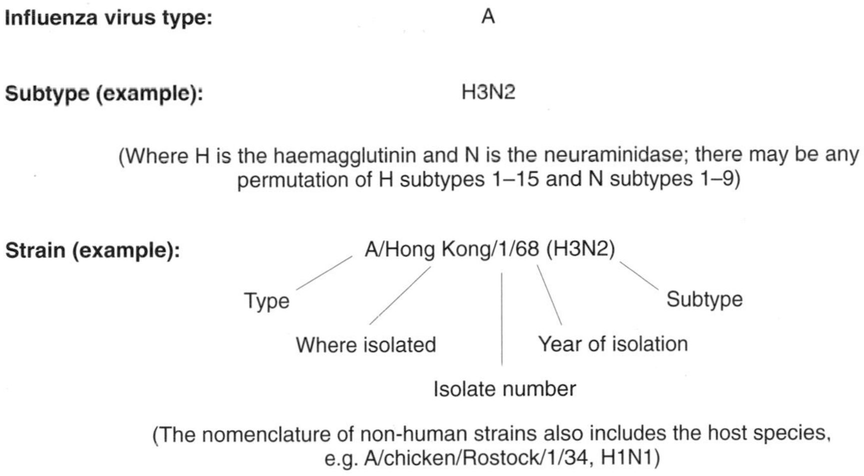

Influenza virus type: A

Subtype (example):   H3N2

(Where H is the haemagglutinin and N is the neuraminidase; there may be any permutation of H subtypes 1–15 and N subtypes 1–9)

Strain (example):   A/Hong Kong/1/68 (H3N2)

Type   Subtype

Where isolated   Year of isolation

Isolate number

(The nomenclature of non-human strains also includes the host species, e.g. A/chicken/Rostock/1/34, H1N1)

Fig. 18.2 Influenza viruses have a formal descriptive nomenclature.

fluid). In elderly people and those with underlying clinical problems of the heart, lungs and kidneys, and in people with diabetes and who are immunosuppressed, it can be life threatening. Immunity to an influenza virus is effective at preventing reinfection by that same strain. The viral antigens relevant to protective immunity are the envelope glycoproteins, haemagglutinin (HA) and neuraminidase (NA—see Fig. 3.16), and immunity is mediated by virus-specific antibody. Why, then, can people suffer several attacks of influenza in their lifetime? The answer is that the HAs and NAs of influenza A and B viruses evolve continuously so that previously acquired immunity is rendered ineffective. How and why this happens are discussed later. In the laboratory, many influenza viruses can be adapted to grow well in 10 day old embryonated chickens' eggs, and this is the culture system used for production of the killed vaccine. Work is now in progress to move to a cell culture system.

Influenza A viruses are classified formally as *types*, *subtypes* and *strains*. This nomenclature is explained in Fig. 18.2. There are 15 subtypes of haemagglutinin and nine of neuraminidase. All 135 permutations are found in nature in wild aquatic birds. By comparison influenza in humans is a side show, and currently only two subtypes, H1N1 and H3N2, are in circulation.

Two mechanisms of evolution

Influenza A viruses undergo two types of change affecting their major surface glycoproteins called antigenic shift and antigenic drift. Since the start of modern virology, there have been four shifts that occurred in 1889, 1918, 1957 and 1968. A shift results in a pandemic (a worldwide epidemic), and is always accompanied by an abrupt change in HA subtype and, apart from the 1968 virus, by a change in NA subtype also. Drift results in epidemics, and is caused by gradual evolution under the positive selection pressure of neutralizing antibody. A new shift virus immediately starts to undergo continuous antigenic drift. Influenza B viruses undergo only drift because they have no major animal reservoir.

The complex evolutionary processes that influenza viruses undergo can be better understood against a background of their biology. These viruses are highly infectious and cause an acutely cytopathogenic infection. Thus, they are a victim of their own success which results in an almost universal immunity among hosts who, given adequate conditions, can live for more than 80 years. Apparently, the rate of production of immunologically naive individuals by new births is not enough to allow the viruses to survive. Thus, influenza viruses have evolved strategies for changing their antigens in order to increase the proportion of susceptible individuals in the population.

The current understanding and implications of shift and drift are discussed below. Information comes from a worldwide network of laboratories, coordinated by the World Health Organization (WHO), which isolate and classify currently circulating influenza viruses. In this way new strains can be quickly spotted and the vaccine changed appropriately.

Antigenic shift

The appearance of shift viruses in humans is shown chronologically in Fig. 18.3. Virus was first cultivated in the laboratory in 1933 by intranasal inoculation of ferrets with nasal secretion from a virologist (Wilson Smith) who had influenza, and was then successfully passaged in embryonated chickens' eggs. However, retrospective information has also been obtained by studying influenza virus-specific antibodies in human sera that had been kept stored in hospital freezers for other purposes. As can be seen, shift has occurred sporadically from the first recorded shift of 1889 at intervals of 11, 18, 39, 11 and >32 years. History of the terrible 1918 pandemic is shown in the box. Shift by definition introduces a virus into a population that has no pre-existing immunity, so that it is nearly always associated with an explosive pandemic with high morbidity and mortality, although these vary between the different shifts. Approximate mortality figures are included in Fig. 18.3.

Until 1977, only one subtype was in circulation at a time and this was replaced completely when the new subtype was introduced. What causes this disappearance is not known. In 1977, during the reign of the H3N2 subtype, an H1N1 virus appeared that was identical to the H1N1 virus that was in circulation in 1950. Thus, this was not a shift but a reintroduction. Where the 1977 virus came from is not known. It had not been infecting the human population because it would have undergone 27 years of antigenic drift (see below). It is as if it had been frozen, and some say that this is literally what happened, but this is conjecture. At the time of writing, the drifted descendants of the 1968 H3N2 and the 1977 H1N1 co-circulate and continue to undergo antigenic drift. However, a new shift could occur at any time.

Also marked on Fig. 18.3 are two instances of abortive shifts that took place in the eastern USA in 1976 and Hong Kong in 1997. These viruses were noted and notable because they caused death in young people, and it was feared that they signalled the start of a new lethal pandemic. Both had their origins in non-human animals. The

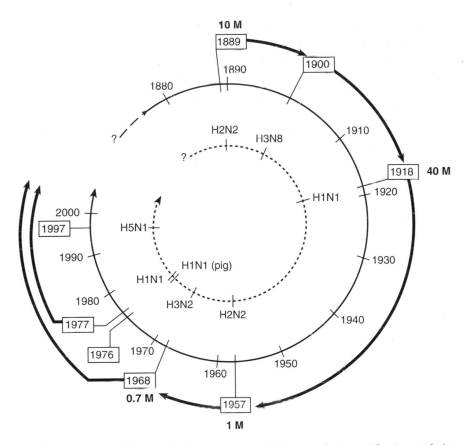

Fig. 18.3 Summary of antigenic shift of influenza A viruses in humans. The time scale in the middle circle is marked with the year of emergence of a new subtype (inner circle), e.g. 1889. See text for an explanation of 1976, 1977 and 1997 viruses. The outer circle indicates the reign of a particular subtype in the population and when it was replaced by another. H3N2 and H1N1 subtypes currently coexist. The approximate worldwide mortality figures for each shift are indicated in millions. The 1900 shift did not cause a serious pandemic.

1976 virus was isolated from local domestic pigs, and the 1997 virus from chickens that had been brought as live birds to the local poultry market. However, neither virus had the ability to spread from person to person, and no more infections were seen.

The mechanism of antigenic shift

In essence shift is thought to occur when virus from a wild bird infects a human being (Fig. 18.4). However, as with most viruses, influenza virus is species specific and does not readily infect other bird species, let alone members of another phylum. Thus, the virus has to adapt by progressive mutation before the jump to humans can occur. To recapitulate, the natural reservoir of influenza A viruses comprises wild aquatic birds, such as ducks, terns and shore birds. These viruses are tropic for the gut (not the res-

The 1918 shift virus was unusual in several ways. It is estimated that in just 1 year it infected 500 million and killed more than 40 million people worldwide. (The enormity of this can be appreciated when compared with the 19 million who have died from AIDS in the 20 years since the HIV-1 epidemic was recognized in the early 1980s.) The highest mortality was in young adults and not the usual high-risk groups mentioned above. Neither its virulence nor its age tropism is understood, and it is greatly feared that a similar virus might appear in a future shift event. The seemingly intractable problem of analysing the virus is being solved by isolating fragments of virus genome from the preserved tissue of victims of the 1918 pandemic. These are either pieces of lung that were preserved in formalin in the US army pathology archives or from the exhumed bodies of people who died and were buried in permafrost in Arctic regions. Virus RNA is being isolated by reverse transcription polymerase chain reaction (RT-PCR). So far the complete sequences of the viral haemagglutinin and neuraminidase, and partial sequences of the NP, M1 and M2 genes, have been obtained, and it may be possible to recreate infectious virus. Needless to say, these experiments will be conducted in a level 4 high-containment laboratory.

piratory tract), cause a subclinical infection and are evolutionarily stable. Many of these birds migrate enormous distances (e.g. from Siberia to Australia) and spread virus as they go through infected faecal deposits.

It is thought that the first link in the chain is the infection of free-range domestic poultry by migrating wild birds. This is an avirulent respiratory infection, although it is not understood how the switch from a gut infection comes about. The infection passes from bird to bird until mutations take place in the protease cleavage site of the HA precursor polypeptide. Cleavage is necessary for virus infectivity. Normally, the HA can be cleaved only by a protease that is present solely in the respiratory tract but, after an increase in basic residues, the HA can be cleaved by another protease that is widely distributed throughout tissues, and this allows the virus to grow throughout the body. At this stage the virus is able to jump from the bird phylum to the mammal phylum, and multiply in humans (Fig. 18.4). In this form it resembles the abortive 1976 or 1997 viruses but, as mentioned above, these viruses lacked the ability to spread to other people. One of the major gaps in modern virology is the understanding of virus transmission, and what gene or genes are responsible for its control. One other variation in the chain of infection from wild birds to humans may be via domestic pigs. In rural areas poultry and pigs are often kept together, giving ample opportunity for the crucial avian-to-mammal adaptation. Amino acid sequence comparisons show that avian, pig and human viruses form distinct family trees, and the 1918 shift virus has an HA sequence that is pig-like in origin.

At this point the virus has two evolutionary options. It can continue to accumulate mutations and become better adapted to humans or it can recombine with an exist-

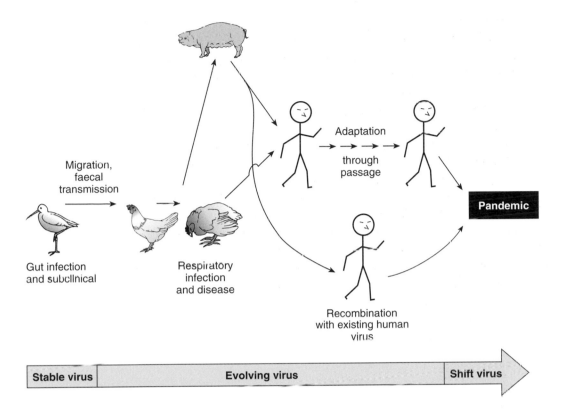

Fig. 18.4 Summary of the events leading up to an antigenic shift of human influenza virus. The evolutionary time scale is not known but probably is several years.

ing human influenza virus strain. In fact it may do both. One essential adaptation is a change in the topography of the viral attachment site, so that the virus can use as a receptor N-acetylneuraminic acid, which has a 2,6 linkage to the carbohydrate moiety on a glycoprotein or glycolipid (as is found in cells of the human respiratory tract), rather than the 2,3 linkage that is most common in birds. The advantage to the virus of recombination is that, at its simplest, there need only be a recombination of the RNA segment that encodes the new HA with the seven existing segments from the human strain. Thus, none of the other genome segments has to adapt to the human environment. This scenario would account for the 1968 shift where an H3N2 virus replaced the H2N2 virus.

Recombination in influenza viruses is very efficient because the genome is segmented—it comprises eight single-stranded, negative-sense RNAs. The process is sometimes called *reassortment* because this describes what happens. The HA and NA proteins are encoded by distinct RNA segments. When a cell is infected simultaneously with more than one strain of virus, newly synthesized RNA segments reassort virtually at random to the progeny (Fig. 18.5). Many, though not all, of the 2^8 (256) possible genetic permutations that can be formed between two viruses are genetically

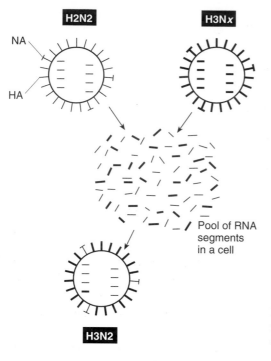

Fig. 18.5 Recombination (reassortment) between an existing human influenza A virus (H2N2) and a new virus from the wild bird reservoir (H3N*x*, where *x* represents an unknown neuraminidase subtype—see text) that gives rise to antigenic shift. The two viruses simultaneously infect a cell in the respiratory tract, and the eight genome segments from each parent assort independently to progeny virions. The example shows a novel progeny virion (H3N2) that comprises the RNA segment encoding the H3 avian haemagglutinin and the seven remaining segments from the existing human virus.

stable. Reassortment occurs readily in cell culture and experimental animals, and in natural human infections between all type A influenza viruses (but never between A and B viruses). Even then more mutational adaptation may be needed before the virus is able to cause a recognized pandemic. Thus it may be that a shift virus is present in the human population, undergoing mutational adaptations, a few years before it causes a pandemic.

Antigenic drift

Influenza A viruses have been isolated every year from humans around the world since their discovery in 1933. Each new isolate is tested serologically with antibody to all other influenza strains. It soon became apparent that the HA and NA of recent isolates were antigenically slightly different from those of earlier strains, and that the difference increased incrementally year by year. This difference is reflected in an amino acid substitution in one of the antigenic sites on the HA or NA. This phenomenon rests on the assumption that, in nature, influenza virus mutants carrying new antigenic determinants that arise are at an evolutionary advantage in the face of an immune response compared with the parent virus. However, although both the HA and NA drift, the HA is the major neutralization antigen and appears evolutionarily the more significant. Thus, year on year new amino acid substitutions—and epitopes—appear. The name 'antigenic drift' is very apt (Fig. 18.6). The nature of the selective force that drives this process is discussed below. In practice a drift variant that can

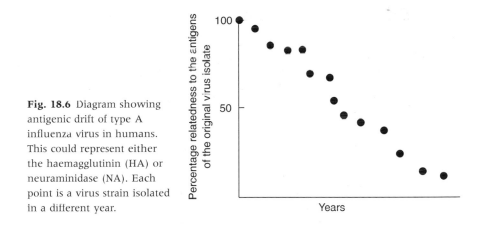

Fig. 18.6 Diagram showing antigenic drift of type A influenza virus in humans. This could represent either the haemagglutinin (HA) or neuraminidase (NA). Each point is a virus strain isolated in a different year.

cause significant disease arises about every 4 years, and this has four or more amino acid substitutions in two or more antigenic sites. What also happens is that the 'old' strain is completely replaced by the 'new' strain. Influenza B viruses also undergo antigenic drift, but this is a slower process and also several different strains co-circulate. The reason for this difference is not known.

Antigenic drift is at least as important in causing human influenza as antigenic shift, and perhaps more so. In the UK, there are 4000–14 000 deaths associated with epidemics of influenza every year, and this extrapolates to 400 000–1 400 000 deaths worldwide. Thus, in the twentieth century, drift viruses have been responsible for approximately 40–140 million deaths worldwide. (In the UK, an influenza epidemic is formally defined as 400 cases of influenza per 100 000 people).

The mechanism of antigenic drift

As soon as a new shift virus appears and infects people, it begins to drift (Fig. 18.7). Drift of influenza A viruses is linear as a result of the dominating effects of favourable mutations (Fig. 18.8). Influenza B viruses undergo less drift than A viruses, and multiple virus lineages coexist. Drift is happening on a global scale.

It can only be theorized how drift takes place because putative drift variants can be assumed to have arisen only from virus that circulated in the previous year. It is generally believed that variants are selected by antibody. The phenomenon of drift can be modelled in the laboratory using a neutralizing monoclonal antibody (MAb) specific for the HA. Virus and MAb are mixed together and inoculated into an embryonated chicken's egg. This is a surprisingly efficient process and the progeny virus is no longer neutralized by the selecting MAb. This 'drift' virus is also known as an escape mutant, and represents the growth to population dominance of an antigenic variant virus that already existed in the inoculum. When sequenced, there is a single amino acid substitution in the expected antigenic site. If the 'drift' virus is subjected to another MAb to a different epitope, then the process is repeated and the progeny virus now has two amino acid changes compared with the original (Fig. 18.9). However, the model falls down when two or more MAbs are mixed together to resemble an

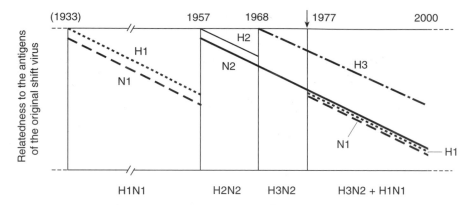

Fig. 18.7 Diagram showing the course of antigenic shift and drift of influenza A viruses in humans. The first virus, isolated in 1933, was H1N1. This arose by antigenic drift from the 1918 virus. Other shift viruses appeared in 1957 (H2N2) and 1968 (H3N2). A 1950 H1N1 virus reappeared in 1977. Drift is shown schematically. The 1957 N2 was acquired by the H3N2 shift virus, and has drifted from 1957 to the present day.

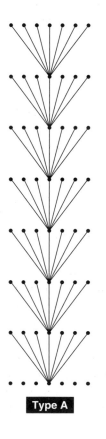

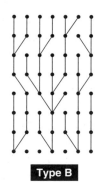

Fig. 18.8 Model of antigenic drift of influenza type A and type B viruses. Points on the same level represent drift variants that arise in the same year. The branch length indicates the relative change in antigenicity from virus in the preceding year. Drift is shown for an arbitrary 7-year period. (See text for further discussion.) (From Yamashita *et al.* 1988.)

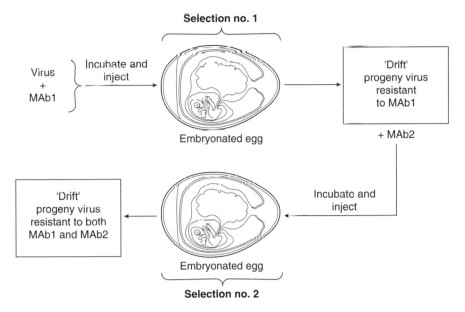

Fig. 18.9 'Antigenic drift' in the laboratory: a single neutralizing monoclonal antibody (MAb) can select a population of influenza virus escape mutants that is no longer neutralized by the selecting MAb. Another round of selection with a second MAb produces virus that now carries two amino acid substitutions, but selection with the two MAbs simultaneously is completely neutralizing (not shown).

antiserum, because no progeny virus—mutant or wild type—is produced at all. Thus it seems that drift can take place only if an antiserum contains antibody to a single epitope. Does this occur in nature?

Influenza viruses have five antigenic sites (although sometimes one is hidden by a carbohydrate group). In H3 viruses these are labelled A–E (see Figs 3.17 and 16.5). Each site comprises around 10 epitopes, so there are 50 epitopes in all, and in theory during infection the immune system should make antibodies to all of these. This was tested by immunizing mice and rabbits with virus and measuring the amount of antibody to individual epitopes. What was found was that one epitope in site B was dominant, and there were only traces of antibody to two other epitopes (Fig. 18.10). The amounts varied between animals. In mice the results were similar but an epitope in site A was dominant. This shows that the immune system responds only selectively to foreign epitopes. In the final test, some of these notionally polyclonal antisera were able to select escape mutants such as an MAb. Others were completely neutralizing. Thus, it seems that the theory for how drift occurs holds water provided that certain individuals have an epitope-biased antibody response. The derivation of the four or more amino acid substitutions in two or more sites could therefore be achieved as shown in Fig. 18.11. It is suggested here that people with different genetic backgrounds have an antibody response biased to different epitopes. Thus the drift

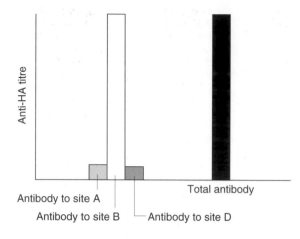

Antibody to site A

Antibody to site B

Antibody to site D

Total antibody

Anti-HA titre

Fig. 18.10 An epitope-biased serum antibody response in a rabbit injected with influenza A virus. Nearly all the haemagglutinin (HA) specific antibody is accounted for by the response to a single epitope in antigenic site B. (From Lambkin and Dimmock 1995.)

variants may arise only in a subpopulation with the relevant antibody response. Furthermore as, over a lifetime, adults suffer repeated influenza infections, they are likely to make a complex antibody response that cannot give rise to drift variants. Thus, the simplest antibody response is likely to occur in children after their first infection, and they may therefore be responsible for the selection of drift variants. The latter is hypothesis, although it can easily be tested.

Vaccines and antivirals for influenza

A killed, split vaccine is currently used to protect against influenza. This is prepared by growing virus in embryonated chickens' eggs. Virus is purified and treated (split) with detergent to release the HA and NA proteins from the lipid membrane of the virus, and then centrifuged again to separate the HA and NA from the other viral proteins which are pyrogenic (cause fever). The vaccine is trivalent and comprises the HA and NA proteins from the currently circulating H1N1 and H3N2 viruses and the current influenza B virus. A single subcutaneous injection is given, but it is recommended only for elderly people, and people with chronic clinical problems of the heart, lungs and kidneys, people with diabetes and immunocompromised individuals. It is not an ideal vaccine because it produces serum antibody but no T-cell-mediated or mucosal immunity, and has to be repeated annually in the autumn, even if the vaccine components have not been changed. It is about 70% effective.

There is continuous global monitoring for new shift and drift virus, which is overseen by the WHO. The shift viruses are the most serious because of the explosive outbreaks and high mortality that they can cause. So far these have all first appeared in winter (June, July) in the Southern Hemisphere. This gives the vaccine manufacturers just 6 months to produce a new vaccine in time to protect people in the Northern Hemisphere (January, February). The correct choice of vaccine strains is vitally important.

For some time the antivirals, amantadine and rimantadine, both inhibitors of M2 ion channel activity, have been available. In 1999, a new generation of tailor-made

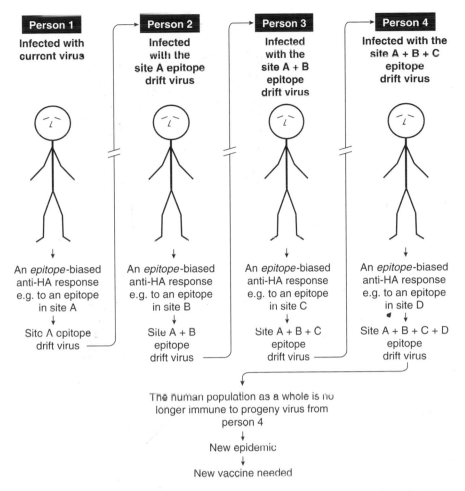

Fig. 18.11 How antigenic drift may occur in nature, bearing in mind that clinically significant drift viruses (that cause epidemics) have four or more amino acid residues changed in two or more antigenic sites present on the haemagglutinin (HA) protein. Note that people with biased antibody responses may be uncommon and that there may be any number of non-selective infections of other individuals occurring between the four people shown here, as indicated by the broken arrow of transmission. A similar process could occur for drift of the neuraminidase (NA) protein.

drugs, zanamivir (Relenza, which is given intranasally), and the orally administered prodrug oseltamivir (see Chapter 16.8), were licensed for the first time by some countries. These antivirals reduce the duration of influenza by 1.5–2.5 days, the amount of secondary bacterial infection and the need for antibiotics. Resistant virus has not been seen. The problem is that these drugs are effective only when taken as the first signs and symptoms appear, and at this stage influenza cannot be distinguished from other respiratory viruses that are not sensitive to the drugs. However, zanamivir and oseltamivir may be a valuable prophylactic against shift viruses.

Unsolved problems relating to influenza virus

Despite considerable knowledge of influenza virus (the complete genome sequence, the atomic structure of the HA, infectious nucleic acid), there are still major areas of ignorance. Most of these concern viral epidemiology, which is immensely difficult to study. Space precludes discussion but the following problems require answering:
- Why is influenza a winter disease?
- How does the same virus appear simultaneously in different places around the globe at the start of an epidemic (i.e. it does not appear to follow any apparent transmission chain)?
- Why is the seasonal restriction of influenza apparently unaffected by the year-round introduction of virus by air travellers incubating influenza? Does this virus not infect people in summer or does it infect them without causing disease?
- How does a clinically significant drift variant travel from northern to southern latitudes in 6 months (and did so before frequent air travel)?
- How does a shift virus (other than the reoccurring H1N1 of 1977) replace the existing virus?
- How does a drift virus replace the existing virus?
- The hypothesis of antigenic drift outlined above needs verifying.
- Why did the 1918 shift virus cause such high mortality?

18.6 Evolution of morphologically similar viruses

An example of a widespread group of viruses are the rhabdoviruses of eukaryotes (see Fig. 3.16). Their 'bullet' shape makes them easy to identify. They have a lipid envelope and single-stranded negative-sense RNA. Rhabdoviruses infect both animals and plants, and are found in invertebrates and cold- and warm-blooded vertebrates (but, like all viruses, most rhabdoviruses infect a very restricted range of host species; rabies virus is the exception because it can infect a wide variety of mammalian species). As these rhabdoviruses are morphologically indistinguishable, it is tempting to speculate that they have arisen once and subsequently spread in a truly remarkable fashion. The rhabdoviruses also present us with an interesting link between viruses of the animal and plant kingdoms, because at least one representative (lettuce necrotic yellows virus) multiplies in both the animal (insect) vector and the plant that the vector feeds on.

18.7 Further reading and references

Black, F. L. (1966) Measles endemicity in insular populations: critical community size and its evolutionary implications. *Journal of Theoretical Biology* **11**, 207–211.

Bush, R. M., Bender, C. A., Subbarao, K., Cox, N. J. & Fitch, W. M. (1999) Predicting the evolution of human influenza A. *Science* **286**, 1921–1925.

Domingo, E. & Holland, J. J. (1997) RNA virus mutations and fitness for survival. *Annual Review of Microbiology* **51**, 151–178.

Domingo, E., Webster, R. G. & Holland, J. J. (1999) *Origin and Evolution of Viruses*. New York: Academic Press.

Fenner, F. & Ratcliffe, F. N. (1965) *Myxomatosis*. London: Cambridge University Press.

Holland, J. J., de la Torre, J. C. & Steinhauer, D. A. (1992) RNA virus populations as quasispecies. *Current Topics in Microbiology and Immunology* **176**, 1–20.

Koonin, E. V. (ed.) (1992) Evolution of viral genomes. *Seminars in Virology* **3**(5), 311–417.

Lambkin, R. & Dimmock, N. J. (1995) All rabbits immunized with type A influenza virions have a serum antibody haemagglutination–inhibition antibody response biased to a single epitope in antigenic site B. *Journal of General Virology* **76**, 889–897.

Oxford, J. S. (2000) Influenza A pandemics of the 20th century with special reference to 1918: virology, pathology and epidemiology. *Reviews in Medical Virology* **10**, 119–133.

Reanny, D. (1984) The molecular evolution of viruses. In: *The Microbe 1984. I Viruses*, Vol. 36, *The Society for General Microbiology Symposium*, pp. 175–196. B. W. J. Mahy & J. R. Pattison (eds). Cambridge: Cambridge University Press.

Taubenberger, J. K. (1999) Gene sequences of the virus from the 1918 influenza pandemic are yielding insights into its origin but little about virulence. *American Society for Microbiology News* **65**, 473–478.

Webster, R. G. (1997) Influenza virus: transmission between species and relevance to emergence of the next human pandemic. *Archives of Virology Supplement* **13**, 105–113.

Webster, R. G. (1998) Influenza: an emerging disease. *Emerging Infectious Diseases* **4**, 436–441.

Yamashita, M., Krystal, M., Fitch, W. M. & Palese, P. (1988) Influenza B virus evolution: cocirculation lineages and comparative evolutionary patterns with influenza A and C viruses. *Virology* **163**, 112–122.

Also check Chapter 22 for references specific to each family of viruses.

Chapter 19
HIV and AIDS

During the past 30 years human immunodeficiency virus (HIV) type 1 has become a common infection of humankind. There is also a second, less common, less virulent, but closely related virus, HIV-2. Both infect mainly T-helper lymphocytes and macrophages using the cell surface CD4 protein as the primary receptor. After a long, essentially symptomless, incubation period, on average 10 years, CD4$^+$ lymphocytes decline to such a low level that the immune system can no longer function efficiently, resulting in the immunodeficiency that gives the virus its name. The consequence of this is that the affected person can no longer restrain certain normally harmless passenger micro-organisms (viruses, bacteria or fungi) which then cause overt clinical disease. A collection of diseases such as this, unrelated except for a common underlying cause, is called a syndrome—hence the name 'acquired immunodeficiency syndrome' or AIDS. HIV infection is almost always lethal and this is typical of viruses that infect a new host (see Sections 18.4 and 21.7).

This chapter discusses the immense progress that has been made in understanding HIV and the reasons why modern science has not, at the time of writing (April 2001), come up with an effective countermeasure. The reader should appreciate that the field is moving with great rapidity and should consult the contemporary literature for an update.

19.1 The biology of HIV-1 infection

The spread of HIV

The HIV/AIDS scenario is tragically grim. The United Nations Programme on AIDS (UNAIDS) estimates that HIV has already infected over 36 million people worldwide (Table 19.1). Nearly 22 million have died of AIDS in the last 20 years, and more than 95% of all new infections arise in the developing world. During 1999, 5.4 million people became infected, equating to 15 000 new infections every day. The epicentre for HIV is sub-Saharan Africa with 1.3 million new infections, with the highest incidence in Botswana, which is estimated to have 50% of its population infected. The predicted expansion into south and south-east Asia has sadly been fulfilled. There is still no cure and no vaccine, but treatment with a combination of two nucleoside analogues and a protease inhibitor (triple therapy—see Section 19.8) has been in use since

Table 19.1 Estimates by the United Nations Programme on AIDS (UNAIDS) of the number of HIV-1-infected adults and children at the end of 2000.

Developed world countries	5 000 000
South and south-east Asia	5 800 000
Sub-Saharan Africa	25 300 000
Worldwide	36 100 000

1994 and is remarkably successful in halting virus replication and the progression to AIDS. This is known as HAART (highly active antiretrovirus therapy) and currently this usually means triple therapy. However, although virus in plasma (blood from which cells have been removed) is undetectable under successful triple therapy, HAART does not clear virus completely. There is latent virus present in resting CD4$^+$ memory T cells, and there may be other reservoirs as well. Treatment must be continued without a break and for the foreseeable future, or the virus rebounds within a few days to its original level. However, HAART is restricted to developed countries because of its expense.

The discovery of HIV

Historically, the disease was recognized before the virus. In 1981, astute clinicians noted an unusual pneumonia caused by *Pneumocystis carinii*, a yeast-like organism, and identified a rare cancer, Kaposi's sarcoma, in homosexual young men in New York and California. Some of those with the sarcoma also had pneumocystis pneumonia. These conditions were then linked to immune deficiency. Later, an association with a variety of common infections, leading to death, was recognized, and AIDS was defined. (Kaposi's sarcoma is actually caused by the ubiquitous human herpes virus (HHV type 8) under conditions of immunodeficiency—see Section 17.5.) Various causes were considered, but it was not until 1984 that it was thought that a virus was involved. In 1985, a virus new to science, HIV-1 (then also known as human T-cell lymphotrophic virus type 3 (HTLV-3) or lymphocyte-associated virus (LAV)) was isolated. Once viral antigens were available as diagnostic reagents, stored sera could be tested retrospectively for the presence of antibody to HIV, and it was demonstrated that there were few HIV infections in humans before 1970. However, as we now know that the incubation period in developed countries averages 10 years (shorter in the developing world), it is clear that HIV had spread in an explosive but silent pandemic throughout that decade (Table 19.2).

Where did it all start?

The history of HIV is surprisingly sketchy, despite much effort to trace its origins. The earliest HIV-1 infections can be traced by serology to people in central parts of sub-Saharan Africa. Either it is a long-established but rare infection in these relatively isolated populations, or a new infection by a virus that has jumped from another species. The latter fits with the generalization that only new infections, not yet in balance with their hosts, cause serious disease. Also supporting this hypothesis is the fact that the closest relative discovered so far is a simian immunodeficiency virus (SIV) from chimpanzees (SIV$_{cpz}$). There have been isolations of three such strains, but the closest to HIV-1 is a strain isolated from a subspecies of chimpanzee (*Pan troglodytes troglodytes*) from central Africa (Fig. 19.1). It is thought that the virus was introduced at least three times into humans. Chimpanzees are a local food source and the infection may have been contracted through cuts and abrasions during butchery and eating. In any event, the factors that caused the virus to spread, first into urban areas in Africa (the earliest reported human virus is 1959), and from there to the USA and the rest of the

Table 19.2 Milestones for AIDS and HIV.

Year	Milestone
1970s	Silent pandemic
1981	AIDS is recognized for the first time
1983	AIDS linked to infection with a new virus
1984	Initial characterization of HIV-1
1994	Triple therapy with inhibitors of reverse transcriptase and viral protease
2000	Still no vaccine or universally affordable treatment

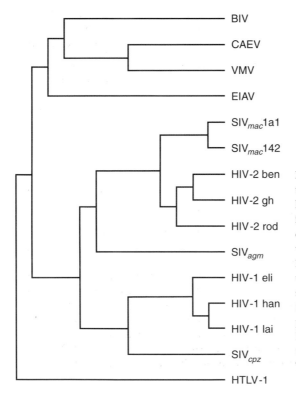

Fig. 19.1 Dendrogram showing the relatedness of different lentiviruses. BIV, bovine immunodeficiency virus; CAEV, caprine arthritis–encephalitis virus; VMV, visna maedi virus; EIAV, equine infectious anaemia virus; SIV, various strains of simian immunodeficiency virus; SIV$_{cpz}$ was isolated from a chimpanzee; HIV, various strains of human immunodeficiency virus type 1 and 2; HTLV-1, human T-cell lymphotrophic virus type 1. (Courtesy of Georg Weiller.)

world, are not known. There are many gaps in the story. HIV-2 is most closely related to SIV$_{sm}$ and there have been at least six independent introductions into humans; again this may have been from sooty mangabeys used for food or possibly kept as pets.

What sort of virus is HIV-1?

HIV-1 is a typical member of the lentivirus genus of the retrovirus family (Baltimore class 6). The name is derived from the Latin *lente*, which refers to the *slow* onset of the disease. However, there is nothing slow about the rate of virus multiplication. There are several well-characterized lentiviruses (Table 19.3), infecting a number of different vertebrate species. Common features of lentivirus infections are listed in Table

Table 19.3 Some typical members of the lentivirus subfamily and their hosts.

Virus	Host	Main target cell	Clinical outcome
Visna-maedi	Sheep	Macrophage	Pneumonia (= maedi) or chronic demyelinating paralysis (= visna); little immunosuppression*
Visna-maedi	Goats	Mammary macrophage	Arthritis; rarely encephalitis; little immunosuppression
Equine infectious anaemia	Horses	Macrophage	Recurrent fever; anaemia; weight loss, little immunosuppression
Bovine immunodeficiency	Cattle	?	Weakness; poor health
Feline immunodeficiency	Cats	CD4$^+$ T cell	AIDS
Simian immunodeficiency†	Monkeys	CD4$^+$ T cell	subclinical or AIDS
Human immunodeficiency type 1	Humans	CD4$^+$ T cell	AIDS
Human immunodeficiency type 2	Humans	CD4$^+$ T cell	AIDS

*The same virus can cause two different diseases in sheep.
†Different SIV strains are named after the species of monkey from which they were first isolated, e.g. SIV$_{man}$ from the mandrill, SIV$_{agm}$ from the African green monkey, SIV$_{sm}$ from the sooty mangabey. These are all Old World African monkeys, and SIV strains cause no disease in their natural host. However, although it does not naturally infect Asian macaque monkeys, SIV$_{sm}$ does do so under experimental conditions and causes AIDS. SIV$_{mac}$ is thought to be SIV$_{sm}$ that accidentally infected a laboratory macaque. (This is a further example of a virus interaction with its natural host being relatively benign compared with that with a new host).

Infection of bone-marrow-derived cells
Persistent viraemia
Life-long infection
Prolonged subclinical infection
Weak neutralizing antibody responses
Continuous virus mutation and antigenic drift
Neuropathology

Table 19.4 Common features of lentivirus infections.

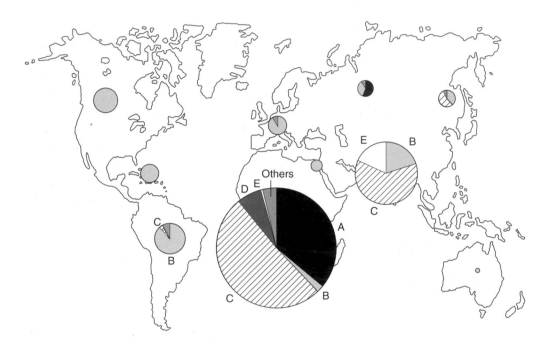

Fig. 19.2 The distribution of HIV-1 clades A–I. The circle size denotes the total number of infections in an area, and the pie charts denote the relative distribution of the major clades. HIV-2 accounts for only a minority of infections and was found originally in West Africa and now also in India. (From UNAIDS.)

19.4. HIV-1 is highly pathogenic and globally dispersed, and causes over 99% of HIV infections. The serologically distinct HIV-2 is much less pathogenic and is endemic in West Africa. HIV-2 is more closely related to SIV than to HIV-1 (see Fig. 19.1).

In fact there is not one HIV-1, but a collection of different evolutionarily linked groups or *clades*. The classification of clades is based solely on sequence, and in the case of HIV-1 by the sequence of the *gag* or *env* genes. This is important because the clades also differ in antigenicity of the envelope gp120–gp41 protein, and this makes vaccine manufacture difficult. Most HIV-1 isolates belong to group M (there are also groups N and O). Group M comprises nine clades—clade A to clade I—and these differ by 25–35% in the sequence similarity of their *env* genes. Their geographical distribution varies enormously (Fig. 19.2). Most of HIV-1 in the developed world belongs to clade B, the virus in sub-Saharan Africa is clade A, C or D, whereas that in south and south-east Asia is clade C, B or E. Compare this distribution with the frequency of the virus in Table 19.1 and it will be apparent that each area will probably need a different vaccine. Racial and ethnic differences in major histocompatibility complex (MHC) haplotypes, which affects the nature of peptides presented to T cells, add to this problem.

19.2 Molecular biology of HIV-1

As with all retroviruses, HIV-1 is diploid and contains two molecules of positive-sense single-stranded (ss) RNA. Its genome conforms to the basic oncovirus plan, described in Fig. 9.8, but is complicated by containing a number of additional regulatory genes (Fig. 19.3). It has the usual retrovirus gene order 5'-*gag–pol–env*-3' and is flanked by the characteristic long terminal repeats (LTRs). One double-stranded DNA (dsDNA) provirus molecule is produced from the RNA template using the virion reverse transcriptase, and the DNA is integrated into the host genome (see Section 8.2). Reverse transcriptase was the first target for antiviral chemotherapy. From the provirus, the simpler oncornavirus synthesizes only two RNAs—the viral genome, which doubles as mRNA for the gag (group antigen) and gag–pol proteins (see Section 9.9), and an mRNA with a single splice, which encodes the envelope protein. Thus, this very simple

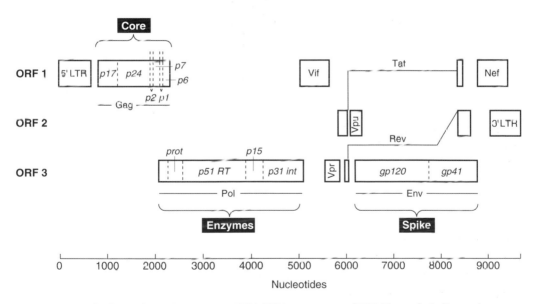

Fig. 19.3 The positive-sense ssRNA 9719 nt. genome of HIV. The scale indicates the genome size in nucleotides. Note that all three reading frames (open reading frames or ORFs) are used, but only simultaneously at around 8500 nucleotides. *Gag, pol* and *env* are transcribed and translated to give polyproteins which are then cleaved by proteases. The nomenclature 'p17' indicates a polypeptide having a relative molecular mass of 17 000, etc. Genes not aligned require a frame shift for expression. Tat and Rev proteins are expressed from spliced RNAs.

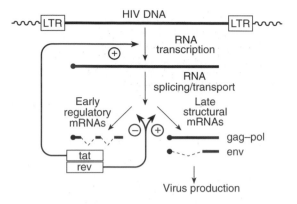

Fig. 19.4 Regulation of HIV-1 gene expression from provirus DNA integrated into the cellular genome. This occurs in two phases; regulatory genes are expressed early and structural genes are expressed late. (From Cullen 1991.)

virus has proteins that have both structural and enzymatic functions. Gene expression is controlled by *cis*-acting (i.e. self) DNA or RNA sequences and by host-coded factors acting in *trans*. HIV-1 imposes on this basic plan a number of proteins (Tat, Rev, Vif, Vpr, Vpu and Nef) with positive or negative regulatory activity (Fig. 19.4). In all, HIV-1 encodes 16 proteins.

Cell transcription factors result in the synthesis of a small amount of full-length viral RNA from integrated HIV-1 DNA, which is then multiply spliced to form small mRNAs encoding Tat and Rev. Tat and Rev are RNA-binding proteins. Tat binds to the so-called TAR element, an RNA sequence just downstream from the transcription start site, and upregulates transcription by increasing processivity of RNA polymerase II. This increases Rev to a certain critical level that downregulates the cytoplasmic accumulation of multiply spliced RNAs and favours the accumulation of a non-spliced and a singly spliced mRNA, which encode the structural proteins Gag and Pol, and Env, respectively (Fig. 19.4). Rev achieves this by binding to an RNA sequence (the RRE) that is present only in non-spliced and singly spliced mRNAs, facilitating their transport from the nucleus. Nef is a complex pleiotropic (multifunctional) regulator. It down-modulates cell surface expression of CD4 and MHC class I proteins, and increases the virus infectivity. In addition, it enhances the spread of infection in the body from infected macrophages by causing them first to chemoattract T cells, and then to activate resting T cells, with the result that they are then able to replicate the virus. Vpr and Vpu are concerned with the morphogenesis of virions and their release from the cell. Vif is associated with virus infectivity. Tat and Rev are essential for replication *in vitro*, whereas the others are not and are consequently called accessory proteins. However, they are presumably needed for the success of the HIV *in vivo*. Much more is still to be learned about all of them.

Virion proteins

Production of structural proteins from the *gag, pol* and *env* genes and their location in the virion is shown in Figs 19.3, 19.4 and 19.5a. In brief, *gag* encodes the virion core proteins, *pol* the virion enzymes and *env* the virion envelope protein. All are derived from polyprotein precursors. Polyproteins from *gag* and *pol* are cleaved by the viral

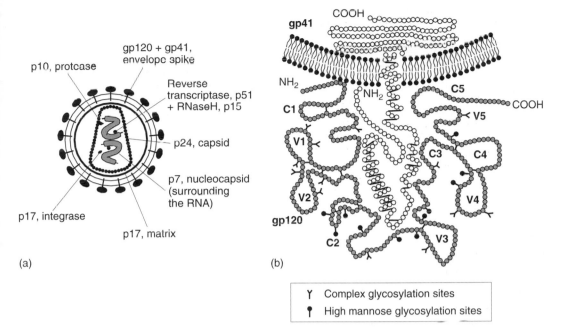

Fig. 19.5 (a) An HIV-1 virion with the major proteins identified. Others (not shown) are accessory proteins (see text). (b) A monomer of the trimeric envelope spike protein adapted from the structures suggested by Leonard *et al.* (1990) and Gallagher *et al.* (1989). Each circle represents an amino acid residue. The gp160 has been cleaved to form gp41 (open circles) with the carboxy-terminal transmembrane anchor, and the distal gp120 (shaded circles). The glycoprotein gp120 contains five antigenically variable (V) regions and five intervening conserved (C) regions; it is heavily glycosylated and the glycosylation sites are indicated in gp120; gp41 is also glycosylated. Intramolecular disulphide bonds in gp120 are also shown.

aspartyl protease (prot), itself a product of *pol*, and the polyprotein from *env* by a cellular protease—hence the importance of aspartyl protease inhibitors in the chemotherapy of HIV. The distal part of the envelope or spike protein, glycoprotein 120 (gp120) comprises five variable (V) loops that tend to vary in sequence. An outline of a monomer is shown in Fig. 19.5b, but the spike is actually a trimer. There are approximately 72 spikes per virion (which is far sparser than on the influenza virus, which has 10 times more spikes on a similar surface area). The glycoprotein gp41 is the transmembrane part of the glycoprotein and has an N-terminal hydrophobic fusion sequence which initiates infection by promoting the fusion of viral and cell plasma membranes. Both gp41 and gp120 have antibody neutralization sites, but much of their surface has evolved to evade antibody binding (see below).

Structure of gp120

A major barrier to progress in understanding antibody neutralization and in designing new antiviral drugs has been the lack of structural information on gp120. For over 10 years it resisted all attempts at crystallization, probably because the molecule was

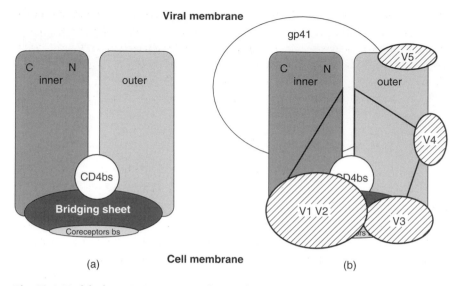

Fig. 19.6 Model of a gp120 monomer of HIV-1 based on the atomic structure determined by Kwong *et al.* (1998). (a) a simplified skeleton structure and (b) with V loops added. The model is orientated with gp41 and the virus membrane at the top of the page. The monomers meet in a trimer and obscure the inner face. The outer face is obscured with carbohydrate. Only the hatched surface of gp120 has epitopes that interact with neutralizing antibody (see text), although there are others on gp41. There is an atomic structure of an isolated gp41 ectodomain, but it is not known how this interacts with gp120. (From Burton & Parren 2000.)

too flexible. This was solved by removing the more flexible parts (parts of V1, V2 and V3, and carbohydrate moieties) and locking up the structure by binding it to a neutralizing Fab and part of the CD4 receptor. Figure 19.6a shows a skeleton version of a gp120 monomer. This forms a U-shaped hairpin with an inner and an outer domain, and the CD4-binding site, the coreceptor binding sites (see Section 19.3) and a bridging sheet at the bend of the 'U'. The CD4-binding site is a conformational region that allows the virus to attach to its T-cell host. In the fleshed-out version (Fig. 19.6b), it can be seen how the virus evades antibody. Loops V1–V3 lie on top, and partly obscure, the two binding sites and the bridging sheet, whereas most of the inner face will be obscured in the trimer. Much of the outer face is obscured by carbohydrate. In all, very little of gp120 is exposed to B cells, so that antibodies are not made to potential neutralization epitopes, and the few antibodies that are made have a very small target to aim at. The V loops that are exposed to antibody are able to sustain many amino acids changes and so virus evades the antibody that is made against them.

19.3 Transmission

CD4 is a protein present on the surface of T-helper cells and those of the monocyte–macrophage lineage. Its normal function is to recognize and interact with major histocompatibility complex (MHC) class II antigens on target cells. It is the main (but not the sole) receptor through which both HIV-1 and HIV-2 initiate infection. Successful HIV infection also requires the presence of the chemokine receptors, CCR5 or CXCR4, which serve as viral coreceptors (Section 19.4), and also bind to specific regions on gp120 (Fig. 19.6). People can contract an HIV infection in three different ways: sexually, via infected blood or by vertical transmission (from mother at birth).

Sexual transmission

The main route of infection is through the transmission of infected CD4$^+$ T lymphocytes or free virus during sexual activity. HIV is spread equally well by heterosexual and male homosexual activity, but is better spread by infected men than infected women. In Africa, HIV is spread predominantly by heterosexual contact. In Western countries, male homosexual contact has been responsible for most infections, but now the greatest percentage increase in new infections is the result of heterosexual activity. Bisexual activity provides a conduit between hetero- and homosexual people.

Transmission through infected blood

Injection using hypodermic needles contaminated with HIV provides a high risk of infection. The virus spreads quickly between injecting drug abusers who share unsterile equipment, and there is a similar risk with hospitals that do not have effective sterilization. Before the virus was recognized, some people with haemophilia were accidentally infected by being given clotting factor VIII prepared from blood contaminated with HIV. Heat sterilization and screening of blood donors have now eliminated this risk.

Vertical transmission

Babies born to HIV-positive mothers may be infected. Fortunately, there is only about a 20% chance of infection, but at this age AIDS develops with a much shorter incubation period of around 2 years. The risk is reduced by application of antiviral therapy (see Section 19.8). The exact route of transmission is not known but breast-feeding increases risk of infection, and is therefore not recommended,

19.4 Course of infection and disease

This section relates to people from the developed world. Although infection in developing countries is essentially similar, the course of infection may be faster as a result of other factors such as nutritional status and burden of infection. However, data are less complete. An average incubation period to AIDS of 10 years means that some will develop disease earlier and some will remain healthy for much longer. The course of infection in Fig. 19.7 refers to HIV antibody-positive people aged 13 or over (as opposed to babies where death ensues in around 2 years). Primary infection is a brief

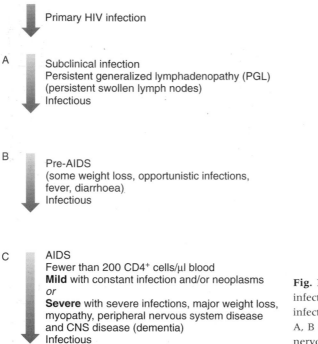

Fig. 19.7 The course of HIV infection, from primary infection through categories A, B and C. CNS, central nervous system.

Table 19.5 Control measures for HIV and AIDS.

Measure	Comment
Avoidance of infection	Available: but how well is it practised?
Diagnosis and detection	Good
Prevention by vaccine	Not available
Chemotherapy	Inhibitors of reverse transcriptase and viral protease: retard the progression to AIDS and reduce virus levels but do not eliminate it; is not a cure. Too expensive to be widely available

influenza-like illness and is followed by category A in which people are asymptomatic but have persistent generalized lymphadenopathy; this is an indication that immune suppression has started and is progressing. Category B people are symptomatic, presenting with a selection of conditions not found in category C, such as candidiasis, fever, or diarrhoea lasting more than 1 month, or more than one episode of shingles. This indicates that that there is a defect in cell-mediated immunity. Category C is AIDS and includes more severe infections or cancers (Table 19.5). The category of HIV infection and AIDS is formally defined by the number of CD4$^+$ T cells in the circulation, so that most of category A have >500 CD4$^+$ T cells/µl or per mm^3 of blood (the normal

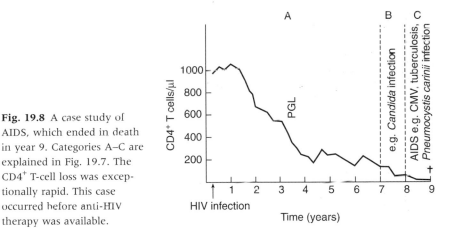

Fig. 19.8 A case study of AIDS, which ended in death in year 9. Categories A–C are explained in Fig. 19.7. The CD4⁺ T-cell loss was exceptionally rapid. This case occurred before anti-HIV therapy was available.

range being 800–1200/µl), most of category B have 200–500 cells/µl, and most of category C have <200 cells/µl. (To calculate numbers per ml, multiply by 1000.) If an AIDS patient does not die from infection with adventitious micro-organisms (Fig. 19.7), HIV progresses to infect CD4⁻ cells and causes disease in the muscles and the peripheral and central nervous systems. 'AIDS dementia' or collapse of brain function is the final stage. The time taken to progress to AIDS, and the symptoms involved, vary greatly between individuals. An example of one case study and the progressive decline in CD4⁺ T cells is shown in Fig. 19.8.

What cells are infected?

Viruses are classified as M tropic and T tropic. M-tropic viruses infect both CD4⁺ macrophages and T cells, and use the CCR5 coreceptor. T-tropic viruses preferentially infect CD4⁺ T-cell lines and use the CXCR4 coreceptor. These properties depend on the gp120 sequence. Most infections take place through the mucosal surface of the genital tract. It is thought that the virus binds to the surface of a dendritic cell, but does not enter or infect it. The dendritic cell then migrates carrying the virus to lymph nodes, where the virus is transferred to, and infects, an activated CD4⁺ T cell. CD4⁺ macrophages can also be infected but these are found infrequently in lymph nodes and blood. Although CD4⁺ T cells are the main target, other cells that do not express the CD4 protein are also infected, which suggests that here the virus can use a different receptor molecule. In the central nervous system, it is thought that HIV infects microglial cells, which belong to the same cell lineage as the macrophage. Infection via the mucosal surface of the rectum, resulting from anal intercourse, may also take place through dendritic cells. Unlike the simple oncornaviruses, HIV can infect resting cells, in this case resting CD4⁺ T cells, because its proviral DNA is transported into the nucleus through associated viral proteins that carry nuclear localization signals. There it is integrated, but no transcription takes place until the cell is activated by antigen.

During the initial/primary infection by HIV, virus-specific CD4$^+$ T cells are stimulated to proliferate by viral antigens. As virus infects and replicates in activated CD4$^+$ cells, these same cells are preferentially destroyed. At the same time, there is a dramatic expansion of virus-specific CD8$^+$ T cells which coincides with the suppression of viraemia. However, CD8$^+$ T-cell proliferation is dependent on CD4 T-cell help. Thus, there is a fine balance between virus destroying CD4$^+$ T cells and leaving enough CD4$^+$ T cells to help produce virus-specific, activated, CD8$^+$ cells. It is suggested that this balance determines the plasma virus load (concentration) at the end of the acute primary infection, and it is this virus load that determines the rate of progression to AIDS. It is further suggested that application of antiviral drug therapy (see Section 19.8) during the primary infection helps preserve HIV-specific CD4$^+$ T-cell function and hence reduce the ensuing plasma virus load.

During the intermediate stages of infection, there is relatively little virus and few infected cells in the plasma, and this led to the mistaken view that there was little virus replication at this time. Later it was found that most of the infected cells were in lymphoid tissue and large amounts of virus were being continuously released into the circulation. However, this was almost balanced by its efficient removal. In fact a healthy HIV-infected person produces a colossal 10^{10} virions per day. In addition there is a small, but vitally important, reservoir of latently infected, resting, CD4$^+$ memory T cells (10^3–10^4/person), and possibly other reservoirs, which cannot be shifted by drug therapy (see Section 19.8).

19.5 Immunological abnormalities

Although loss of CD4$^+$ T cells is the major result of HIV infection, there are many other alterations to immune cells and functions, some of which are listed in Table 19.6. Those in the first category are always found in AIDS patients and are therefore diagnostic. Others are found irregularly.

What causes the immunological abnormalities?

Immunological abnormalities result from one or more of the following:
1 The virus kills the infected cell.
2 Envelope protein expressed on the surface of an infected cell attaches to CD4 molecules on non-infected CD4$^+$ T cells and causes the cell membranes and cells to fuse and form a single cell (a syncytium); this is a lethal process and fused cells die.
3 Infected cells suffer no direct cytopathology but viral proteins expressed on the cell surface are recognized by antibody, and are then attacked by complement or phagocytic cells bearing Fc receptors (see Fig. 14.3).
4 Infected cells proteolytically process viral proteins and present peptide–MHC I or II protein complexes on their surface. These cells are then attacked by CD8$^+$ or CD4$^+$ cytotoxic T cells (see Chapter 14).

Table 19.6 Some immunological abnormalities in
AIDS*.

Diagnostic abnormalities

CD4+ T-cell deficiency
Reduction in levels of all lymphocytes
Lowered cutaneous delayed-type hypersensitivity
Non-specific elevation of immunoglobulin concentration
   in serum

Other abnormalities

Decreased proliferative responses to antigen
Decreased cytotoxic responses to all antigens
Decreased response to new immunogens
Decreased CD8+ T-cell cytotoxic response to HIV
Decreased macrophage functions
Production of autoantibody
Decreased natural killer cell activity
Decreased dendritic cell number and activity
Loss of lymph node structure

*Note that the effects of HIV on the immune system
extend far beyond the CD4+ T cell and vary greatly
between individual patients. Delayed-type
hypersensitivity is a T-cell-mediated reaction.

5 Non-infectious virus or free envelope protein that is released from infected cells
can attach to non-infected CD4+ cells and render those cells liable to immune attack,
either by antibody or, if the antigen is processed and presented by MHC proteins, by
cytotoxic T cells, as described above.

19.6 Why is the incubation period of AIDS so long?

The onset of AIDS is defined by CD4+ T-cell numbers dropping below a concentration
of 200 cells/μl of plasma. Thus the question can be rephrased in terms of the time
taken—10 years on average in the developed world—to reduce CD4+ cells to this level.
It is now clear that this is not caused by viral latency, because an HIV-positive person
is infectious at all times (see above). Figure 19.9 shows a simplified version of the
interrelationship of the circulating virus, T cells and overt clinical disease. After the
primary infection, the virus-specific, CD8+ T-cell response increases and virus reaches
a stable low-level plateau. After an initial fall brought about by infection, the CD4+ T-
cell population recovers. At some point years later, the onset of AIDS is announced
by a dramatic rise in circulating virus and fall in circulating T cells. The key event is
the reduction in CD4+ T cells to a level that makes them unable to sustain the immune
system. This permits the adventitious infections that will eventually kill the patient
(see Table 19.6). These data, and evidence from people who survive far longer than
the average 8 years of infection (long-term non-progressors) and from those on triple
therapy (see Section 19.8) indicate that the key to survival is the maintenance of a
low plasma virus concentration.

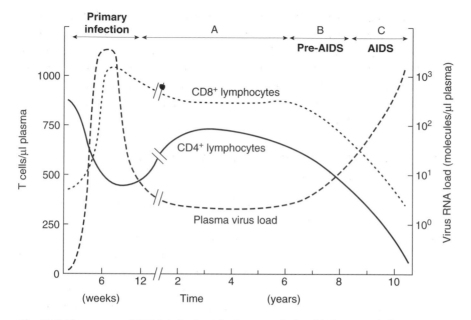

Fig. 19.9 The course of HIV-1 infection, the inverse relationship between infectious virus load/overt disease and T-cell concentrations in plasma. A, B, C refer to the categories of infection (see Fig. 19.7 and text).

Hypothesis 1: activation of virus and death of CD4⁺ cells is limited by the rate at which CD4⁺ cell clones are stimulated by cognate antigen

Important here is the fact that HIV multiplies only in dividing CD4⁺ T cells. In nature, T cells divide only when they are stimulated by the antigen that their unique T-cell receptor recognizes (termed the 'cognate antigen'), and many T cells never divide because they never meet the relevant cognate antigens, e.g. HIV will be latent in a T cell specific for antigen *x*, until the cell is stimulated to divide by contact with antigen *x*. Antigen *x*-specific cells will then produce a burst of virus and die by one of the mechanisms already discussed. Many more CD4⁺ cells will be infected by the new progeny HIV and the cycle repeats itself. Thus, the time taken to reduce the number of CD4⁺ cells to the level required for the definition of AIDS depends on the time it takes to come into contact with enough antigens to react with all those T cells. Taking this hypothesis to logical absurdity, if an HIV-positive person met no CD4⁺ cell-specific antigens, he or she would not develop AIDS.

Hypothesis 2: autoimmunity

The envelope protein of HIV-1 has a small sequence homology with the MHC II protein. Normally, the immune system recognizes MHC antigens as self and mounts no immune response to them or to the epitope that is mimicked by HIV. The idea here is that the long incubation period reflects the number of years that it takes for the self-tolerance mechanism to break down and the immune system to respond to the MHC II-like epitope on the HIV-1 envelope. Once that response is in place, the immune response will then react with the cross-reactive MHC II epitope, and cells bearing this antigen are then killed. CD4$^+$ T cells, CD8$^+$ T cells and B cells all express MHC II proteins constitutively, whereas other cells express them only after exposure to interferon-γ (see Section 14.5).

Hypothesis 3: evolution of the envelope protein in the infected individual

Sequencing of the gene encoding the envelope protein from isolates of HIV from different infected individuals shows that it can be divided into variable and constant regions (see Fig. 19.5b). These isolates also differ antigenically. Of particular importance is that the envelope protein evolves in infected *individuals*, and this is probably driven by the immune response, in a manner similar to antigenic drift in influenza virus (see Section 18.5). Evolution of viral proteins during the lifetime of an infected individual is also found in other lentiviruses. One view of the long incubation period is that infected cells are continuously removed by the immune system, and that the virus eventually gains the upper hand when it evolves a variant to which the immune system cannot respond. This virus then kills cells by one or more of the mechanisms discussed above. (This evolutionary plasticity of the envelope protein also gives rise to problems in regard to vaccines, which will be discussed later.)

19.7 Death and AIDS

The key failure in immune responsiveness is the loss of the helper function of the CD4$^+$ lymphocyte. This cell is the pivotal part of the immune system, because most antibody responses are T-helper-cell dependent and helper cells also assist in the maturation of T-cell effectors, such as CD8$^+$ cytotoxic T cells, and can have cytotoxic activity in their own right. One of the functions of T cells is to control a variety of micro-organisms (viruses, bacteria, fungi, protozoa—Table 19.7), a selection of

Table 19.7 Common infections associated with AIDS.

Organism	Infection
Viral	
Cytomegalovirus, Epstein–Barr virus	Generalized infection
Bacterial	
Mycobacterium tuberculosis	Tuberculosis
Fungal	
Candida albicans	Candidiasis (thrush)
Pneumocystis carinii (a yeast-like organism)	Pneumonia

which is carried by all individuals. The immune system is normally unable to evict these micro-organisms, but ensures that they remain subclinical and harmless. However, when the T cells are compromised by HIV infection, the brakes are released and the micro-organisms multiply at full speed. Infections are much more severe and prolonged than normal because of the immunosuppression. Different micro-organisms are activated in AIDS patients in different parts of the world. Tuberculosis is a common complication in developing countries. This period of chronic adventitious infection may last for several years, but eventually becomes overwhelming and death ensues.

19.8 Prevention and control

No other virus has been subjected to such intense scrutiny as HIV but, as yet, there is little sign of any effective vaccine. At first sight, this is surprising, because the small-pox vaccine has been in use for over 200 years and, in the last 30 years, some very effective vaccines have been devised and used. The problem with HIV is complex. Part of it lies in the fact that a micro-organism is sensitive only to one particular arm of the immune system (e.g. it might be the CD8$^+$ cytotoxic T cell), and other parts of the immune system are ineffective. A different micro-organism might be sensitive to, say, antibody. The rules governing immunity are not understood. Thus, all vaccines are made empirically. By definition, a successful vaccine stimulates the required immune response, along with many other irrelevant responses. However, we do not know which element of the immune response is required for protection in any disease or how to stimulate it to order. Further, there are a number of poor vaccines that have defied attempts to improve them (e.g. against cholera and typhoid). HIV is simply an extreme case. We need to know much more about the immune system in general before the rational design of vaccines becomes a reality. That day cannot come soon enough for HIV. The following sections deal initially with the vaccine problem and then discuss chemotherapy.

Prevention of HIV infection

In the absence of any vaccine, avoiding contact with HIV is paramount (see Table 19.5). The main risk, already discussed, is through sexual contact. Education programmes have been promulgated worldwide, with advice to have as few sexual partners as possible, and to practise safer sex (whether heterosexual or homosexual) by avoiding the exchange of, and contact with, body fluids and using condoms as a barrier to infection. However, there are immense social problems in dealing with any sexually transmitted disease, and these vary according to local conventions around the world. Historical parallels are not encouraging, because it is well documented that many were undeterred by the risk of contracting syphilis at a time when no treatment was available. The involvement of male and female prostitutes is a particular problem, because, in some areas of the world, a high proportion are infected with HIV.

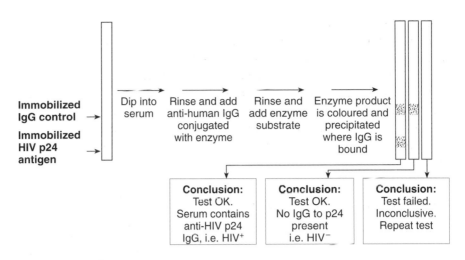

Fig. 19.10 The dip-stick test for antibody to HIV.

Diagnosis and detection

Control of any disease needs quick, cheap and reliable methods for detecting signs of infection. These are in place. Possession of serum *antibody* to the HIV internal antigen, p24, is the main diagnostic criterion, because this protein provokes the strongest and most reliable antibody response. This test decides if a person is 'HIV positive'. It is used as a 'dip-stick' in an enzyme-linked immunosorbent assay (ELISA) format (Fig. 19.10). The test is not carried out until 2 months after the suspected infection, to allow antibody to form and avoid a falsely negative diagnosis. With babies born to HIV-positive mothers, a period of 3–6 months has to elapse to allow the decay of maternal HIV antibody in the infant's circulation.

It is not easy to isolate HIV-1, although many strains can be grown in CD4$^+$ cells in the laboratory. This is a slow and labour-intensive process and has the intrinsic problem of all such techniques, that of selecting variants from the infected person which happen to grow well in culture but which are unrepresentative of the original virus population. However, this can be circumvented by use of RT-PCR (reverse transcription polymerase chain reaction) directly on viral genomes obtained from a patient. A positive result is the amplification of a DNA molecule of a known size, as seen after gel electrophoresis. This amplified DNA can be sequenced directly to give accurate information on the virus present in the body.

Chemotherapy

Chemotherapy of HIV infections has two mains aims: (i) to prevent the progression of infection to AIDS, and (ii) to clear the infection completely. The first objective has been achieved through the advent of the treatment known as HAART which now usually means triple drug therapy. This commenced in 1994, and under the best

conditions can reduce plasma virus to undetectable levels. However it does not completely restore the CD4$^+$ T-cell population and function, and does not clear virus from
resting CD4$^+$ memory cells in which the virus is latent (and possibly other reservoirs).
Also even a few days without drugs sees a rapid rebound to normal virus levels.
Rebound virus may be drug resistant and can be passed on with catastrophic results.
Thus, chemotherapy must be continuous. It is calculated that, with the natural
turnover of the pool of resting CD4$^+$ memory cells, it will take 10–60 years for the
latently infected cells to disappear. Triple therapy is prohibitively expensive for all but
the richest countries and is not being used where it is most needed, in Africa and Asia.
The cost in developed countries is approximately $US7 000 per person per year,
whereas the *total* health budget in some African countries is less than $US10 per
person per year.

Triple therapy is so called because a mixture of three anti-HIV-1 drugs is used simultaneously. These inhibit synthesis of proviral DNA from viral RNA or inhibit the action
of the virus protease (see Fig. 19.3). The former are either nucleoside analogues or
non-nucleosides, such as nevirapine. They target different sites on the enzyme. The
best known is the deoxythymidine analogue, AZT, also known as zidovudine. This is
3′-azido-2′,3′-dideoxythymidine (Fig. 19.11). Its antiretrovirus properties were dis-

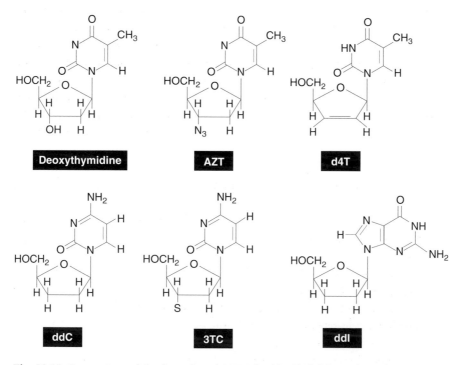

Fig. 19.11 Comparison of the formulae of AZT (3′-azido-2′,3′-dideoxythymidine) and
deoxythymidine, and the formulae of some other nucleoside analogues that are also
inhibitors of HIV DNA synthesis (see text). Note that all analogues shown lack the 3′-
hydroxyl group, and so act as chain terminators. All require phosphorylation by cellular
enzymes before being incorporated into DNA. (See text for commercial names.)

covered before the HIV emergency, but at that time there was no known human retrovirus infection. The nucleoside analogues are incorporated into viral DNA and, because they all lack a 3'-hydroxyl group (see deoxythymidine, Fig. 19.11), no further base can be added and DNA synthesis ceases. Other analogues with similar function (such as stavudine (d4T; deoxythymidine), lamivudine (3TC) and zalcitabine (ddC; deoxycytidine), and didanosine (ddI; deoxyguanosine)—see Fig. 19.11) have since been discovered. The protease inhibitors were tailor-made after the atomic structure of the viral protease had been determined. Examples are saquinavir, indinavir, ritonavir and nelfinavir. Triple therapy usually uses two nucleoside inhibitors and one protease inhibitor. The rationale for using triple therapy to avoid the appearance of resistant mutants was covered in Section 16.8. An additional advantage of triple therapy is that the inhibitors are synergistic, meaning that the overall protection is greater than the sum afforded by the drugs individually. Thus, the amount of each drug can be decreased, reducing both the inevitable toxicity and the very high cost.

Vaccines

Animal model systems

As it is ethically not possible to test the efficacy of a vaccine or a treatment by infecting people with a potentially lethal agent, vaccine trials can only be made in people whose lifestyle places them at high risk from infection (see below) or in model systems. The latter have limitations, as only primates have a CD4 protein sufficiently closely related to the human CD4 protein to permit infection with HIV, but almost no primate infected with HIV-1 progresses to AIDS (Table 19.8). The use of higher primates, such as chimpanzees, is fraught with many problems and, aside from ethics, they are exceedingly expensive to maintain and there will never be enough for statistically meaningful experiments. Many experiments use simian immunodeficiency virus (SIV)

Key points on triple therapy for HIV infection

- Uses three drugs simultaneously; these bind to different parts of the viral reverse transcriptase or protease molecules.
- Reduces virus load in plasma.
- Restores lost immune functions.
- Retards progression to AIDS.
- Has no effect on latently infected CD4$^+$ memory cells.
- Lapses result in rebound to normal virus levels.
- Is not tolerated by all people.
- Has to be taken for the foreseeable future (up to 30 pills each day).
- Is expensive (approximately $US7000 per person per year).
- Has only been in use since 1994.
- Can it be tolerated for the foreseeable future?

Table 19.8 Some animal models for HIV infection and AIDS, and their limitations.

Virus	Animal	Comment
HIV-1, -2	Chimpanzee	Infection but no AIDS; used to test vaccines
HIV-2	Rhesus monkey	Infection and lymphadenopathy— similar to early stages of AIDS
SIV	Macaque monkey	Infection and AIDS
SHIV	Macaque monkey	Infection and AIDS
FIV	Cats	Infection and AIDS
HIV-1	SCID-hu mouse	Infection, no AIDS

SIV, simian immunodeficiency virus; SHIV, a simian virus that has been engineered to express the envelope protein of HIV-1; FIV, feline immunodeficiency virus; SCID-hu mouse, mice with severe combined immunodeficiency, i.e. in their homozygous form, lack both B- and T-cell responses. These can be reconstituted with human (hu) lymphocytes (see text).

and Asian monkeys as a model of the human situation. (SIV is subclinical in its native African monkey host.) Closer to HIV is SHIV, an SIV in which the envelope gene has been replaced by that of HIV and which retains the capacity to infect and cause disease in monkeys. This allows the activity of HIV envelope-specific immunity to be assessed.

A major breakthrough came when immunologists decided to take mice that were congenitally deficient in both B and T cells (known as severe combined immunodeficient (SCID) mice) and reconstitute them with human lymphocytes. By conventional immunological wisdom, this experiment was a waste of time, because the transplanted cells should have mounted an immune response against mouse antigens in a classic 'graft-vs-host' reaction. For some reason, not understood, this did not happen and it is now possible to study infection of human lymphocytes and their reaction to experimental vaccines. However, human T cells do not repopulate the mice in their normal proportions and the mice do not develop AIDS. The alternative approach is to study other non-human immunodeficiency lentivirus–host systems (see Table 19.3) and hope to apply lessons learned from these to HIV infections.

Immunity to HIV

Infected people and infected animals mount both antibody and T-cell responses that are antiviral, and the appropriate laboratory assays can demonstrate the presence of anti-HIV neutralizing antibody and cytotoxic T cells. Why, then, do HIV-positive people (i) not clear the virus and (ii) go on to develop AIDS? The first question has many precedents, e.g. hepatitis B and herpes simplex viruses (already discussed in Sections 15.5 and 15.6), and viruses have a variety of ways of evading immune responses. It needs to be remembered that antibody and T cells cannot get into cells and thus act only on antigens that are exposed on the outside of the cell membrane. Also, although antibody can recognize almost any molecule, T cells recognize only peptide fragments that are complexed with MHC proteins and expressed on the cell surface. Antibodies

Table 19.9 Some of the ways in which the HIV-1 virion evades the stimulation and action of the antibody response.

Both gp120 and gp41 are extensively glycosylated, and this masks much of the viral protein

The gp120 has hypervariable (V) loops that vary under immunological pressure without losing essential functions

The variable loops, V1 and V2, partly mask the binding site for the primary virus receptor, CD4, and with V3 mask the coreceptor binding site

As well as varying, the V3 loop attracts the immune response away from other epitopes that are less able to vary, and has been described as a 'decoytope'

The CD4 binding site is a recessed cavity which makes it difficult for an antibody paratope to contact it

The binding sites are invariant regions that are intrinsically unable to sustain amino acid replacements. However, the CD4bs cavity is lined with hydrophilic residues that make no contact with CD4 and are subject to mutation

The infected cell releases large amounts of monomeric gp120, which has highly immunogenic surfaces that are occluded in the mature trimer by monomer–monomer interactions. Antibodies made to these surfaces are non-neutralizing, and monomeric gp120 is regarded as an immunological decoy. Thus, the mature HIV-1 envelope protein is non-immunogenic, and stimulates antibodies of a specificity that poses no danger to the virus

---

See Fig. 19.6 for details of the gp120–gp41 envelope protein.

which bind to epitopes that mediate neutralization can render the virus non-infectious. Both antibodies and activated cytotoxic T lymphocytes (CTLs) can kill infected cells bearing their cognate antigens in their different ways (see Chapter 14). However, the HIV-infected person presumably does not make antibody to the right epitope, or enough of this antibody, or antibody of high enough affinity to deal with the infection. There is the same sort of problem with T cells. HIV-1 has a large number of ways in which it evades the host's immune responses, and some of these that relate to antibody are listed in Table 19.9.

The discussion so far focuses on people who are already infected. Although therapeutic vaccines aim to boost the immune responses of those already infected, the main aim of immunization is to put in place immune responses that will prevent HIV from initiating infection. Regrettably, this has not so far proved to be possible.

Failure of experimental vaccines to protect against HIV infection

All vaccines in use to date have been produced empirically, i.e. they have been made without knowledge of the exact immune response(s) that were needed to control a particular infectious disease. In the case of HIV, many of the types of vaccine discussed

Table 19.10 Different types of experimental HIV vaccine.

Type of vaccine	Comment
Killed (inactivated) vaccine	Work ongoing on the problem of totally inactivating virulent virus; more work on immune responses is needed. Also problem of cost and no production of non-structural proteins
Live attenuated vaccine (mostly with SIV with deletions in *nef*)	Attenuated for adult monkeys but not for neonates. Induces variable protection in adults against infection and/or disease after challenge by virulent virus. Infected animals have a persistent infection, which eventually reverts to virulence, usually by repair/partial repair of the deletion
Soluble recombinant gp120	Simulates antibodies but these are poorly neutralizing and not crossreactive; non-neutralizing epitopes are dominant
Soluble recombinant gp120 trimers	New in 2000; should expose only external neutralizing epitopes; not yet tested for induction of immune responses
Expression of gp120 by recombinant vaccinia virus	Often used in prime-boost mode (inject with immunogen and boost with recombinant virus); has given protection of primates
Peptides and various protein expression systems	Poor neutralizing antibody and T-cell responses
DNA immunization with plasmids encoding gp120 and other proteins	Newest concept. Systemic and mucosal immune responses are being evaluated; has protected chimpanzees

in Sections 16.1–16.7 have been investigated: inactivated whole virus, purified viral protein, recombinant protein, recombinant protein expressed in a vaccinia virus vector, peptides and DNA immunization; there has also been intensive work on attenuated vaccines, but so far only with SIV (Table 19.10). All stimulate some part of the immune system, but none appears reliably to stimulate an immunity—or enough of it—that gives protection. Responses are often transient. There is individual variation in immune responses, and antibody responses are rarely crossreactive with other HIV-1 strains. It is putting it mildly to say that the scientific world is surprised by this universal lack of success—it was expected that at least one of the preparations would have been protective. Vaccinologists are now, for the first time, forced to follow a logical approach, to determine what types of immune response(s) protect against HIV and how these can be stimulated, but it is taking a long time to do this.

Production of a protective vaccine is one of the major goals of the worldwide campaign to prevent AIDS, and an international collaborative scientific effort, with large amounts of scientific time and money, is bent on achieving this goal. It is generally agreed that the aim is to stimulate both virus-specific antibody and T-cell responses. Antibody work has concentrated on the envelope protein, which carries the only neutralization epitopes, and the best way of presenting it to the immune system. T-cell immunity can, however, be directed against peptides from any HIV protein, but it is not yet clear which is the best candidate. One experimental vaccine consists of a number of known T-cell peptide epitopes from different HIV proteins joined together by chemical synthesis.

Early on it was realized that systemic immunity (immunity in the blood and the body core) would not protect against HIV infection of the genital or rectal mucosae. Such naked epithelial surfaces, which are used as the route of entry for many different pathogens, have a local mucosal immune system that has to be stimulated by the direct application of the vaccine (see Fig. 14.2). Local cognate B and T lymphocytes then migrate to the draining lymph node and, after undergoing activation, return to mucosal surfaces, and form a barrier to infection. Mucosal vaccines often stimulate systemic immunity but not vice versa. There is increasing evidence that mucosal immunity may be crucial in preventing HIV infection.

Various human HIV vaccine trials have been conducted in HIV-negative people, and are continuing to be undertaken (see box). By the end of 1998, there have been 25 phase I trials, 4 phase II trials, and 2 phase III trials in the USA and Thailand; others are planned and include Africa. The cost and logistic problems are enormous, and many doubt that there is sufficient evidence to suggest that the vaccine candidates in use are likely to be successful in phase III trials. However, the political pressure to get 'real vaccine work' under way cannot be underestimated.

Key points on clinical trials for HIV vaccines or drugs

Phase I: safety and immunogenicity testing; 8–12 months duration; 10–30 people with a low risk of contracting HIV infection.

Phase II: further safety and immunogenicity testing, with variation of dose, route of administration and sample population; 18–24 months duration; 50–500 people with low to higher risk of contracting HIV infection.

Phase III: efficacy trial for the prevention of infection in high-risk volunteers (protective vaccine) in which progress is compared with people who receive a placebo vaccine; takes a minimum of 3 years but it is difficult to determine the end-point; therapeutic vaccines for the treatment of HIV patients could eventually be tested. Several thousand people. Cost of one trial about $US25 million.

HIV is susceptible to neutralizing antibody

Many hundreds of HIV-1-specific monoclonal antibodies have been produced and, although many do not neutralize infectivity, some neutralize quite well and a few neutralize very well indeed. The reader will already be familiar with the fact that the gp120 monomer has a preponderance of non-neutralizing epitopes (see Table 19.10), but there are other complications lying in wait in the neutralization of HIV. Virus that has been adapted to grow in laboratory T-cell lines is relatively easy to neutralize, but primary virus strains (isolated from an infected person) requires 100-fold higher concentrations of the same antibody for significant neutralization. It is well known that adaptation of any virus to growth in cells in the laboratory is associated with selection of the best-suited virus variants, and this is well illustrated here. The other problem is clade variation (see Section 19.1), because the immunity stimulated by any future vaccine should protect against as many of the different clades of HIV-1 as possible. In all, there are just three monoclonal antibodies (MAbs) so far described that neutralize primary viruses from different clades: these are the gp120-specific MAbs b12 and 2G12, and the gp41-specific MAb 2F5. All are human antibodies, which shows that our immune systems do have the ability to produce such antibodies, so now we only have to learn how to tell the immune system to do this. However, each has been isolated only once from infected individuals, and none of the three antibodies can be raised by immunization of humans or any animal even by non-infectious preparations of envelope protein, which carry the cognate epitope in a recognizable form. It is not known why.

Neutralization, although important, is a laboratory phenomenon, but what is vital is the ability to protect the whole animal or person. This can be tested by injecting preformed antibody into an animal or a person. Here HIV and chimpanzees or SCID mice reconstituted with human lymphocytes, and SHIV or SIV and monkeys have been used. (This is called *passive transfer* of antibody or *passive immunity* because the animal does not actually make the antibody.) The animals are then challenged with infectious virus by intravenous or genital routes, and in this way it was shown that the MAbs were capable of protecting the animal against HIV-1 infection. This was an important experiment as it demonstrated that the *right sort of immunity* was able to prevent HIV infection.

Key points about HIV-specific antibodies

- Primary virus isolates are intrinsically difficult to neutralize.
- There are at least nine virus clades (A–I) that differ in sequence and antigenicity of their gp120–gp41 envelope protein, hence…
- Different clades have very few epitopes in common that mediate efficient neutralization.
- Three monoclonal antibodies (MAbs) are known that neutralize primary viruses from different clades.
- The same three MAbs protect against infection in model systems *in vivo*.
- Mucosal immunity (locally produced) may be essential for preventing infection.

Any successful vaccine would need to stimulate all three of the above antibodies or antibodies similar to them, otherwise neutralization escape mutants would soon be selected if only one was stimulated. Escape mutants arise with the same frequency as any other mutant, and the arguments above for using triple therapy apply equally to antibodies. One advantage of having all three MAbs present is that they neutralize synergistically and hence less of each is needed, again as described for chemical inhibitors. Presumably, the binding of one antibody alters the conformation of the protein so that a second antibody can bind more easily.

Future vaccines

Today, the vaccine problem is being attacked on a number of different fronts. Empirical work is still in progress and investigates the almost infinite permutations of immunization protocols in terms of type of vaccine, amount inoculated, site(s) of inoculation, adjuvant, number of doses and the interval between them. The main aim, however, is directed at achieving a fundamental understanding of the immune system. This is a huge endeavour and inevitably slow, but will in the end provide the information necessary to devise an anti-HIV vaccine as well as establishing principles that will enable vaccines to be made to many other problem diseases. A future vaccine will probably consist of a mixture of different immunogens, each of which stimulates a different part of the immune response, and this will surely include both antibody and T-cell responses. New adjuvants are being used and the ability of certain cytokines to enhance immune responses is being evaluated. Mucosal responses are likely to be essential in preventing the establishment of infection. Whatever the form of any new vaccine, it will need to be inexpensive so that, unlike current drug therapy, it will be available to developing countries where it is so desperately needed.

19.9 The cost of HIV infection

Every country has a finite amount to spend on health care, and AIDS is making substantial inroads into this budget at a time when healthcare costs are generally rising. Cost is a major issue in any chronic disease where treatment has to be provided over a number of years, and the cost in human terms of any AIDS case is incalculable. As a result of the high incidence of infection, the immense loss of life in central and east Africa is devastating the countries of those regions. Demographic changes are predicted in a manner not seen since the time of the Black Death in Europe in the fourteenth century. There is every indication now that HIV is advancing through Asia on a scale similar to that in Africa, and will result in social and economic disaster unless preventive measures are rapidly put in place. This is not helped when some countries refuse to admit that they have an HIV problem.

19.10 Further reading and references

http://www.unaids.org/epidemic_update

Ada, G. (2000) HIV and pandemic influenza virus: two great infectious disease challenges. *Virology* **268**, 227–230.

Aldrovandi, G. M., Feuer, G., Gao, L., Jamieson, B., Kristeva, M., Chen, I. S. Y. & Zack, J. A. (1993) The SCID mouse as a model for HIV-1 infection. *Nature (London)* **363**, 732–736.

Balfour, H. H. (1999) Antiviral drugs. *New England Journal of Medicine* **340**, 1255–1268.

Birmingham, K. (2000) UN acknowledges HIV/AIDS as a threat to world peace. *Nature Medicine* **6**, 117.

Collins, K. L. & Baltimore, D. (1999) HIV's evasion of the cellular immune response. *Immunological Reviews* **168**, 65–74.

Cullen, B. R. (1991) Human immunodeficiency virus as a prototypic complex retrovirus. *Journal of Virology* **65**, 1053–1056.

Essex, M. (1999) Human immunodeficiency virus in the developing world. *Advances in Virus Research* **53**, 71–88.

Gallagher, W. R., Ball, J. M., Garry, R. F., Griffin, M. C. & Montelaro, R. C. (1989) A general model for the transmembrane proteins of HIV and other retroviruses. *AIDS Research and Human Retroviruses* **5**, 431–440.

Garcia-Blanco, M. A. & Cullen, B. R. (1991) Molecular basis of latency in pathogenic human viruses. *Science* **254**, 815–820.

Gotch, F. & Hardy, G. (2000) The immune system: our best antiretroviral. *Current Opinion in Infectious Diseases* **13**, 13–17.

Gotch, F., Hardy, G. & Inami, N. (1999) Therapeutic vaccines in HIV-1 infection. *Immunological Reviews* **170**, 173–182.

Kwong, P. D., Wyatt, R., Robinson, J., Sweet, R. W., Sodroski, J. & Hendrickson, W. A. (1998) Structure of an HIV gp120 envelope glycoprotein in complex with the CD4 receptor and a neutralizing human antibody. *Nature (London)* **398**, 648–659.

Landau, N. R. (1999) Recent advances in AIDS research: genetics, molecular biology and immunology. *Current Opinion in Immunology* **11**, 449–450.

Lehner, T., Bergmeier, L., Wang, Y., Tao, L. & Mitchell, E. (1999) A rational basis for mucosal vaccination against HIV infection. *Immunological Reviews* **170**, 183–196.

Leonard, C. K., Spellman, M. W., Riddle, L., Harris, R. J., Thomas, J. N. & Gregory, T. J. (1990) Assignment of intrachain disulfide bonds and characterization of potential glycosylation sites of the type 1 recombinant human immunodeficiency virus envelope protein (gp120) expressed in chinese hamster ovary cells. *Journal of Biological Chemistry* **265**, 10373.

Levy, J. A. (1998) *HIV and the Pathogenesis of AIDS* (2nd edn). Herndon, VA: ASM Press.

McMichael, A. (1998) Preparing for HIV vaccines that induce cytotoxic T lymphocytes. *Current Opinion in Immunology* **10**, 379–381.

Parren, P. W. H. I., Moore, J. P., Burton, D. R. & Sattentau, Q. (1999) The neutralizing antibody response to HIV: viral evasion and escape from humoral immunity. *AIDS* **13**, S137–S162.

Pillay, D., Taylor, S. & Richman, D. D. (2000) Incidence and impact of resistance against approved antiretroviral drugs. *Reviews in Medical Virology* **10**, 231–253.

Ruprecht, R. M. (1999) Live attenuated AIDS viruses as vaccines: promise or peril? *Immunological Reviews* **170**, 135–149.

Sattentau, Q. J., Moulard, M., Brivet, B., Botto, F., Guillemot, J.-C., Mondor, I., Poignard, P. & Ugolini, S. (1999) Antibody neutralization and the potential for vaccine design. *Immunology Letters* **16**, 143–149.

Stevenson, M., Bukrinsky, M. & Haggerty, S. (1992) HIV-1 replication and potential targets for intervention. *AIDS Research and Human Retroviruses* **8**, 107–117.

Stott, E. J., Hu, S.-L. & Almond, N. (1998) Candidate vaccines protect macaques against primate immunodeficiency viruses. *AIDS Research and Human Retroviruses* **14**, S265–S270.

Tudor-Williams, G. & Lyall, E. G. H. (1999) Mother to infant transmission of HIV. *Current Opinion in Infectious Diseases* **12**, 27–52.

Also check Chapter 22 for references specific to each family of viruses.

Chapter 20
Prion diseases

20.1 Introduction

This chapter deals with an intriguing group of diseases that affects a variety of animals, including humans. These diseases are known collectively as *spongiform encephalopathies* because of the characteristic pathology that they display in regions of the brain. Historically, these were known as slow virus diseases after the demonstration of an infectious basis for the sheep and goat disease, *scrapie*, the first identified disease of this type, and in recognition of the fact that the disease course was prolonged. However, it has proved impossible to identify any nucleic acid-containing entity associated with this infectivity and instead there is a considerable body of evidence which ascribes infectivity in these diseases to a protein. This notion, the *prion hypothesis* (which includes the concept of a replicating protein), is contrary to the standard dogma of molecular biology but has now gained widespread, although not universal, acceptance. Its principal champion, Stanley Prusiner, has received the Nobel prize for his work on this subject. As the agents causing scrapie and other related diseases are probably not viruses, it could be argued that they do not belong within the scope of this text. However, these diseases are the subject of widespread interest and concern as threats to public health after an epidemic of *bovine spongiform encephalopathy* (BSE) in the UK in the 1980s and 1990s, and cases of BSE in other European countries, and so merit some consideration.

20.2 The spectrum of prion diseases

A number of diseases of animals and humans are now believed to be caused by prions. These are summarized in Table 20.1. All show a characteristic pathology in the central nervous system, with parts of the brain becoming vacuolated or spongy, in many cases also with extensive deposits of extracellular protein fibrils or plaques, but with no sign of inflammation, i.e. no invasion of immune cells. Different brain regions are affected by the various spongiform encephalopathies (Fig. 20.3 shows an example). For most of the diseases listed, the presence of infectivity has been shown by transmission of disease to an experimental animal (typically a mouse or hamster) by intracranial injection of a homogenate of diseased tissue and subsequent serial passage to further mice or hamsters. Hence, these diseases have been termed *transmissible spongiform*

Table 20.1 Possible prion diseases in the date order that transmissibility was demonstrated.

Disease	Occurrence	Host species	Date
Scrapie	Common in several countries throughout the world	Sheep, goats	1936
Transmissible mink encephalopathy (TME)	Very rare, but adult mortality rate nearly 100% in some outbreaks	Mink	1965
Kuru	Once common among the Fore-speaking people of Papua New Guinea; now rare	Humans	1966
Creutzfeldt–Jakob disease (CJD)	Occurs in iatrogenic, familial and sporadic forms. The latter has uniform worldwide incidence of 1 per million per annum	Humans	1968
Gerstmann–Sträussler–Scheinker (GSS) syndrome	An inherited disease; <0.1 per million per annum	Humans	1981
Chronic wasting disease (CWD)	Colorado and Wyoming, USA	Mule-deer, elk	1983
Bovine spongiform encephalopathy (BSE)	Europe, principally UK	Cattle	1988
Feline spongiform encephalopathy (FSE)	UK	Domestic cat	1991
Fatal familial insomnia	An inherited disease; very rare	Humans	1995
Variant Creutzfeldt–Jakob disease (vCJD)	UK, France	Humans	1997

encephalopathies (TSEs). Characteristically, all the TSEs have long incubation periods (e.g. 1 year for a mouse; 4–6 years for a cow). For the human TSEs, the age at onset of symptoms depends on the disease type. Sporadic classic Creutzfeldt–Jakob (CJD) symptoms start at a mean age of 60 years whereas symptoms of inherited forms emerge somewhat earlier (45 years). One of the defining features of variant CJD (vCJD) (see Section 20.5) is its early age at onset, typically <30 years. All TSEs are invariably fatal.

20.3 The prion hypothesis

As noted in the introduction, the infectious agents in the TSEs are widely held to be proteins. What evidence has led many scientists to accept this *prion hypothesis*?

First, there are data that argue that the TSE agents cannot be conventional viruses. Although the scrapie agent passes through filters that admit only viruses and can be titrated in mice (often reaching concentrations of $>10^7$ infectious units/ml), it is much more resistant than typical viruses to inactivation by heat, radiation and chemicals such as formaldehyde. By measuring the rate at which infectivity is destroyed by radiation, the size of any nucleic acid genome present in the agent has been estimated as equivalent to at most 250 nucleotides (nt.) of single-stranded (ss) nucleic acid or 125 nt. of double-stranded (ds) nucleic acid. For comparison, the genome of a plant pathogenic viroid comprises 359 nt. of ssRNA and encodes no protein, and some defective–interfering animal viruses have single-stranded genomes of only 250 nt. and also encode no protein. From this information, the agent could at best be argued to be viroid like, encoding none of its own proteins. This is the *virino hypothesis*. However, other data (discussed below) make it unlikely that this hypothesis is correct.

Second, the agent is not susceptible to nucleases and no unique nucleic acid molecule, even of the minimal size defined by radiation inactivation, has been found to co-purify with infectivity. These results argue against both the virus and the virino hypotheses.

Third, there are positive data supporting the prion hypothesis. Infectivity co-purifies with a protein, PrPsc, which forms the protein plaques seen in some TSE-affected brain tissues, and from its sequence this protein is clearly a host-encoded, abnormal form of the product of the *prnp* gene (or the *sinc* gene—scrapie *inc*ubation time—in mice). In uninfected brain (and other tissues), the normal protein PrPc is present as a glycoprotein on the cell surface. PrPsc has the same sequence as PrPc, but a very different tertiary structure which renders it relatively resistant to protease digestion. Small amounts of PrPsc can catalyse the conversion of PrPc molecules into the PrPsc structure in the test tube. This result offers a replication mechanism for prions in an infected animal (Fig. 20.1), where the substrate for the production of new infectious agent (prion replication) is provided by the host, and its conversion into the infectious form is driven by an incoming molecule already in the altered structural form. In short, the infectious aberrant form is thought to act as a template on which existing normal protein molecules are converted into new altered structures, which precisely resemble the original infectious agent and are themselves infectious. Supporting this model, transgenic mice that have been engineered to lack all PrPc expression are completely resistant to experimental TSE. Further support comes from the properties of mutant *prnp* genes (see Section 20.4). However, it should also be noted that *in vitro* conversion of PrPc to PrPsc has not yet been shown to generate any new infectivity—a key prediction of the model.

Fourth, this novel replication mechanism has gained support from the study of two genetic traits in yeast—*PSI*+ and *URE3*—which show all the properties that would be expected from being determined by prions, i.e. the particular phenotypes result from the presence of aberrant conformations of two normal proteins—Sup35p and Ure2, respectively. Interestingly, a molecular chaperone (a protein that catalyses changes in the conformation of other proteins) has been shown to be involved in generating the *PSI*+ trait in a previously negative (*psi*−) cell population, and in subsequent curing of the trait from that population. Mutant forms of the normal proteins have also been shown to increase the rate of conversion to give the abnormal phenotypes.

Given this evidence, why is the prion hypothesis not universally accepted? Aside from natural scepticism about such a novel concept, there are some observations that the prion hypothesis struggles to accommodate. The most significant of these is scrapie

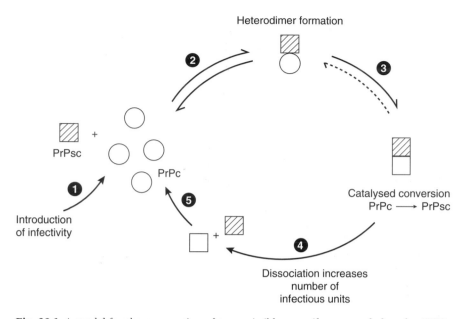

Fig. 20.1 A model for the propagation of transmissible spongiform encephalopathy (TSE) infectivity according to the prion hypothesis. PrPc (circles) and PrPsc (squares) represent the normal and an abnormal conformation, respectively, of the *prnp* gene product. The initiating infectivity is distinguished by shading, but is proposed to have the same structural features and properties as the progeny PrPsc molecules (although see discussion of the species barrier, Section 20.6). Steps 2 and 3 in the propagation cycle may be reversible. Alternative models differ principally in suggesting that polymerization of the altered structural form is important to its potential for catalysing further structural conversions.

strain variation. The scrapie agent has been isolated in mice, and several different strains have been defined by their very precise, but different, incubation periods, and by the severity and distribution of the lesions that they cause in the central nervous system (CNS). Mouse-adapted scrapie strains can then be passaged on to hamsters where, after crossing the species barrier (see Section 20.6), they again show reproducible and discrete incubation periods (Fig. 20.2). The incubation periods in mice are also host strain specific and are controlled by the *sinc* gene (see above). Thus, in a single inbred mouse strain, multiple scrapie strains can be serially passaged while retaining distinct biological properties. If the scrapie agent had a nucleic acid component, this phenomenon would be easily explained as the result of sequence differences between strains of the agent. For the prion hypothesis to explain these data, it must be argued that the protein product of the *sinc* gene of a particular mouse strain, which has a specific and defined amino acid sequence, can be caused to adopt multiple distinct aberrant conformations by encounter with molecules of different scrapie strains. Moreover, the conformation produced in each case must be a replica of the infecting prion, so that the unique pathogenic properties of that agent are faithfully passed on to the progeny molecules, which will mediate infection upon

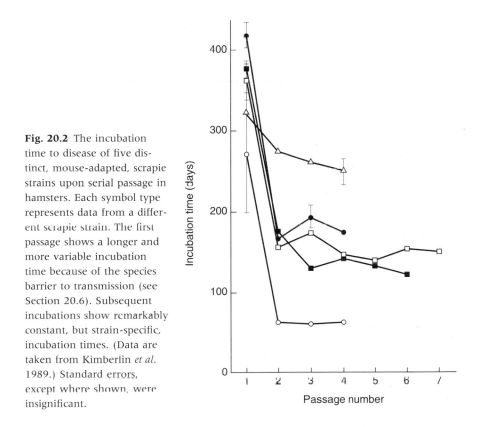

Fig. 20.2 The incubation time to disease of five distinct, mouse-adapted, scrapie strains upon serial passage in hamsters. Each symbol type represents data from a different scrapie strain. The first passage shows a longer and more variable incubation time because of the species barrier to transmission (see Section 20.6). Subsequent incubations show remarkably constant, but strain-specific, incubation times. (Data are taken from Kimberlin *et al.* 1989.) Standard errors, except where shown, were insignificant.

subsequent passage. This idea is difficult to reconcile with our understanding of how stable protein folding is achieved by a nascent polypeptide, although the properties of human *prnp* mutants (see Section 20.4) suggest that multiple PrP conformations are possible.

20.4 The aetiology of human prion diseases

The first TSE to be identified was scrapie, a natural infection of sheep, which takes its name from the tendency of diseased animals to scrape themselves against fence posts, presumably to relieve itching of the skin. The disease is usually spread maternally but can also be spread horizontally and its infectious basis was further confirmed by transmission into experimental animals (see Section 20.3). There is also clear evidence for an infectious basis to other prion diseases of animals, such as BSE (see Section 20.5).

Among the human prion diseases listed in Table 20.1, kuru clearly has an infectious basis. It was spread by ritual cannibalism and, since its identification, its incidence has declined along with this practice. *Iatrogenic* (meaning related to medicine) CJD also shows infectious aetiology. It has arisen through various medical treatments involving transfer of certain tissues/tissue extracts between individuals, e.g. the treatment of dwarfism caused by pituitary growth hormone deficiency with hormone extracted from cadavers led to a number of cases of CJD, through inadvertent use of material from individuals harbouring inapparent prion disease at the time of death.

On the other hand, Gerstmann–Sträussler–Scheinker syndrome (GSS), fatal familial insomnia (FFI) and familial CJD are autosomal dominant inherited genetic disorders whereas sporadic CJD, as the name suggests, occurs at a random low frequency with no known cause. How can these be accounted for within the prion hypothesis? For the inherited diseases, the explanation lies in the discovery of specific mutations in the *prnp* gene, each associated with a distinct disease phenotype, e.g. replacement of glutamic acid by lysine at position 200 results in CJD whereas replacing proline with leucine at position 102 results in GSS. There are also other specific mutations that can be associated with these diseases in other individuals. In addition to these pathogenic mutations, the *prnp* gene also shows polymorphism, most notably at position 129, where alleles may encode either methionine or valine with either giving a normal phenotype. Interestingly, this polymorphism has a dramatic influence on disease in the context of pathogenic *prnp* mutations. Thus, almost all the disease resulting from the specific CJD and GSS mutations just mentioned occurs in individuals in whom methionine residues are found at position 129 on both the normal and the mutated *prnp* alleles. Even more remarkably, replacing aspartic acid at position 178 with asparagine causes FFI when linked with methionine 129, but CJD when linked with valine 129 (Fig. 20.3) and in each case, disease onset and progression are exacerbated when the normal allele carries the same amino acid at the polymorphic position 129 as does the mutated allele. All this suggests that subtle differences in the

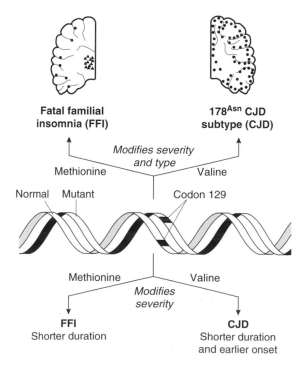

Fig. 20.3 The interaction between pathogenic mutations at position 178 of human PrP and the amino acid present at the polymorphic position 129 on either the mutant or normal allele which determine disease pathology, time of onset and rate of progression. ●, regions of spongiform degeneration; ▲, regions of neuronal loss and replacement with astrocytes and glial cells. See text for further details. (Reproduced from Gambetti 1996, Fig. 1; with permission from Springer-Verlag.)

structure of PrP can affect the time course and nature of disease, as the prion hypothesis predicts.

Changes in the sequence of PrPc that are linked with disease are proposed to destabilize the protein so that spontaneous conversion into a PrPsc form inevitably occurs, after which replication of the aberrant structure can take place as already described. Again, the precise nature (shape and structure) of this altered form must determine the disease phenotype. Supporting this idea, mice made transgenic for the human GSS mutant *prnp* gene spontaneously develop disease. Following the same reasoning, sporadic CJD is thought to occur when a molecule of the normal PrPc undergoes this same spontaneous conversion to initiate replication of the relevant PrPsc form. In the absence of a destabilizing mutation, this spontaneous conversion would be a rare event, accounting for the low frequency of sporadic CJD.

20.5 Bovine spongiform encephalopathy (BSE), scrapie and vCJD

Origins of the BSE epidemic

BSE (popularly known as 'mad cow disease') appeared suddenly in dairy cattle in the UK in the 1980s, and was formally recognized in 1986. It developed into a large-scale epidemic, with serious economic repercussions, and did not show any sign of abating until late 1993, cases continue to appear at the time of writing (2001). There are two related theories to explain this epidemic. Either BSE resulted from the scrapie agent of sheep adapting to cattle, or an unrecognized sporadic case of BSE was propagated into the British cattle herd. In either case, transmission of the infection was made possible by the practice of giving cattle artificial food concentrates that contained certain tissue residues of slaughtered sheep and cattle. Scrapie has been endemic in sheep in the UK for nearly 300 years and the practice of feeding animal protein to cattle was not new, so if scrapie was the progenitor of BSE, what was the cause of its sudden transmission to cattle? It is likely that this resulted from an abrupt change in the treatment of animal carcass waste products between 1980 and 1983, during which the proportion of meat-and-bonemeal cattle food produced by extraction of tallow with organic solvents decreased by nearly 50%. This treatment had involved processing abattoir products for 8 h at 70°C and then the application of superheated steam to remove traces of solvent. Both would have been effective in inactivating scrapie, or indeed unrecognized BSE, infectivity. Such extreme measures are needed because TSE infectivity is immensely stable, far more so than other infectious agents (see Section 20.3). So the BSE epidemic probably had a single origin in either a scrapie-infected sheep carcass or a cattle carcass infected with unrecognized sporadic BSE that was incorporated into cattle feed. Infection was then propagated by the continued recycling of infected cattle residues from the abattoir in cattle food concentrates.

BSE has now been transmitted experimentally into mice and different isolates appear to be remarkably homogeneous with regard to incubation period in a given mouse strain (Fig. 20.4a), i.e. they behave as a single strain. Each of four mouse strains

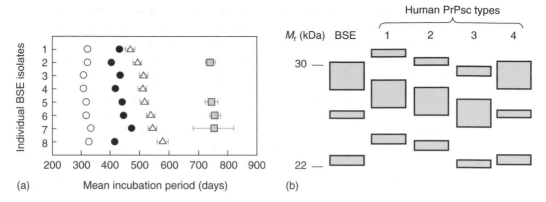

Fig. 20.4 The biological and molecular properties of different bovine spongiform encephalopathy (BSE) isolates transmitted to mice are very similar to each other, and to those of variant Creutzfeldt–Jakob (vCJD) isolates. (a) The incubation times of BSE isolates from cattle on transmission to mice. Each line on the figure represents a different BSE isolate and the four types of symbol show the mean incubation time for that isolate in four different strains of mouse. (Reprinted with permission from *Nature* (Bruce *et al.* 1997). Copyright 1997 Macmillan Magazines Limited.) (b) Molecular signatures of TSE isolates. A schematic representation of an immunoblot analysis of polypeptides from the brains of mice infected with different transmissible spongiform encephalopathy (TSE) isolates that have been separated by sodium dodecylsulphate polyacrylamide gel electrophoresis (SDS-PAGE). The bands represent protease-resistant core fragments of PrPsc, differing in length and glycosylation pattern, which have been detected with antibody to PrPsc. The thickness of each band represents its intensity in the original analysis. Human PrP types 1, 2 and 3 are seen in various sporadic and iatrogenic CJD isolates; the type 4 pattern is seen in variant CJD isolates and is very similar to the pattern consistently generated by BSE isolates in this type of analysis. (Reprinted from *Current Opinion in Genetics and Development* **9**, Wadsworth *et al.*, Molecular biology of prion propagation, pp. 338–345, Copyright 1999, with permission from Elsevier Science; using original data from Collinge *et al.* 1996.)

tested shows a different characteristic BSE incubation time. Different BSE isolates also show similar brain pathology on experimental transmission and similar molecular features (number and location of glycosylation sites and the size of the protease-resistant core of PrPsc), which give each isolate a similar molecular 'signature' (Fig. 20.4b). In contrast, scrapie strains differ widely in all these properties. Although most scrapie isolates have signatures that distinguish them from BSE, one that shows similar properties has now been identified.

BSE and the emergence of vCJD

Recognition of the BSE epidemic led to changes in practice, i.e. a ban on using meat and bonemeal in feed to prevent further spread in cattle and a ban on the human consumption of specified bovine offals (brain, spinal cord and lymphoid tissues—see Section 20.6) that were considered potentially infectious. However, it was recognized that people had been eating potentially contaminated meat products for a number of years before the recognition of BSE as a disease and the implementation of these precautions. There were also other risks to be considered, e.g. calf serum is widely used in the pharmaceutical industry for the growth of cultured cells to make virus vaccines and other cattle byproducts are used in cosmetics.

The risk of transmission of BSE through the food chain was considered to be low (but not zero) because scrapie had been endemic in the UK sheep population for nearly three centuries and there was no evidence of transmission to humans, despite the number of sheep consumed. Experimental evidence from studies of scrapie at the time also showed that transmission between animals of different species was very much harder than between two members of the same species, the so-called *species barrier*. Nevertheless, the disease transmissible mink encephalopathy (TME—see Table 20.1) had previously emerged in captive mink through feeding them scrapie-contaminated sheep and the emergence of feline spongiform encephalopathy (FSE) in domestic cats in the early 1990s, presumably as a result of consuming contaminated food, suggested that this species barrier could also be breached by the BSE or scrapie agent.

This view of the risk to humans from BSE was changed dramatically in 1996 by the recognition of a novel human disease, now known as variant CJD. This differed somewhat from previously described CJD (however acquired) in its clinical characteristics, but strikingly also in the age of its victims. Variant CJD patients are typically much younger (< 30 vs. average 60 years) than are sporadic CJD patients. The evidence that vCJD represents infection by the BSE agent comes from studies of transmission to mice. Transmission from either BSE or vCJD samples occurred with very similar incubation times and with similar characteristic pathogenic signatures in the brain, features that were very different from transmissions of classic CJD carried out at the same time. The molecular signatures of the BSE and vCJD agents are also similar (Fig. 20.4b).

So far, more than 100 people have been diagnosed with vCJD. All the cases so far have been homozygous for methionine at the polymorphic position 129 in their *prnp* genes. This polymorphism is already known to influence disease in the context of pathogenic *prnp* mutations (see Section 20.4). The annual incidence of new cases of vCJD, after being stable for several years, has perhaps begun to show a gradual increase. However, because the average incubation time for the disease is not known, it is not yet possible to say what the eventual size of the epidemic will be. There may also be differences in the susceptibility of individuals to infection. Only about 40% of the population are homozygous for methionine 129, the group so far found to be susceptible, although whether other genotypes are completely resistant to infection by BSE or simply show a longer incubation time cannot yet be determined.

20.6 TSE pathogenesis

If we accept the prion hypothesis, there are some very intriguing issues that arise about the pathogenesis of the various TSEs. Sporadic and inherited TSEs might arise directly from pathogenic conversion of PrP being initiated within the CNS (see Section 20.4). Some cases of iatrogenic CJD, which have arisen through direct introduction of infectivity into the CNS, e.g. by grafting of contaminated dura mater (the covering of the brain) during brain surgery or grafting of contaminated corneas, could similarly arise through propagation of infectivity from the site of exposure through the CNS. Experiments with PrP^0 transgenic mice (i.e. they make no endogenous PrP) carrying a graft of PrP$^+$ brain tissue show that expression of PrP in the CNS is essential if pathogenesis is to be seen, leading to the idea that, within the CNS, infectivity is propagated through the tissue from the site of infection by the progressive conversion of normal PrP to a pathogenic form.

Other cases of iatrogenic CJD have come from peripheral exposure to infectivity, e.g. intramuscular injection of contaminated pituitary growth hormone, whereas kuru, BSE and vCJD are, with varying degrees of certainty, known to have been transmitted through ingestion of contaminated food. How, in all of these varying circumstances, does infectivity reach the CNS? Information on this subject has again come mostly from studies of parenteral infection in various transgenic mouse lines. These show that elements of the immune system are crucially important to amplify the infectivity and transmit it on to the CNS. Indeed the spleen has long been known as a site of replication of these agents (Fig. 20.5). Both B cells and follicular dendritic cells (FDC) have been implicated in TSE agent peripheral replication and, possibly, different types of TSE require different immune cell types for the agent to penetrate the CNS. Amplification of infectivity in FDC seems to be crucial if disease is to be manifested in the brain and the FDC must, as predicted from the prion hypothesis, be expressing PrP for them to fulfil this role.

A further aspect of pathogenesis concerns the species barrier. Experimentally, this is observed as a very much greater difficulty (lower frequency of transmission and longer duration to overt disease) in transmitting infection between species than within a species (see Fig. 20.2). This has been thought to reflect the need for structural similarity (ideally identity) between the infecting prion and the endogenous PrPc for efficient structural conversion of the latter protein to a prion. Thus, human prions

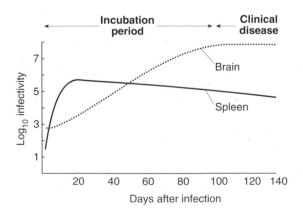

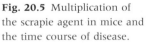

Fig. 20.5 Multiplication of the scrapie agent in mice and the time course of disease.

transmit disease to standard laboratory mice very poorly, more easily to mice carrying transgenic copies of the human *prnp* gene in addition to their mouse *prnp* genes, and most easily to mice carrying the human gene but not the mouse gene. The disease-enhancing effects of homozygosity at position 129 (see Section 20.4) may reflect a similar effect. However, animals that have failed to 'take' infection in cross-species transfers can harbour considerable infectivity, albeit without apparent disease, and this calls into question this simple molecular view of the species barrier.

20.7 Conclusions

The recent epidemic of BSE, and its transfer into the human population as vCJD, has ignited an explosion of research interest in the TSEs. However, there remain several significant unanswered questions about them. First, it is unclear why the accumulation of a protein in an aberrant conformation should cause the mass death of neurons and spongiform degeneration of the brain. Second, it is unclear how the different pathologies associated with each human prion disease, which include different effects on specific brain areas in each case, can be generated through the structural conversion of the same normal protein. In other words, why does conversion of PrP to one form damage area A of the brain and spare area B, whereas conversion to another form damages area B but not A when the substrate for conversion, normal PrPc, is present in both areas? Finally, what is the normal function of PrP, aberrations in which seem to have such disastrous consequences? Prevention and cure for TSEs will surely depend on a much fuller understanding of these diseases.

20.8 Further reading and references

Andrews, N. J., Farrington, C. P., Cousens, S. N., Smith, P. G., Ward, H., Knight, R. S. G., Ironside, J. W. & Will, R. G. (2000) Incidence of variant Creutzfeldt–Jakob disease in the UK. *Lancet* **356**, 481–482.

Belay, E. (1999) Transmissible spongiform encephalopathies in humans. *Annual Review of Microbiology* **53**, 283–314.

Bruce, M. E., Will, R. G., Ironside, J. W. *et al.* (1997) Transmissions to mice indicate that 'new variant' CJD is caused by the BSE agent. *Nature* **389**, 498–501.

Büeler, H., Aguzzi, A., Sailer, A., Greiner, R.-A., Autenreid, P., Aguet, M. & Weissman, C. (1993) Mice devoid of PrP are resistant to scrapie. *Cell* **73**, 1339–1347.

Chesebro, B. (1998) BSE and prions: uncertainties about the agent. *Science* **279**, 42–43.

Collinge, J., Sidle, K. C., Meads, J., Ironside, J. & Hill, A. F. (1996) Molecular analysis of prion strain variation and the aetiology of 'new variant' CJD. *Nature* **383**, 685–690.

Dormont, D. (1999) Bovine spongiform encephalopathy and the new variant of Creutzfeldt–Jakob disease. In: *Prions: Molecular and cellular biology*, Harris, D. A. (ed.), pp. 177–191. Wymondham: Horizon Scientific Press.

Fraser, H., Bruce, M. E., Chree, A., McConnell, I. & Wells, G. A. H. (1992) Transmission of bovine spongiform encephalopathy and scrapie to mice. *Journal of General Virology* **73**, 1891–1897.

Gambetti, P. (1996) Fatal familial insomnia and familial Creutzfeldt–Jakob Disease: a tale of two diseases with the same genetic mutation. *Current Topics in Microbiology and Immunology* **207**, 19–25.

Hill, A. F., Antoniou, M. & Collinge, J. (1999) Protease-resistant prion protein produced *in vitro* lacks detectable infectivity. *Journal of General Virology* **80**, 11–14.

Kimberlin, R. H., Walker, C. A. & Fraser, H. (1989) The genomic identity of different strains of mouse scrapie is expressed in hamsters and preserved on reisolation in mice. *Journal of General Virology* **70**, 2017–2025.

Mabbott, N. A., Farquhar, C. F., Brown, K. A. & Bruce, M. E. (1998) Involvement of the immune system in TSE pathogenesis. *Immunology Today* **19**, 201–203.

Pattison, J. (1998) The emergence of bovine spongiform encephalopathy. *Emerging Infectious Diseases* **4**, 390–394.

Prusiner, S. B. (1996) Molecular biology and pathogenesis of prion diseases. *Trends in Biochemical Sciences* **21**, 482–487.

Prusiner, S. B., Scott, M. R., DeArmond, S. J. & Cohen, F. E. (1998) Prion protein biology. *Cell* **93**, 337–348.

Tuite, M. F. (2000) Yeast prions and their prion-forming domain. *Cell* **100**, 289–292.

Wadsworth, J. D. F., Jackson, G. S., Hill, A. F. & Collinge, J. (1999) Molecular biology of prion propagation. *Current Opinion in Genetics and Development* **9**, 338–345.

Weissman, C. (1999) Molecular genetics of transmissible spongiform encephalopathies. *Journal of Biological Chemistry* **274**, 3–6.

The UK Creutzfeldt–Jakob Disease Surveillance Unit web site: www.cjd.ed.ac.uk

The UK Government BSE web site: www.maff.gov.uk/animalh/bse (this site may shortly become part of the site: www.defra.gov.uk as a result of UK government changes)

The World Health Organization TSE web pages: www.who.int/emc/diseases/bse

Chapter 21
Trends in virology

What will be the trends in virology in the near future? Clearly, prevention of virus disease in humans, domestic animals and crops is the major aim of virologists. For human and animal virus diseases, the priorities are improvements in vaccines and the search for antiviral compounds. Our relative success in treating human immunodeficiency virus (HIV) in the developed world has underlined our impotence in combating the HIV pandemic in the developing world, and the need for affordable treatments and vaccines. Plant viruses are no less important, because there are serious diseases of virtually all crops, whether grown for fruit, vegetable, grain, beverage or fibre. Such viruses often do not kill the affected host but they can reduce its vigour, leading to reduced yield.

There are interesting developments involving the insertion of disease 'resistance' genes through DNA technology, which extend the traditional breeding programmes to obtain virus-resistant cultivars. Meanwhile, the provision of virus-free seeds and plants through elimination of persistent infection remains an important weapon in the armoury of the grower. Recombinant DNA technology and the polymerase chain reaction (PCR) that followed soon after, and the advent of monoclonal antibodies (MAbs) very rapidly fulfilled expectations and truly changed the face of virology, as well as other branches of the biological sciences and medicine. All continue to reach out into both applied and academic aspects of virology. We are continually surprised by the speed and precision with which these technologies are able to tackle a variety of fundamental and fascinating problems, particularly in allowing investigation of non-cultivable viruses. Other trends will see the application of modern methods to the analysis of virulence and this, no doubt, will feed back to help in the preparation of better preventive measures. Further, there is now concern that the appearance of new virus diseases is an inevitable consequence of evolution, replacing the mood of self-congratulation for the successful conquest of smallpox and containment of measles and poliomyelitis. The immense problem of HIV has already been discussed in the preceding chapter. Just as serious is the ever-present possibility of a new, possibly lethal pandemic of influenza. Meanwhile other emerging virus diseases continue to make their appearance.

21.1 Virulence—the major unsolved problem

There is greater ignorance about virulence than there is of any other area in the virology of eukaryotes. Although we can describe what happens during a virus disease, we have very little notion of the molecular events that distinguish otherwise identical virulent and avirulent strains. There can be no doubt that, whenever their nucleic acids are compared by sequence analysis, differences will be found (see below), but as there is more than one nucleotide change the essential problem is to correlate which of these controls virulence, and then to understand in molecular terms how the virulent virus strain interacts with its host to produce a violent end-result, whereas the avirulent strain does not. The analysis is complicated because virulence is the end-product of the very complex reactions of the animal or plant host to the infectious agent and this, by its essence, makes experiments done in cultured cells irrelevant. This is a monumental problem but progress is slowly being made.

Reversion of poliovirus vaccine to virulence

There are three poliovirus serotypes and immunity to each is required for protection against poliomyelitis. Accordingly the vaccine is trivalent. The avirulent type 3 component of poliovirus live oral vaccine is known to revert to virulence through mutation, with the result that there is a very, very low incidence of vaccine-associated paralytic poliomyelitis (approximately one case per 10^6 doses of vaccine for the first dose and diminishing to zero by the third dose). Viruses isolated from such patients are neurovirulent when inoculated into monkeys in the test normally applied to ensure the safety of new batches of vaccine. Sequencing the 7431 nucleotides of poliovirus has enabled the research groups headed by Jeffrey Almond, then at Leicester, and Philip Minor in London to determine which changes in sequence correlate with the reacquisition of virulence.

In all, there were just seven nucleotide changes, one in each of the terminal non-coding regions and five in the coding regions; of the latter, one is silent (at position 6034) and four result in amino acid changes in structural proteins (Fig. 21.1b). Are all or only a few of these mutations responsible for reversion to virulence? Two approaches to answering this problem were used. First, the wild-type Leon strain of type 3 poliovirus, from which the vaccine strain was derived by Albert Sabin, was cloned and sequenced (Fig. 21.1a). This gave the interesting information that only 10 mutations had occurred in the attenuation process (which, incidentally, was achieved empirically by adaptation of the virus to growth in cell culture at 31°C). There were two mutations in the 5′ non-coding region, one in the 3′ non-coding region, three leading to amino acid changes and four silent mutations in the coding regions. However, the main point is that there was only a single nucleotide change common to the acquisition of avirulence and the subsequent reversion to virulence, at residue 472 in the 5′ non-coding region from cytosine to uridine and back to cytosine. The implication that this was the change responsible for reacquisition of virulence was strongly backed up by the finding that the 472 U → C change occurred in all known cases of vaccine-associated illness.

This is exciting stuff, but the final story has not yet been told, because, first, residue 472 is in the non-coding region so it is not obvious how its change affects virus function. However, it is known that the 472 U → C change greatly alters the secondary

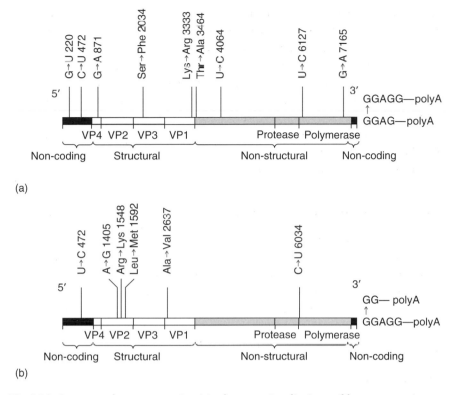

Fig. 21.1 Sequence changes occurring (a) when type 3 poliovirus wild-type neuroviru-lent Leon strain was attenuated to form the current Sabin vaccine, and (b) when the vaccine strain reverted to the neurovirulent 119 isolate. (Reproduced with permission, from the *Annual Review of Microbiology*, Volume 41 ©1987 by Annual Reviews www.AnnualReviews.org.)

structure of the viral RNA. Exactly how this affects virulence is not known, but virus with 472 U has reduced replication and translation, and decreased interaction of its RNA with a cellular RNA-binding protein in neuronal cells, but not in cells of non-neuronal origin. Second, the mutation at 472 increases but does not fully restore neu-rovirulence to that of the original Leon strain. It seems that one or more other residues contributes to virulence. In anatomical terms, the avirulent vaccine virus multiplies solely in the epithelial cells of the small intestine, whereas virulent virus invades the central nervous system (CNS). It remains to be explained how the 472 mutation is responsible for this difference in cell tropism.

For further information see Almond (1987), Racaniello (1988) and Gutiérrrez *et al.* (1997).

21.2 Subtle and insidious virus–host interactions

When it infects an animal, a virus is faced with a variety of differentiated cells which display a great range of structures, physiological functions and biochemical equipment.

However, an individual virus infects only certain cells and it is the specificity of this interaction that can be unexpected and interesting.

Destruction of specialized cells after infection

In Section 12.1, viruses are defined as acutely cytopathogenic if they kill the cells that they have infected. However, *in vivo*, viruses only infect (and only sometimes kill) cells of their target organ or target tissue. This can be exquisitely specific, as demonstrated by those viruses that have an affinity for the β cells of the pancreas, which produce insulin, the hormone responsible for the storage of glucose as glycogen. If the β cells are destroyed, the animal becomes diabetic. In humans, diabetes is common (occurring at the rate of 1–2/1000) and is mainly an autoimmune disease. However, the condition of 'juvenile diabetes' occurs suddenly in previously healthy children, and it is tempting to speculate that virus infection of the β-pancreatic cells is responsible. Unfortunately, several different viruses are suspected of being able, on occasion, to cause such effects (e.g. in animal models, representatives of the picornaviruses and reoviruses), so the prospects of reducing the incidence of diabetes by immunization are remote. In practice, it is found that repeated infections have an additive effect on the destruction of β cells and are necessary to produce clinically apparent diabetes.

Loss of cellular 'luxury' functions following infection

In Sections 12.1 and 12.2, viruses are classified as acutely cytopathogenic or persistent, with the tacit assumption that the latter coexist peacefully with their host cell. Recent data show this to be an over-simplification, because viruses can alter the production of specialized cellular products (termed 'luxury' functions), while leaving the everyday 'housekeeping' functions of the cell unharmed. In one experimental system, Michael Oldstone and colleagues from La Jolla, California, demonstrated the ability of the arenavirus, lymphocytic choriomeningitis virus (LCMV), to stunt the growth of newborn mice. Fluorescent antibody staining located viral antigen in many tissues, but, significantly, it was present in the pituitary, where growth hormone is synthesized. The pituitary showed no sign of pathology when examined microscopically, but a link between LCMV infection and growth retardation was established when virus was found only in those pituitary cells that make growth hormone. Growth hormone has major effects on growth and glucose metabolism, and infected mice die prematurely of severe hypoglycaemia within 3 weeks after infection. Transplantation into infected mice of growth hormone-producing cells ensures both their survival and their normal development (Fig. 21.2). Although it is not known how the virus causes the reduction of growth hormone, it is more likely to be affecting its synthesis than its structure or function. Michael Oldstone has found that the LCMV can also switch off insulin without damaging the β cells of the pancreas. Armed with this knowledge, it seems profitable to re-examine human conditions characterized by hormone, cytokine or neurotransmitter deficiencies to see whether viruses are affecting the activity of relevant cell types. Such cells would appear normal in all respects save for the loss of a

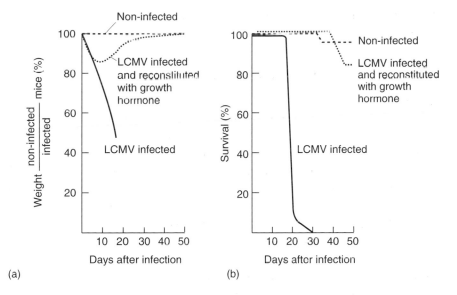

Fig. 21.2 Stunting of growth (a) and death (b) in newborn mice infected by the Armstrong 1371 strain of lymphocytic choriomeningitis virus, and reversal of the trend by transplantation of cells secreting growth hormone. (b) Without hormone replacement, the mice die. (Reprinted with permission from *Nature* (Oldstone *et al.* 1984a). Copyright 1984 Macmillan Magazines Limited.)

'luxury' function—luxury for the cell, maybe, but not for the unfortunate person, of course.

For further information see Notkins (1984), Oldstone *et al.* (1984a,b), Oldstone (1989).

Viruses and behavioural changes

There are examples of viruses causing changes in the normal behaviour of infected animals and none is more striking than rabies virus. This rhabdovirus is usually transmitted by bites, which transfer infected saliva, but also by inhalation of virus in airborne particles. After infection, the virus slowly ascends the peripheral nerves on its way to the spinal cord and brain. The time between infection and the appearance of signs of disease can be up to several months and is proportional to the distance that the virus has to travel. Getting bitten in the face gives the shortest incubation period. The initial stages of disease in feral animals are characterized by their loss of fear of humans, and this progresses through violent rabid behaviour, during which infectious virus can be transferred by biting, to death.

Much more subtle are the effects of Borna disease virus, a single-stranded (ss) negative-sense RNA virus and the type member of the new bornavirus family, which produces slowly progressive disease in a number of vertebrate model systems. Infection of the tree shrew (*Tupaia glis*), a primitive primate, causes a variety of behavioural changes, including hyperactivity and spatial and temporal disorientation. In breeding pairs, the normal division between the active aggressive male and more passive female behaviour becomes blurred. The normal pattern of sociosexual activity, which is a necessary preamble to successful mating, is so disturbed that breeding no longer occurs, although the animals remain in good physical health. Such changes are all consistent with alterations to neurotransmitters, already referred to above. Animals are permanently affected and can thus be rendered effectively sterile as a result of infection with Borna disease virus.

More insidious is the evidence that Borna disease virus is present in the human population. At first the evidence depended on the presence of virus-specific antibodies or antigens, then the isolation of virus-specific sequences by reverse transcription polymerase chain reaction (RT-PCR), but now the infectious virus itself has been isolated—although it is technically difficult. The virus has also been isolated from people with psychiatric disorders, but this does not necessarily mean that the virus is the causal agent. Nevertheless, the parallel with non-human disease suggests for the first time that a virus may alter human behaviour.

The examples given illustrate some of the subtle interactions that occur between host and virus and leave wide open for investigation the possibility that viruses may be the prime causative agent in a variety of human diseases of unknown aetiology. Suspicion that progressive pathological conditions affecting the CNS, such as multiple sclerosis (MS—see Section 21.8) and Alzheimer's disease, involve virus infections has too often raised false hopes, but there is still no reason to deny the possibility.

For further information see Lipkin *et al.* (1995), Bode and Ludwig (1997) and Nakamura *et al.* (2000).

21.3 Cofactors in virus diseases

We have learned something of the complexity of virus–host interactions in Chapters 14 and 16, but there are an increasing number of situations where infections or diseases involve a third party, an agent that modulates the interaction between virus and host.

Immunosuppression by malaria potentiates viral carcinogenesis

It has long been suspected that viruses require the presence of a cofactor to cause a particular disease, but evidence has come to hand only recently. Epstein–Barr virus (EBV) plays a crucial role in Burkitt's lymphoma by immortalizing B lymphocytes. These form a tumour of the lymphoid tissue of the jaw, which is found in high incidence in children, particularly boys, aged 6–9 in tropical Africa and New Guinea. However, EBV occurs throughout the world and causes infectious mononucleosis (glandular fever), and the presence of EBV-specific antibody in nearly all young people by the age of 20 indicates the ubiquitous nature of the infection. Thus there was *de facto* evidence that a cofactor was responsible for the lymphoma, and epidemiological evidence from Dennis Burkitt himself suggested that it was malaria. How do malaria and EBV interact? EBV, as all herpesviruses, causes a lifelong persistent/latent infection which is kept in check by the immune system, including neutralizing antibody and cell-mediated immunity. People who are immunocompromised cannot control the virus and develop a fatal proliferation of EBV-immortalized cells. It has been demonstrated that malaria drastically lowers the control of EBV-infected cells by EBV-specific T lymphocytes, by decreasing the proportion of helper T cells in relation to suppressor T cells. One attack by the malarial parasite, *Plasmodium falciparum*, is not likely to be sufficient, but repeated attacks in areas where malaria is endemic reduce the immune response to the level needed to establish a tumour. Exactly how malaria achieves this immunosuppression is not known, and an added complication is a likely genetic predisposition of the unfortunate victim; in addition boys, as already mentioned, have a higher incidence of Burkitt's lymphoma than girls.

For further information see Epstein (1984).

Enhanced transmission of viruses by mosquitoes infected with nematodes

Rift Valley fever virus (a bunyavirus) is an important pathogen of cattle and sheep, which is transmitted by the mosquito *Aëdes taeniorhynchus*. The virus is ingested with a blood meal from a viraemic animal, but is not passed on unless it gets to the insect's salivary gland. From there, it is injected with salivary anticoagulant into the blood of the next victim. Normally, this occurs in only 5% of mosquitoes but, if they feed on animals which, in addition to Rift Valley fever virus, are infected with a microfilarian nematode (a microscopic parasitic 'worm'), transmission of virus is increased to 30%. It appears that the microfilariae burrow out of the gut and enhance the passage of virus to the salivary gland. In addition, the presence of microfilariae allows mosquitoes exposed to very low titres of virus to transmit infection.

As many other bunya-, alpha- and flaviviruses are transmitted by biting arthropods, it will be important to discover whether microfilariae promote infection in other situations or there are other examples of synergism between viral and other parasites.

For further information see Turell *et al.* (1984).

21.4 Virus infection can be an evolutionary advantage

Certain parasitic wasps (of the *Ichneumonidae* and *Braconidae*) are infected with polydnaviruses which replicate in the ovary. The female wasp deposits eggs by injection into caterpillars, which are simultaneously infected by virus carried along with the eggs. Normally, the eggs hatch out and the wasp larvae feed on the living caterpillar until it dies, whereupon the wasp larvae pupate and emerge in due course as new adult wasps. If eggs from the female wasp are separated from virus by centrifugation and injected artificially into caterpillars they fail to develop. On examination, the eggs are found to have been overcome by the caterpillar's immune system. However, if purified virus is injected together with eggs, normal development takes place (Fig. 21.3).

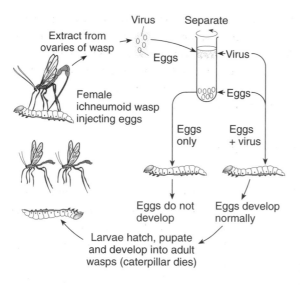

Fig. 21.3 Eggs of the ichneu-moid wasp develop in their host caterpillar only when injected together with a polydnavirus.

Thus, virus is needed to suppress (by some unknown mechanism) the immune responses of the caterpillar, and we have the situation where a virus has become essential to the successful life cycle of its eukaryotic host. We do not know if other viruses bestow evolutionary advantages upon their hosts, but this example stimulates us to turn the normal virus–host relationship on its head and look for positive aspects of infection. Who knows, one day it may be proved that some virus infections are good for you!

For further information, see Turrell *et al.* (1984).

21.5 The eradication and control of viral diseases

The euphoria that followed the successful elimination of smallpox (variola) virus has been replaced by a realization that other viruses may not yield so easily (see Section 16.4). Although eradication remains the ultimate goal, it is apparent that not all viruses (e.g. those with non-human hosts as well) lend themselves to this process. Containment by the better application of existing resources is perhaps more realistic, at least in the short term. However, even in the wealthy countries, there are still enormous problems, ranging from devising a new vaccine when none currently exists (e.g. against HIV), to improving existing vaccines (e.g. against influenza) and, even where there is an effective vaccine, to persuading people to use it. For instance, in the UK in recent years the take-up rates in children for measles vaccine have been 50–60%, rubella vaccine 80–90% and poliomyelitis vaccine 70–80%. In the developing world, problems of finance, communication, education and priorities all limit the effectiveness of virus immunization campaigns, and we should not forget that malnutrition alone is responsible for most of the very high measles mortality. Fortunately there is help from the World Health Organization (WHO), the World Bank, governments and relief organizations. Nor should we lose sight of the fact that in some countries other serious diseases, such as malaria, schistosomiasis and non-viral diarrhoea (which can be fatal in infants), have priority for already limited resources. Aciclovir and the anti-HIV drugs are a welcome sign that effective antiviral chemotherapeutic agents do in fact exist, but there is no easy solution to the discovery of others.

For further information, see Sections 16.3, 16.4 and 16.9.

21.6 Emerging virus infections

Emerging viruses are usually first identified by the diseases that they cause. There are four categories: (i) the invasion of a new host population by a known virus; (ii) a totally novel virus; (iii) the isolation of a new virus to a well-known disease; and (iv) the increase of clinical severity caused by an existing virus. Table 21.1 shows that most emerging viruses are in the first category, and the list, which is far from complete and centred on just one animal species, is worryingly long. The reasons why a virus switches species are manifold, but central to this is the degree of contact between humans and the main virus reservoir (see zoonoses—Section 14.12). Switching species is a rare event, and it is even rarer for the virus to become established in the new host. Switching of viruses between the great apes, the closest relatives of humans, and humans occurs relatively readily in both directions because of the great genetic homologies, but this facility is lost with the lower primates such as monkey. Of even greater potential for infection of humans is the vast reservoir of viruses that are specific to every other animal species.

However, viruses that switch species will not necessarily cause the same disease, and a virus in a new host can be infinitely more pathogenic than it is in its normal host, e.g. herpes B virus, which causes an asymptomatic infection of monkeys, is usually lethal when contracted by humans. HIV-1 infects but does not cause AIDS in chimpanzees. Presumably, these all reflect an upset in the normally favourable virus–host balance that has evolved over a long period of time in the natural host species. However, this is not a universal argument because HIV-2 from more distantly related sooty mangabey monkeys is less pathogenic than HIV-1. In either case it is likely that any virus that becomes established in the human population and transmits person to person will undergo considerable mutation and adapt to the new host. In addition hosts, whose reproductive potential is affected by the virus, will be selected for their resistance to the virus. Together such adaptations would be expected to reduce pathogenicity, as observed with the infection of the European rabbit with myxoma virus in Australia and Europe (see Section 18.3). Some of the examples from Table 21.1 are discussed below—some of the others have been covered in earlier chapters.

Novel paramyxoviruses

The dramatic and distressing appearance of large numbers of dead seals around the North Sea and other areas of north-west Europe at the end of the 1980s heralded the

Table 21.1 Some emerging infectious diseases that affect humankind.

Year recognized	Disease	Causative agent and comment	Category*
Already present	Influenza	New influenza A viruses arise at irregular intervals through antigenic shift which creates a new mix of genome segments and a new virus (see Section 17.4)	D
Already present	Influenza	Influenza A and B viruses evolve continuously through antigenic drift so that in 4 years a 'new' virus is made that is no longer affected by immune responses to the parent virus (see Section 17.4)	D
1997–8	Influenza	Hong Kong; H5N1 chicken virus jumped the species barrier; caused great concern with 40% lethality but did not spread (see Section 17.4; Cohen 1997; Shortridge 1999)	A
1981	A silent pandemic in the 1970s with AIDS recognized in 1981	Caused by HIV-1; origin probably chimpanzees; 36 million already infected, 22 million dead and increasing (see Chapter 19)	A
1986	AIDS with a slower course	Caused by HIV-2; originated from the sooty mangabey monkey (pets, food?) on >six occasions	A
1993	Hantavirus pulmonary syndrome	Caused by a virus of deer mice in the USA that jumped the species barrier; a bunyavirus called sin nombre virus	A
Already present	Non-A, non-B hepatitis	This is now known to be caused by hepatitis viruses C–G, none of which can be cultivated *in vitro*	B
Already present	Chronic hepatitis and liver cancer	Caused by hepatitis C virus (flavivirus); no vaccine available	B
Already present	Hepatitis with mortality during pregnancy	Hepatitis E virus with a single molecule of single-stranded positive-sense RNA cloned in 1990; possibly from a zoonosis; 20% mortality rate during pregnancy	A or B
1995	Psychiatric illness?	Borna disease virus in horses known for more than two centuries; infects a variety of animals and alters behaviour patterns; now also isolated from patients (see Section 21.3)	A or B
1996	Postpolio syndrome	Previously known as the prototype causative agent of acute infections in humans, poliovirus somehow persists in some who were paralysed by the original infection and causes a second round of neuromuscular problems (see text)	C
Already present	Myalgic encephalomyelitis (ME)	Some cases are associated with aberrant picornavirus (enterovirus) RNA or Epstein–Barr virus DNA in muscle (see text)	C
1998–9	Encephalitis	Malaysia. Probably a virus of fruit bats that moved to pigs to man; a paramyxovirus called Nipah, spread by *Culex* mosquitoes (see text)	A

Continued

Table 21.1 *(Continued)*

Year recognized	Disease	Causative agent and comment	Category*
1994	Encephalitis	Tropical Australia. Horses and humans; a paramyxovirus of fruit bats called Hendra, that moved to horses and humans (see text)	A
1986	BSE	UK; a neurotropic disease of cattle; probably a mutated form of scrapie of sheep; not a virus; never been purified (see Section 20.5)	A
1996	Variant CJD	UK; almost certainly from infection with BSE; 60 deaths by August 2000; an unknown number of people infected probably by eating infected meat/meat products (see Section 20.5)	A
1999	Encephalitis	In New York City, USA. Killed many wild birds and horses, and 7 of 56 people. Caused by West Nile virus, a *Culex* mosquito-borne flavivirus (Rappole *et al.* 2000)	A
1991	Exanthem subitum/sixth disease	Universal infection of children: fever and rash. Infection known long before virus (human herpesvirus type 6, HHV-6) was isolated. Infects T cells and other cells. Persistent in salivary gland. Latent in monocytes and bone marrow cells. A problem during immunosuppression in transplantation and AIDS; causes a fatal pneumonitis	B
1995	Kaposi's sarcoma	A rare cancer known since 1872 that came to notice as an early correlate of AIDS; associated with human herpesvirus type 8: HHV-8 antibodies occur in 25% of the normal population, thus cancer must depend on other unknown cofactors (see Biberfield *et al.* 1998; Schultz 1998)	B

*A, a virus that invades a new host species; B, the isolation of a novel causative virus to a well-known disease; C, an increase in clinical severity/change in clinical picture caused by an existing virus; D, a totally novel virus.
BSE, bovine spongiform encephalopathy; CJD, Creutzfeldt–Jakob disease.

first indication that previously unknown paramyxoviruses were capable of causing significant disease. It was rapidly established that the virus causing the seal deaths is closely related to the well-known virus that causes distemper of dogs, and is a member of the morbillivirus genus of the paramyxovirus family, which includes measles virus. Following the seal outbreak, a similar disease was reported in dolphins, particularly those found in the western Mediterranean Sea. Retrospective analysis of samples from marine mammals has shown that an outbreak had occurred along the Atlantic coast of the USA, predating the European experience. With no intervention from humans these diseases have declined naturally, possibly as a result of an increase in immunity in the host animals, but recent reports have described infections in the Gulf of Mexico and the Pacific Ocean. The origin of these viruses remains unknown. Since the appearance of the morbillivirus of sea mammals, similar viruses have been identified as causing infection, and frequently death, in a number of animals in various parts of the world. Some also represent a risk to humans.

In the mid-1990s lions in the Serengeti desert area began to succumb to a respiratory infection of such severity that many died, putting this unique population at risk. Analysis to identify the cause of the disease quickly implicated a virus and the agent responsible was shown to be canine distemper virus (CDV). The data indicated that CDV had moved from its normal canine host, in which infection is frequently fatal, to a new host species. The absence of protective immunity in the lions and the pride structure has permitted a very rapid spread of the disease.

In September 1994 the first reports of a serious respiratory disease in horses in a suburb of Brisbane, Australia, appeared in the international media. The reason for the interest was not only the high level of rapid mortality in the affected horses but also the associated illness of the trainer from the stable. After a short period in hospital the trainer died. At the same time a stable hand was also affected. The causative agent had the hallmarks of a virus transmitted by the respiratory route, by which all paramyxoviruses are transmitted, but there was no similarity to any known virus. Subsequent analysis confirmed that a virus, structurally similar to paramyxoviruses, was responsible and that it had never before been encountered. The causative agent was named Hendra virus, after the area in Brisbane where the outbreak had occurred. A small number of outbreaks affecting horses has occurred since the first, but only one additional human death has been reported. Molecular analysis of the Hendra virus genome confirmed that it is a morbillivirus, related to but distinct from measles virus. An extensive search for possible animal reservoirs of Hendra virus has shown that the natural hosts are pteropid fruit bats, commonly known as flying foxes, which are migratory animals indigenous to Australia and much of south-east Asia. The most likely explanation of the outbreaks is that the virus spreads from aborted foetuses and birth material from infected bats which come into direct contact with susceptible hosts. Infection of humans does not appear to be direct and is limited to those in very close contact with animals that have acquired the infection from bats. Since the description of Hendra virus, two additional, newly discovered, closely related viruses have appeared with devastating consequences.

A rapid increase in the number of stillborn piglets suffering from severe degeneration of the brain and spinal cord was noted at a piggery in New South Wales, Australia, in 1997. This was shown to be the result of infection of the pigs with a paramyxovirus named Menangle virus. Two staff at the piggery were also infected and suffered a severe, but not life-threatening, febrile illness. Determination of the complete nucleotide sequence of the genome of the virus shows that it is a close relative of Hendra virus. As with Hendra virus, the natural reservoir was shown to be several species of pteropid fruit bats.

The most dramatic recent outbreak of infection associated with a previously unknown virus began at the end of 1998 and continued into 1999. During this outbreak, 276 people in Malaysia and Singapore contracted viral encephalitis and 106 died. It was quickly shown that the causative agent was a virus closely related to, but distinct from, Hendra virus. It was named Nipah virus after the area in which it first appeared. A consistent feature of the disease was that the affected individuals were all pig farmers, or closely associated with pig farmers, and the virus was shown to be present in pigs in the affected areas. Such was the severity of the situation that government agencies sanctioned the slaughter of almost the entire pig population in the affected areas, amounting to more than a million animals. The natural host and reservoir of Nipah virus are pteropid fruit bats of the same species that host Hendra and Menangle viruses. The presence of these three viruses was not suspected until the outbreaks of disease in other species, and this serves to emphasize our ignorance of viruses found in nature and the need for continued study and vigilance.

For further information see Murray *et al.* (1995), Philbey *et al.* (1998), Chua *et al.* (2000) and Halpin *et al.* (2000).

Myalgic encephalomyelitis (ME) or postviral fatigue syndrome

Myalgic encephalomyelitis (ME) is a distressing condition in which patients suffer extreme fatigue of muscles after moderate exertion, and this may be prolonged for a year or more, although sometimes with partial remissions. Described as a syndrome (meaning a collection of different diseases with a common expression), it is defined by exclusion of other causes of chronic fatigue. ME has been recognized for over 30 years, although until very recently there has been considerable uncertainty about its cause—whether it was the result of a persistent or latent infection or it was psychosomatic or plain malingering. Now, new sensitive methods of detection have shown the presence of viral genomes in muscle biopsies of some cases of ME.

The breakthrough came through the discovery of the presence of viral DNA. A biopsy needle was used to remove a small core of muscle, from which nucleic acid was extracted. This was then tested, as described in Table 21.2. A highly significant proportion of people with ME had enterovirus RNA in their muscle; a smaller proportion had EBV DNA instead. None had both. Later work showed no active involvement of human herpesviruses 6 or 8. An important question is how enterovirus RNA persists, when these viruses normally cause an acute infection. Data from Len Archard's group in London show that viral RNA synthesis is abnormal: whereas cells infected with a cytopathogenic virus contain 99% positive-sense RNA (mRNA and virion RNA) with very little negative-sense RNA, muscle biopsies from ME patients have similar amounts of positive- and negative-sense RNA. It will be of interest to see whether full-length viral RNA is present, and whether it is mutated in regions controlling RNA synthesis.

These data are in good agreement with earlier serological evidence of enterovirus infection. Here, 51% of ME patients had virion protein 1 (VP1)–antibody complexes circulating in serum and 20% excreted virus–antibody complexes in faeces. Finally, Drs J. W. Gow and W. M. H. Behan have reported from Glasgow that electron microscopy shows the presence of abnormal mitochondria in muscle of ME patients.

Table 21.2 Evidence for the presence of enterovirus and Epstein–Barr virus (EBV) (a herpesvirus) nucleic acid in muscle biopsy samples from patients with myalgic encephalopathy (ME).

Technique	Primer/probe	ME	Controls
RT-PCR	Enterovirus sequence from the 5′ non-translated region	32/60 (53%)	6/41 (15%)
Hybridization	Common cDNA from the enterovirus polymerase gene	34/140 (23%)*	0/152 (0%)
Hybridization	cDNA from EBV nuclear antigen 1 gene	8/89 (9%)*	0/48 (0%)

*Same set of biopsies tested; none was positive for both viruses.
cDNA, complementary DNA; RT-PCR, reverse transcription and polymerase chain reaction.

Obviously, any defect in energy production by the mitochondria would make muscle prone to fatigue. We now have some of the pieces of the ME jigsaw puzzle, but they have yet to be fitted together.

The opposing views on enterovirus involvement in ME can be found in Melchers *et al.* (1994), and Muir and Archard (1994).

Postpolio syndrome

Poliovirus conducts the archetypal hit-and-run infection and is well known to be acutely cytopathogenic both *in vitro* and *in vivo*, but recent data suggest that it can cause rare persistent or latent infections which manifest in clinical disease after 30–40 years. In all, 25–28% of postpolio syndrome patients suffer clinical disease, and 60% are female. The average age is 57. Before immunization, poliovirus caused common childhood infections, although >99% were subclinical. Of the fewer than 1% people who experienced clinical disease some were left with permanent paralysis caused by destruction of motor neurons. That should have been the end to the virus, but it has been reported that some of these people were experiencing further neuromuscular problems, and that persistence of the original virus was suspected. These highly controversial findings have now been vindicated because RT-PCR has revealed the presence of viral RNA sequences in the cerebrospinal fluid of 50% (5 of 10) postpolio syndrome patients compared with 0 of 23 controls. Such virus contains mutations compared with wild-type virus, but the significance of these has not yet been established and nothing is known of the mechanism of the persistence/latency. These findings may mean that any virus can cause a persistent infection, provided that particular conditions are met.

For further information see Leparc-Goffart *et al.* (1996) and Pavio *et al.* (1996).

Non-A, non-B hepatitis viruses

The hepatitis viruses are joined in name by virtue of their common target tissue and so far all belong to different families. All cause a malfunction of the liver (hepatitis) recognized at the acute stage by jaundice. This is a yellowing of the skin and eyeballs due to the presence of large amounts of bilirubin, a pigment that is formed from the normal breakdown of haemoglobin from old red blood cells and which is normally excreted in bile but is not metabolized properly in the infected liver. Infectious hepatitis A virus (HAV) is a picornavirus whereas hepatitis B virus (HBV) is a hepadnavirus. Until recently it was known only that there were other viruses termed collectively 'non-A, non-B' viruses. These are now identified as hepatitis viruses C, D, E, F and G. At best, hepatitis viruses (apart from the picornavirus, hepatitis A) grow poorly in cell culture and there is no small animal model. All infect primates and all biological work, vaccine and drug testing have to be done in this system with its attendant difficulties. There is most information on their genomes, which has come about only since the advent of recombinant DNA technology.

Hepatitis C virus (HCV) is a flavivirus that appears to be on the increase. It causes a massive amount of infection, estimated as 200 million (or 3% of the world's population). Around 80% of these become life-long chronic infections and lead to complications, including liver failure and primary hepatocellular carcinoma, in this respect very similar to HBV (see Sections 15.5 and 17.7). HCV is notable as the only true RNA virus that is carcinogenic, and we have no knowledge of its oncogenic properties. The genome of approximately 9500 nucleotides (nt.) was sequenced in 1989, and there are six major and 12 minor genotypes. Clearance of infection depends on the genome diversity generated during infection: with little diversity the virus is cleared, with much diversity coupled with a weak immune (possibly Th2) response the infection becomes chronic. Treatment with interferon-α gives a poor recovery rate of 4–20% but, in combination with the antiviral drug ribavirin, this rises to 33–70%. However, the treatment takes 6–18 months and is expensive ($US15 000–20 000). Genotype 1b is relatively more resistant to treatment.

Hepatitis D virus (HDV) is a very unusual animal virus because it is a satellite virus of HBV, and hence only found as a dual infection. Satellite viruses depend for their multiplication on a specific helper virus, and differ from defective–interfering (DI) virus as their sequences are completely unrelated. In this relationship HBV provides HDV with envelope protein. HDV has a covalently closed, circular, negative-sense, ssRNA genome of around 1700 nt. It is about 70% base-paired, similar to a viroid genome (see Sections 4.7 and 7.5). There are three genotypes. Replication is nuclear and possibly involves the cellular pol II. The RNA has a single open reading frame (ORF) that encodes the core protein. The precise role of HDV in liver disease is not known, but it appears to exacerbate the pathogenic effects of HBV. Treatment of HBV (with interferon-α) concomitantly treats HDV, and prevention of HBV infection by the HBV vaccine prevents HDV at the same time.

Hepatitis E virus (HEV) has a single molecule of plus sense ssRNA of 7500 nt., which resembles the genome of caliciviruses. In the USA, clinical disease is rare but in some inner city areas there is a 20% incidence of HEV-specific antibodies. Antibodies are widespread in wild rats and domestic pigs, and it may be that there is a zoonosis with virus being passed from rat to person early in life, followed by successful virus clearance. HEV can cause severe complications during pregnancy, with 20% fetal mortality if contracted in the third trimester. Neutralizing antibodies that protect monkeys from infection and disease have recently been isolated from a random combinatorial DNA library (see Section 21.8) obtained from an infected chimpanzee, and may lead the way to an antibody-based therapy.

Little is known about hepatitis G virus, which has a flavivirus-like genome (cloned and sequenced in 1996) and is a minor cause of human disease; the reality of hepatitis F virus remains to be confirmed.

For further information see Purcell *et al.* (1994), Harrison (1996), Karayiannis *et al.* (1998), Spector (1999), Anonymous (2000), Flint and McKeating (2000) and Liang and Hoofnagel (2000).

21.7 Viruses and multiple sclerosis

Multiple sclerosis (MS) is a disease of the CNS (the brain and spinal cord). About 10% of the cells in the CNS are neurons and the rest are support cells, called glia. About one-third of glial cells are oligodendrocytes, which insulate the electrical activity of nerve cell axons by surrounding them with a cytoplasmic 'Swiss roll' containing myelin in the plasma membrane. MS is associated with a number of focal areas in the CNS in which oligodendrocytes have degenerated, and this results from autoimmune attack by myelin-specific CD4$^+$ T cells. The disease is chronic, affects adults and involves mainly the limbs, sight and incontinence. MS either progresses continuously or is characterized by periods of remission of a few months to a year or more, when there is at least partial recovery, alternating with reappearance of the MS symptoms. During remission, myelin sheaths regrow around the neuronal axons and neuronal function is restored. The incidence of MS is about 1 per 800 in the UK and the USA, but in other areas of the world this may be higher or lower.

The major question is what initiates the antimyelin autoimmunity. MS has both genetic and environmental origins, e.g. with one parent with MS the incidence in children rises to 1 in 200, although with an affected non-identical twin it is 1 in 33, and with an affected identical twin it is 1 in 4. Largely because of the lack of any other predisposing factor, a virus or viruses are suspected to be the environmental factor. Many viruses have been proposed but on examination none was shown to be the cause. One of the latest suspects is human herpesvirus type 6 (HHV-6), the causative agent of a mild fever and rash called exanthem subitum, but the evidence is not strong. Another is a human retrovirus that is related to an endogenous retrovirus sequence. The problem is that everyone harbours viruses as harmless passengers, especially members of the herpesvirus family which remain latent or persistent for life. It is very difficult and expensive to prove that a virus is *not* involved. The jury is still out.

One attractive hypothesis is that MS involves enveloped viruses in general and results not from infection by one virus but by successive infections with several different viruses. It is postulated that each virus enters the CNS asymptomatically and incorporates myelin components into its envelope; these are then presented to the immune system, together with the foreign viral antigens. Myelin components in this context are seen by the immune system as foreign, and autoimmunity results. Continued presence of the virus(es) is not required once the antimyelin T cells have been activated. The genetic component would then include differences in the ability of T cells to respond to the autoantigen.

For further information see Steinmann (1996, 2000).

21.8 Antibodies: human monoclonal antibodies from DNA libraries and microantibodies

Since the initial technology for the production of MAbs was devised in 1975 by George Köhler and César Milstein at Cambridge, the face of virology has been transformed by their multifarious uses. We see MAbs in the forefront of new analytical, preparative and diagnostic technologies, and nowhere have they had more impact than in the creation of a new sphere of employment in the commercial sector. Now, it is difficult to imagine how virology functioned without them.

Humanizing mouse MAbs

There is, however, one important area where the use of MAbs has been frustrated by technical limitations and this is in their therapeutic use. The problem is that MAbs are made by immunizing a mouse or rat, extracting primed B cells from the spleen and then immortalizing the B cell by fusion with a mouse myeloma (B cancer) cell. Although the system works for mice, it is not applicable to humans and no human myeloma cell has proved effective. However, there has been some success by immortalizing human B cells with Epstein–Barr virus. The problem of treating people with a mouse MAb is that their immune system very rapidly reacts to the mouse antigens: at best, this destroys the antibody and, at worst, initiates a potentially lethal anaphylactic response. One answer is to humanize a mouse MAb and to use recombinant DNA technology to replace all mouse sequences with sequences from human antibody, except those sequences responsible for binding to antigen (the complementarity-determining regions (CDRs) of the hypervariable parts of the variable (V) regions of the light (L) and heavy (H) chain—see Fig. 14.4). By minimizing the amount of foreign (mouse) protein, the problem of rejection is overcome. The first humanized MAb is now available and licensed for use against respiratory syncytial virus.

Human antibodies from random combinatorial libraries

Another alternative does not use mammalian cells for antibody production at all. Instead mRNA is isolated from antibody-synthesizing cells and subjected to reverse transcription (RT) and PCR (Fig. 21.4). Such cells are withdrawn, using a syringe, either from the bone marrow or from blood. In the first reaction, specific primers are

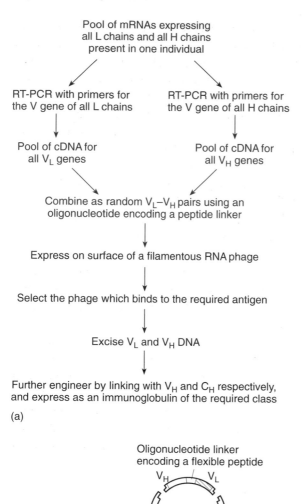

Pool of mRNAs expressing
all L chains and all H chains
present in one individual

RT-PCR with primers for
the V gene of all L chains

RT-PCR with primers for
the V gene of all H chains

Pool of cDNA for
all V_L genes

Pool of cDNA for
all V_H genes

Combine as random V_L–V_H pairs using an
oligonucleotide encoding a peptide linker

Express on surface of a filamentous RNA phage

Select the phage which binds to the required antigen

Excise V_L and V_H DNA

Further engineer by linking with V_H and C_H respectively,
and express as an immunoglobulin of the required class

(a)

Oligonucleotide linker
encoding a flexible peptide

V_H V_L

Attachment
protein gene

(ii)

(i)

(iii)

Attachment
protein

(b)

Fig. 21.4 (a) Scheme outlining the random combinatorial method for the cloning of human antibody Fv fragments. Its success lies in the ability to select rare recombinant phages expressing the required antibody activity (about 1 in 10^6 phages). (b) Construction of the fd phagemid containing the fused V_L–linker–V_H gene. This is transformed into *E. coli*. The bacterium is then infected with phage (fd) and progeny phage are produced having a single recombinant V_L–linker–V_H polypeptide (i). This folds to form a functional antibody-binding site (ii). The 'phage antibody' requires at least one molecule of non-modified attachment protein (iii) to be infectious.

used to amplify all V_L regions present and, in the second reaction, all V_H regions are similarly amplified. One V_L and one V_H are then linked together at random, using an oligonucleotide encoding a flexible peptide linker made of 15 amino acid residues (Gly$_4$–Ser)$_3$, so that, when expressed, the resulting polypeptide folds up to form an antigen-binding site. Any antibody of interest will be present at a very low frequency in the population of expressed hybrid molecules, but screening has been solved in a novel fashion by expressing the V_L–linker–V_H polypeptide on the surface of a filamentous bacteriophage (fd or M13). These phages encode an attachment protein, expressed in low numbers—about four copies per virion. This binds to the F pilus of *Escherichia coli*. The V_L–linker–V_H is expressed as a fusion protein with the phage attachment protein. The phage is still infectious and the V_L–V_H is expressed in a form that is able to bind to antigen. This is called a 'phage antibody'. Selection of the required antibody activity is achieved by panning a phage population that expresses a myriad of antibody specificities on to the required antigen. Non-bound phage is washed away and bound phage eluted and grown up on a lawn of *E. coli*. Plaques are picked and retested. DNA encoding the V_L–V_H gene can be excised and V_L joined to C_L and V_H to C_H, and reconstituted as FAbs or as a complete antibody molecule. The phage technique allows 1 in 10^6 antibody specificities to be selected. This type of procedure has been used recently to obtain powerful neutralizing antibody from an HIV-positive individual. The system is known as the random combinatorial method of making antibody, because it allows the formation of antibody from any pair of V_L and V_H genes, even those that may not exist as such in the donor.

Therapy with recombinant mouse–human monoclonal antibodies is not intended to replace the stimulation of immunity with vaccines, but, rather, to provide immunity in circumstances in which no other treatment is possible or effective. Examples are congenitally immunodeficient individuals, or those whose systems are compromised by accident or in the course of transplantation surgery or cancer chemotherapy, cases of acute life-threatening disease where there is no time to immunize, and in the protection of the newborn from infection by the mother. It is being considered for use with HIV infections.

For further information see Burton and Barbas (1994).

Microantibodies

The ultimate in downsizing an antibody is the microantibody. This is a peptide with a sequence derived from one of the six CDRs (complementarity-determining regions)—highly variable sequences of approximately 8–23 amino acid residues, three on each of the V_L and V_H, that are folded together to form the paratope or epitope-binding site (see Fig. 14.4). The surprise is that such a peptide has sufficient of the sequence and conformation to act like an antibody. The best example is a microantibody derived from a monoclonal IgG that neutralizes HIV-1. It comprises 17 amino acid residues from the CDR-H3. Work in the laboratory of one of the authors showed that the microantibody recognizes exactly the same epitope as the IgG from which it derives, and binds to the gp120 envelope protein and neutralizes infectious HIV-1. Such a small structure can be chemically synthesized and so obviates the problems of contamination by infectious agents such as endogenous viruses which can exist in hybridoma cell lines. Its small size will permit the study of the poorly understood neutralization process through alteration of individual amino acid residues. The still unsolved problem is to identify microantibody peptide sequences from other antibodies. So far only two neutralizing microantibodies are known.

For further information see Jackson *et al.* (1999).

21.9 Applications of recombinant viruses

As well as providing potent and constantly evolving threats to the health of humans, animals and plants, viruses also offer the possibility of being used beneficially through the construction and appropriate use of *recombinant viruses*. This term refers to any virus where the genetic material has been altered in a planned way by experimental manipulation.

The genetic material of a virus, once isolated, is amenable to all the same techniques as any other RNA or DNA molecule. Thus, DNA virus genomes may be cloned directly whereas RNA virus genomes may be cloned as cDNA. Many transcription regulatory elements and replication origins in common use in the molecular biology laboratory today have come from cloned viral sequences. Once cloned, viral DNA/cDNA molecules may be modified by site-specific mutation or, more drastically, larger segments may be removed and replaced with foreign DNA sequences. A recombinant virus must then be created by rescuing these cloned, manipulated sequences into infectious particles. This is not yet possible for all virus types; some of the double-stranded RNA viruses remain refractory to all attempts at recombinant virus rescue. The practical details of recombinant virus construction lie outside the scope of this text. Useful sources for further information on this topic are listed at the end of the chapter.

Recombinant viruses have been invaluable in the study of virus replication cycles, and much of the information in the earlier chapters of this book has come from their analysis. They are also of value as vehicles for the introduction of specific nucleic acid sequences into cells, either in cell culture or *in vivo*. In the former context, they are powerful tools for heterologous protein expression, whereas in the latter context they have been valuable as vaccines, e.g. against rabies, and are under extensive development as gene therapy agents.

Recombinant viruses for heterologous protein expression

Although many viruses have been experimented with for the purpose of expressing heterologous gene sequences in a culture system, the most widely used examples are the poxvirus, vaccinia and the insect baculovirus, *Autographa californica* nuclear polyhedrosis virus (AcNPV). Insect cells infected with AcNPV provided a very superior expression system. Recombinant baculoviruses containing an inserted influenza virus type A haemagglutinin (HA) or β-galactosidase gene have produced up to 5 mg and 400 mg protein/L, respectively. Unfortunately, baculoviruses are lytic for the host cell, so only batch production is possible. None the less, baculoviruses currently outperform any other expression system, whether bacterial, yeast or higher animal cell, and the commercial potential is attracting industrial interest and finance. Vaccinia virus recombinants actually have many other applications (see below). Like AcNPV, they can be used to express a protein in batch culture. In the laboratory, vaccinia virus recombinants have been particularly useful in dissecting the immune response to specific antigens, despite the powerful immune system-modulating factors that they express. Other viruses that have seen development as expression vectors include the positive-sense RNA genome alphaviruses.

Recombinant viruses as vaccines

We have already discussed the new generation of recombinant vaccines centred on vaccinia virus, in which relevant foreign antigens are expressed as if they were proteins native to the vaccinia virus (see Fig. 16.8). Essentially, any DNA sequence may be incorporated into the vaccinia virus genome and expressed as protein upon infection. There is considerable interest in veterinary circles, too, because poxviruses have representatives native to many animal species and offer the same advantages as a cheap, multiple, one-dose, live vaccine as vaccinia virus does for humans. Other viruses, such as adenovirus, have also been shown to express foreign antigens that elicit immune responses *in vivo*, although they have not been widely used to date.

Recently, the ability to perform 'reverse genetics' on negative-strand RNA viruses, such as measles, has opened the way to creating recombinant attenuated viruses as possible vaccines. By altering the order of coding regions on the genome, or by deleting certain functions, viruses may be dramatically attenuated for growth. In essence, this process mirrors the *ad hoc* selection of attenuated virus by serial passage—a traditional method for vaccine strain isolation—but here attenuated virus is produced in a planned way rather than arising through chance genetic change. Such experiments have, however, to be carried out under close containment because it is possible to imagine viruses of increased rather than decreased virulence resulting, although in practice this is not observed.

Recombinant viruses for gene therapy

The concept of gene therapy is simply the introduction of a potentially therapeutic or otherwise beneficial DNA sequence into the cells of a patient. Originally conceived as a way of restoring normal function to people with specific inherited gene defects, the concept has now grown to encompass a variety of applications, with the greatest number of clinical trials being in the field of cancer therapy. Viruses are potentially valuable in carrying therapeutic DNA efficiently into cells. Those that have seen extensive work for this purpose to date include murine retroviruses, human adenovirus and the parvovirus, AAV (see Section 9.8).

Retrovirus vectors have all their normal coding sequences removed. The recombinant virus has then to be grown in the laboratory in a special cell line which provides the necessary viral proteins *in trans*. When used therapeutically, the virus particles reverse transcribe and integrate a copy of their genome randomly into the host cell genome, after which no further events of the infectious cycle can occur. Hence, the foreign gene that the virus has carried into the cell is present for the life of the cell.

The principal problem with retroviral vectors has been the low titres of particles produced, which has meant that they have typically been used only to transduce cells taken from a patient, which must then be reimplanted appropriately.

As adenoviruses are more complex than retroviruses, it has not been possible yet to make packaging cells that would allow the growth of vectors in which all the viral genes had been deleted. Some vectors of this type have been grown using a second virus to provide the necessary viral proteins, but most applications to date have used vectors where most of the viral genes are retained. Only selected genes, the loss of which can be complemented in a cell line, are deleted and replaced by foreign DNA. The major problem that has been encountered has been the host immune response, which effectively eliminates cells that have been transduced. This, however, may be an advantage in the context of cancer therapy if the virus can be targeted specifically to tumour cells.

AAV has only a small genome, so even by removing all its genes its capacity for foreign DNA is limited. Its principal advantage is its ability to integrate its genome specifically into human chromosome 19. This gives the same benefits as retroviral vector integration, but greater safety because there is not the same risk of random insertional mutagenesis (see Section 17.6). Herpes simplex virus has also been considered as a vector, particularly for gene delivery to neurons where it naturally establishes latency.

A factor that limits the use of all viral vectors for gene therapy at present is the lack of ability to target gene delivery to specific target cell populations. What is ideally required is to target viral attachment only to the desired cell type, so improving effectiveness of gene delivery and limiting the scope for potential side effects. Although this is not yet possible, considerable progress has been made in recent years in the re-targeting of viral infectivity on to novel receptors. This is a key area for future development.

For further information see Ali *et al.* (1994), Kovesdi *et al.* (1997), Leppard (1999) and Primrose *et al.* (2002).

21.10 Educating the general public

Even among the quality press and media sources, virologists are constantly saddened and irritated by fundamental misconceptions about viruses. If these professionals make mistakes, what hope is there for the general public? The answer, perhaps, lies in all virologists (and especially future virologists) helping to inform those around them. With greater understanding will come, for example, the willingness to accept vaccines, because, as mentioned earlier, fear and apathy make acceptance rates abysmally low. Every medical treatment carries a finite risk of personal damage, and vaccines are no exception. However, it is to be hoped that the public will appreciate that not taking a vaccine exposes them to far greater risk. By understanding more about infectious agents, sensible precautions against infection can be taken, and the hysteria that has in turn accompanied increases in hepatitis B virus, genital herpes viruses and HIV will be replaced by more objective reactions. We are confident that the increasing numbers of students taking courses in virology will have a positive impact on the understanding of the public at large, and on the general adverse reaction to genetically manipulated organisms in particular, and the problem that this poses for realizing the benefits of recombinant viruses.

21.11 Further reading and references

Ali, M., Lemoine, N. R. & Ring, C. J. A. (1994) The use of DNA viruses as gene therapy vectors. *Gene Therapy* **1**, 367–384.

Almond, J. W. (1987) The attenuation of poliovirus neurovirulence. *Annual Review of Microbiology* **41**, 153–180.

Anonymous (2000) Predicting the course of HCV infection. *Nature Medicine* **5**, 512.

Biberfeld, P., Ensoli, P., Sturzl, M. & Schultz, T. F. (1998) Kaposi sarcoma-associated herpesvirus/human herpes-virus 8, cytokines, growth factors and HIV in pathogenesis of Kaposi's sarcoma. *Current Opinion in Infectious Diseases* **11**, 97–105.

Bode, L. & Ludwig, H. (1997) Clinical similarities and close genetic relationship of human and animal Borna disease virus. *Archives of Virology* **S13**, 167–182.

Burton, D. R. & Barbas, C. F., III (1994) Human antibodies from combinatorial libraries. *Advances in Immunology* **57**, 191–280.

Chua, K. B., Bellini, W. J., Rota, P. A. *et al.* (2000) Nipah virus: a recently emergent deadly paramyxovirus. *Science* **288**, 1432–1435.

Cohen, J. (1997) The flu pandemic that might have been. *Science* **277**, 1600–1601.

Epstein, M. A. (1984) Clues to the role of malaria. *Nature (London)* **312**, 398.

Fazakerley, J. K., Amor, S. & Nash, A. A. (1997) Animal model systems in multiple sclerosis. In: *Molecular Biology of Multiple Sclerosis*, pp. 255–273. Russell, W. C. (ed.). Chichester: John Wiley & Sons Ltd.

Flint, M. & McKeating, J. A. (2000) The role of the hepatitis C virus glycoproteins in infection. *Reviews in Medical Virology* **10**, 101–117.

Gutiérrez, A. L., Denova-Ocampo, M., Racaniello, V. R. & del Angel, R. M. (1997) Attenuating mutations in the poliovirus 5' untranslated region alter its interaction with polypyrimidine-tract binding protein. *Journal of Virology* **71**, 3826–3833.

Halpin, K., Young, P. L., Field, H. E. & MacKenzie, J. S. (2000) Isolation of Hendra virus from pteropid bats: a natural reservoir of Hendra virus. *Journal of General Virology* **81**, 1927–1932.

Harrison, T. J. (1996) New agents of viral hepatitis. *Reviews in Medical Virology* **6**, 71–75.

Jackson, N. A. C., Levi, M., Wahren, B. & Dimmock, N. J. (1999) Mechanism of action of a 17 amino acid microantibody specific for the V3 loop that neutralizes free HIV-1 virions. *Journal of General Virology* **80**, 225–236.

Karayiannis, P. (1998) Hepatitis D virus. *Reviews in Medical Virology* **8**, 13–24.

Koprowski, H. & Lipkin, W. I. (1995) Borna disease. *Current Topics in Microbiology and Immunology* **190**, 1–140.

Kovesdi, I., Brough, D. E., Bruder, J. T. & Wickham, T. J. (1997) Adenoviral vectors for gene transfer. *Current Opinion in Biotechnology* **8**, 583–589.

Leparc-Goffart, I., Julien, J., Fuchs, F., Janatova, I., Aymard, M. & Kopecka, H. (1996) Evidence of the presence of poliovirus genomic sequences in cerebrospinal fluid from patients with postpolio syndrome. *Journal of Clinical Microbiology* **34**, 2023–2026.

Leppard, K. N. (1999) Mutagenesis of DNA viruses. In: *DNA Viruses—a practical approach*, pp. 47–81. Cann, A. J. (ed.). Oxford: Oxford University Press.

Liang, T. J. & Hoofnagel, J. H. (eds) (2000). *Hepatitis C.* New York: Academic Press.

Lipkin, W. I., Schneemann, A. & Solbrig, M. V. (1995) Borna disease virus: implications for human neuropsychiatric illness. *Trends in Microbiology* **3**, 64–69.

Melchers, W., Zoll, J., van Kuppveld, F., Swanink, C. & Galama, J. (1994) There is no evidence for persistent enterovirus infections in chronic medical conditions in humans. *Reviews in Medical Virology* **4**, 235–243.

Muir, P. & Archard, L. C. (1994) There is evidence for persistent enterovirus infections in chronic medical conditions in humans. *Reviews in Medical Virology* **4**, 245–250.

Murray, K., Selleck, P., Hooper, P. *et al.* (1995) A morbillivirus that caused fatal disease in horses and humans. *Science* **268**, 94–97.

Nakamura, Y., Takahashi, H., Shoya, Y. *et al.* (2000) Isolation of Borna disease virus from human brain tissue. *Journal of Virology* **74**, 4601–4611.

Notkins, A. L. (1984) Diabetes: on the track of viruses. *Nature (London)* **311**, 209–220.

Oldstone, M. B. A. (1989) Viruses can cause disease in the absence of morphological evidence of cell injury: implication for uncovering new diseases in the future. *Journal of Infectious Diseases* **159**, 384–389.

Oldstone, M. B. A., Rodriguez, M., Daughaday, W. H. & Lampert, P. W. (1984a) Viral perturbation of endocrine function: disordered cell function leads to disturbed homeostasis and disease. *Nature (London)* **307**, 278–281.

Oldstone, M. B. A., Southern, P., Rodriguez, M. & Lampert, P. (1984b) Virus persists in β cells of islets of Langerhans and is associated with chemical manifestations of diabetes. *Science* **224**, 1440–1443.

Pavio, N., Buc-Caron, M. H. & Colbère-Garapin, F. (1996) Persistent poliovirus infection of human fetal brain cells. *Journal of Virology* **70**, 6395–6401.

Philbey, A. W., Kirkland, P. D., Ross, A. D. *et al.* (1998) An apparently new virus (family Paramyxoviridae) infectious for pigs, humans, and fruit bats. *Emerging Infectious Diseases* **4**, 269–271.

Primrose, S. B., Twyman, R. & Old, R. (2002) *Principles of Gene Manipulation* (6th edn). Oxford: Blackwell Science.

Purcell, R. H. (1994) Hepatitis viruses: changing patterns of human disease. *Proceedings of the National Academy of Sciences of the United States of America* **91**, 2401–2406.

Racaniello, V. R. (1988) Poliovirus neurovirulence. *Advances in Virus Research* **34**, 217–246.

Rappole, J. H., Derrickson, S. R. & Hubálek, Z. (2000) Migratory birds and spread of West Nile virus in the Western hemisphere. *Emerging Infectious Diseases* **6**, 319–328.

Richt, J. A., Pfeuffer, I., Christ, M., Frese, K., Bechter, K. & Herzog, S. (1997) Borna disease virus infection in animals and humans. *Emerging Infectious Diseases* **3**, 343–352.

Schultz, T. F. (1998) Kaposi's sarcoma-associated herpesvirus (human herpesvirus-8). *Journal of General Virology* **79**, 1573–1591.

Shortridge, K. F. (1999) Poultry and the H5N1 outbreak in Hong Kong, 1997: abridged chronology and virus isolation. *Vaccine* **17** (Suppl. 1), S26–S29.

Spector, S. (ed.) (1999) *Viral Hepatitis: Diagnosis, therapy and prevention*. Totowa: Humana Press.

Steinman, L. (1996) Multiple sclerosis: a coordinated immunological attack against myelin in the central nervous system. *Cell* **85**, 299–302.

Steinman, L. (2000) Multiple approaches to multiple sclerosis. *Nature Medicine* **6**, 15–16.

Turell, M. J., Rossignol, P. A., Spielman, A., Rossi, C. A. & Bailey, C. L. (1984) Enhanced arboviral transmission by mosquitoes that concurrently ingested microfilariae. *Science* **225**, 1039–1041.

Also check Chapter 22 for references specific to each family of viruses.

Chapter 22

The classification and nomenclature of viruses

What follows is a list of the major groups of viruses, together with a brief description and a sketch of each type of virus particle (although not drawn to scale). The principles by which viruses are classified are still being decided, and some viruses have not yet been assigned to a family or even a genus. The list is divided for convenience into: viruses that infect animals, plants, fungi and bacteria; satellite viruses and satellites; and viroids. Some families cross these boundaries, with members infecting animals or plants, etc., and the family name will be found in both categories. In this chapter, we have italicized family, subfamily and genus names, but not those of viral species. The list is further ordered according to the revised Baltimore scheme (see Section 4.2) and alphabetically by family. At present most is known about viruses that cause diseases of humans, domestic animals and crops. However, every species has its own broad range of viruses, and these will eventually be discovered. Thus the list is undergoing continual revision. The current total of virus families by host is shown in Table 22.1. What is remarkable is the skewed distribution, with classes 1, 4 and 5 predominant in animals, class 4 in plants, and class 1 in bacteria. Table 22.1 does not, however, take into account the success of individual families or even virus species. There are no class 1 viruses in plants, few class 5 viruses in hosts other than animals, one family of class 6 viruses in animals, and one family of class 7 viruses in animals and another in plants.

At least one specific reference to each virus group is cited as a means of providing further information and references, and a list of general references can be found at the end of the chapter. Lipid envelopes are drawn with a wavy line to distinguish from purely protein structures.

22.1 Viruses of vertebrate and invertebrate animals

Some vertebrate viruses have an invertebrate vector; some of these also multiply in the vector whereas others are carried passively on its mouthparts.

Table 22.1 Current count of virus families according to host and class in the revised Baltimore scheme.

Class	Animal	Plant	Algae, fungi and protozoa	Bacteria*
1	10	0	2	9
2	2	2	0	2
3	2	2 (1)†	3	1
4	9	7 (20)†	2	1
5	7	2 (2)†	0	0
6	1	0	0	0
7	1	1	0	0

*Includes *Archaea, Mycoplasma* and *Spiroplasma*.
†In parenthesis, genera not yet assigned to a family.

Viruses with double-stranded DNA genomes (class 1)

Family: *Adenoviridae*

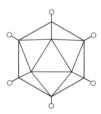

Non-enveloped icosahedral particles of 60–90 nm with a fibre protein at each vertex. Contain one molecule of linear double-stranded DNA (dsDNA) of 25 000–42 000 base-pairs (bp). A protein is covalently linked to the 5′ end and forms a pseudo-circular genome by non-covalently linking to the 3′ end. Replicate and are assembled in the nucleus. Infect vertebrates but *one genus infects fungi* (see Section 22.3).

Genus: *Mastadenovirus*. Infect mammals, e.g. human adenovirus type 2 or type 5.
Genus: *Aviadenovirus*. Infect birds, e.g. fowl adenovirus type 1.
Genus: *Atendovirus*. Bovine adenoviruses.
Genus: *Siadenovirus*, e.g. frog adenovirus 1, turkey adenovirus 3.

Doefler, W. & Böhm, P. (eds) (1995) The molecular repertoire of adenoviruses. *Current Topics in Microbiology and Immunology* **199**, Parts 1–3.

Family: *Ascoviridae*

Enveloped allantoid particle, 400 nm × 130 nm, with complex symmetry and one molecule of linear dsDNA of 100 000–180 000 bp. Replication nuclear. Infect insects, mainly *Lepidoptera*, e.g. *Spodoptera frugiperda* ascovirus 1a.

Cheng, X. W. *et al.* (1999) Circular configuration of the genome of ascoviruses. *Journal of General Virology* **80**, 1537–1540.

Family: *Asfarviridae*

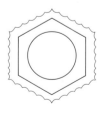

This group has properties of both *Iridoviridae* and *Poxviridae*. Enveloped 180 nm icosahedral particle containing an isometric 80 nm core surrounded by lipid. One molecule of linear dsDNA of 170 000–190 000 bp. Virion contains enzymes for mRNA synthesis. Cytoplasmic. Infect vertebrates and spread by ticks. African swine fever virus.

Mebus, C. A. (1988) African swine fever. *Advances in Virus Research* **35**, 271–312.

Family: **Baculoviridae**

Enveloped rod-shaped particles each with a nucleocapsid of 300 nm × 30–60 nm with one molecule of circular dsDNA of 80 000–180 000 bp. Virion may be occluded in a protein inclusion body (which makes it very stable). A large and diverse group, common in over 50 species of *Lepidoptera*; also infect *Coleoptera*, etc.

Genus: *Granulovirus*. May be occluded in a polyhedron containing one particle, e.g. *Cydia pomonella* granulovirus.

Genus: *Nucleopolyhedrovirus*. Virions may be occluded in a polyhedron containing many particles, e.g. *Autographica californica* multiple nucleopolyhedrosis virus (three such particles are illustrated).

Kurstak, E. (1993) *Viruses of Invertebrates*. New York: Marcel Dekker Inc.
Miller, L. K. (ed.) (1997) *The Baculoviruses*. New York: Plenum.
Volkman, L. E. (1997) Nuclearpolyhedrosis viruses and their insect hosts. *Advances in Virus Research* **48**, 313–348.

Family: **Herpesviridae**

Enveloped 120–220 nm particle with spikes, enclosing successively a tegument and an icosahedral nucleocapsid of 125 nm. One molecule of linear dsDNA of 120 000–240 000 bp. May bud from nuclear membrane. Very large family that infects vertebrates; latency is a common feature of the life cycle. Some are oncogenic.

Subfamily: *Alphaherpesvirinae*
Genus: *Simplexvirus*, e.g. human herpes (simplex) virus type 1 and 2—HHV-1 and -2.
Genus: *Varicellovirus*, e.g. HHV 3 or varicella–zoster (chickenpox) virus.

Subfamily: *Betaherpesvirinae*
Genus: *Cytomegalovirus*, e.g. HHV-5 or human cytomegalovirus.
Genus: *Muromegalovirus*, e.g. mouse cytomegalovirus.
Genus: *Roseolavirus*, e.g. HHV-6, HHV-7.

Subfamily: *Gammaherpesvirinae* (lymphoproliferative viruses; oncogenic)
Genus: *Lymphocryptovirus*, e.g. HHV-4 or Epstein–Barr virus.
Genus: *Rhadinovirus*, e.g. ateline herpesvirus 2, herpesvirus saimiri, HHV-8.
Unclassified: Marek's disease virus, turkey herpesvirus.

Davison, A. J. (1991) Varicella zoster virus. *Journal of General Virology* **72**, 475–486.
Davison, A. J. (1993) Herpesvirus genes. *Reviews in Medical Virology* **3**, 237–244.
Jones, C. (1999) Alphaherpes virus latency: its role in disease and survival of the virus in nature. *Advances in Virus Research* **51**, 81–133.
Medveczky, P. G. & Friedman, H. (eds) (1998) *Herpesviruses and Immunity*. New York: Plenum.

Family: *Iridoviridae*

Icosahedral isometric particle of 120–200 nm which may be surrounded by a lipid envelope. Purified virus pellets are iridescent blue or green. One molecule of linear dsDNA of 140 000–300 000 bp, but virions may contain one or more copies. Contain several enzymes. Cytoplasmic. Includes viruses of insects, fish (flounder, dab and goldfish) and many frog species, e.g. frog virus 3.

Kurstak, E. (1993) *Viruses of Invertebrates.* New York: Marcel Dekker Inc.
Williams, T. (1996) The iridoviruses. *Advances in Virus Research* **47**, 345–412.

Family: *Papillomaviridae*

Non-enveloped, 55 nm icosahedral particle with 72 capsomers in a skewed (*T* = 7) arrangement. One 8000 bp molecule of double-stranded, covalently closed, circular DNA. Replicates and assembles in the nucleus. Over 60 human papillomaviruses (HPVs) and others from many species. Oncogenic.

McCance, D. J. (ed.) (1998) *Human Tumor Viruses.* Herndon, VA: ASM Press.
Turek, L. (1994) The structure, function, and regulation of papillomaviral genes in infection and cervical cancer. *Advances in Virus Research* **44**, 306–356.

Family: *Polydnaviridae*

Enveloped particles, variable in size with nucleocapsids of about 300 nm × 80 nm. Bud successively from nuclear and plasma membranes. Complicated complement of DNA with multiple copies of circular dsDNA with or without some linear host DNA. Genome estimated as 2000–31 000 bp, but may be up to 300 000 bp. Infect insects.

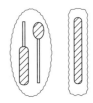

Genus: *Ichnovirus.* Double-enveloped 330 nm × 85 nm particle (illustrated right) with one nucleocapsid. Infect parasitic ichneumoid wasps.
Genus: *Bracovirus.* Single-enveloped, drop-shaped virions 30–150 nm long × 40 nm, containing 1–12 nucleocapsids (illustrated left). Infect parasitic braconid wasps.

Gruber, A. *et al.* (1996) Polydnavirus DNA of the braconid wasp *Chelonis inanitus* is integrated in the wasp's genome and excised only in later pupal and adult stages of the female. *Journal of General Virology* **77**, 2873–2879.
Kurstak, E. (1993) *Viruses of Invertebrates.* New York: Marcel Dekker Inc.

Family: *Polyomaviridae*

Non-enveloped 40 nm icosahedral particle with 72 capsomers in a skewed (*T* = 7) arrangement. One 5000 bp molecule of covalently closed, circular dsDNA. Replicates and assembles in the nucleus. Includes murine polyomavirus and simian virus type 40 (SV40). Oncogenic.

McCance, D. J. (ed.) (1998) *Human Tumor Viruses.* Herndon, VA: ASM Press.

Family: *Poxviridae*

Enveloped brick-shaped virions of 220–450 nm × 140–260 nm × 140–260 nm thick. Complex structure enclosing one or two lateral bodies and a biconcave core, with all enzymes required for mRNA synthesis. One molecule of linear dsDNA of 130 000–375 000 bp. Cytoplasmic. Infect mostly vertebrates but also insects.

Subfamily: *Chondropoxvirinae* (viruses of vertebrates).

Genus: *Orthopoxvirus*, e.g. vaccinia and related viruses.

Genus: *Avipoxvirus*, e.g. fowlpox and related viruses.

Genus: *Cupripoxvirus*, e.g. sheeppox and related viruses.

Genus: *Leporipoxvirus*, e.g. myxoma of rabbits and related viruses; spread passively by arthropods.

Genus: *Molluscipoxvirus*, e.g. molluscum contagiosum virus.

Genus: *Parapoxvirus*, e.g. orf virus/milker's node virus and related viruses.

Genus: *Suipoxvirus*, e.g. swine pox virus.

Genus: *Yatapoxvirus*, e.g. yaba/tanapox and related viruses of monkeys.

Subfamily: *Entomopoxvirinae* (viruses of insects).

Genera: *Entomopoxvirus* A, B and C.

Moyer, R. W. & Turner, P. C. (eds) (1990) Poxviruses. *Current Topics in Microbiology and Immunology* **163**, 1–211.

Niemialtowski, M. G. *et al.* (1997) Orthopoxviruses and their immune escape. *Reviews in Medical Virology* **7**, 35–47.

Viruses with ssDNA genomes (class 2)

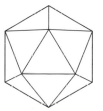

Family: *Circoviridae*

Non-enveloped small icosahedral particles with circular ssDNA. **Another genus infects plants** (see Section 22.2).

Genus: *Circovirus*. A 20 nm virion with one molecule of DNA of 1800–2300 nucleotides (nt.), e.g. chicken anaemia virus; probably also TT (transfusion-transmitted) virus of humans.

Gibbs, M. J. & Weiller, G. F. (1999) Evidence that a plant virus switched hosts to a vertebrate and then recombined with a vertebrate-infecting virus. *Proceedings of the National Academy of Sciences of the United States of America* **96**, 8022–8027.

Family: *Parvoviridae*

Non-enveloped icosahedral particles of 18–26 nm which contain no enzymes. One 5000 nt. linear molecule of *either* negative- *or* positive-sense ssDNA per particle. On extraction, these form an artefactual double strand. Replication is nuclear.

Subfamily: *Parvovirinae* (viruses of vertebrates)

Genus: *Parvovirus*, e.g. canine and feline parvovirus, minute virus of mice.

Genus: *Erythrovirus*. B19 virus which causes fifth disease of children.

Genus: *Dependovirus*. **These are satellite viruses** (see Section 22.5).

Subfamily: *Densovirinae* (viruses of insects).

Collett, M. S. & Young, N. S. (1994) Prospects for a human B19 parvovirus vaccine. *Reviews in Medical Virology* **4**, 91–103.

Hsieh, S.-Y. *et al.* (1999) High prevalence of TT virus infection in healthy children and adults and in patients with liver disease in Taiwan. *Journal of Clinical Microbiology* **37**, 1829–1831.

Parrish, C. R. (1995) Autonomous animal parvoviruses. *Seminars in Virology* **6**, 269–355.

Viruses with dsRNA genomes and a virion-associated RNA-dependent RNA polymerase (class 3)

Family: *Birnaviridae*

Non-enveloped icosahedral 60 nm particle with a 45 nm core containing two segments of linear dsRNA of 2800 and 3200 nt. Has a VPg covalently linked to 5′ end of each segment. Cytoplasmic.

Genus: *Aquabirnavirus*, e.g. pancreatic necrosis virus of fish.
Genus: *Avibirnavirus*, e.g. infectious bursal disease of chickens, turkeys, ducks.
Genus: *Entomobirnavirus*, e.g. *Drosophila* X virus.

Kurstak, E. (1993) *Viruses of Invertebrates*. New York: Marcel Dekker Inc.

Family: *Reoviridae*

Large family with members found in vertebrates, insects and *plants* (see Section 22.2). Non-enveloped 60–80 nm icosahedral particle containing an isometric nucleocapsid with 10–12 segments of linear dsRNA of 18 000–30 000 bp. Cytoplasmic. Within a genus RNA segments in a mixed infection readily assort to form genetically stable hybrid virus.

Genus: *Orthoreovirus*. Infect vertebrates (humans, dogs, cattle, birds). Ten dsRNA segments, e.g. reoviruses of humans.
Genus: *Orbivirus*. Viruses that replicate in vertebrates and their insect vectors. Ten dsRNA segments, e.g. bluetongue virus of sheep.
Genus: *Rotavirus*. The name comes from its wheel-like particle with spokes. Causes life-threatening diarrhoea in very young vertebrates of many species, including humans. Eleven dsRNA segments.
Genus: *Aquaerovirus*. Viruses of fish and *Crustacea*. Eleven dsRNA segments.
Genus: *Coltivirus*. Colorado tick fever virus of vertebrates. Twelve dsRNA segments.
Genus: *Cypovirus*. Cytoplasmic polyhedrosis viruses of insects (including *Lepidoptera*, *Hymenoptera* and *Diptera*). Ten dsRNA segments.

Chiba, S. *et al.* (eds) (1996) *Viral Gastroenteritis (Archives of Virology,* Suppl. **12**). Vienna: Springer-Verlag. (Rotaviruses)

Ramig, R. F. (ed.) (1994) Rotaviruses. *Current Topics in Microbiology and Immunology* **185**, 1–371.

Roy, P. & Gorman, B. M. (eds) (1990) Bluetongue viruses. *Current Topics in Microbiology and Immunology* **162**, 1–200.

Tyler, K. L. & Oldstone, M. B. A. (eds) (1998) Reoviruses. *Current Topics in Microbiology and Immunology* **233** (Parts 1 & 2).

Viruses with positive-sense ssRNA genomes (class 4)

Family: *Arteriviridae*

Reclassified from the *Togaviridae*. With the *Coronaviridae* forms the order *Nidovirales*. A 45–60 nm enveloped particle containing an icosahedral nucleocapsid with one molecule of linear positive-sense ssRNA of 13 000 nt. Cytoplasmic; has a nested set of subgenomic mRNAs with a common leader sequence.

Genus: *Arterivirus*, e.g. equine arteritis virus, lactate dehydrogenase-elevating virus.

Snijder, E. J. & Meulenberg, J. J. M. (1998) The molecular biology of arteriviruses. *Journal of General Virology* **79**, 961–979.

Family: *Astroviridae*

Genus: *Astrovirus*. Non-enveloped icosahedral 30 nm star-like particle with five or six points. Contain one molecule of linear positive-sense ssRNA of 7000–8000 nt. Cytoplasmic. Include human, bovine and duck astroviruses.

Chiba, S. *et al.* (eds) (1996) *Viral Gastroenteritis (Archives of Virology*, Suppl. **12**). Vienna: Springer-Verlag.
Willocks, M. M. *et al.* (1992) Astroviruses. *Reviews in Medical Virology* **2**, 97–106.

Family: *Caliciviridae*

Non-enveloped icosahedral 35–40 nm particle with 32 calyx-like (cup-shaped) depressions. Contain one molecule of linear positive-sense ssRNA of 7500 nt. The 5′ end has a VPg or a cap structure (hepatitis E virus). Cytoplasmic with subgenomic mRNA. Infect many species, e.g. human calicivirus, Norwalk virus, hepatitis E virus, rabbit haemorrhagic fever virus; swine exanthema virus.

Clarke, I. N. & Lambden, P. R. (1997) The molecular biology of caliciviruses. *Journal of General Virology* **78**, 291–301.

Family: *Coronaviridae*

With the *Arteriviridae*, forms the order *Nidovirales*. Enveloped particles of 120–160 nm with club-shaped sparse protein spikes. Contains a helical nucleocapsid with one molecule of positive-sense ssRNA of 28 000–31 000 nt. One of the largest RNA genomes. Cytoplasmic replication with a nested set of subgenomic mRNAs with a common leader sequence. Buds from Golgi apparatus and endoplasmic reticulum.

Genus: *Coronavirus*, e.g. avian infectious bronchitis virus and human coronavirus OC43.
Genus: *Torovirus*. Biconcave or toroidal or doughnut-shaped particles. Berne virus of horses and Breda virus of cattle.

Enjuanes, L. *et al.* (eds) (1997) *Coronaviruses and Arteriviruses*. New York: Plenum.
Koopmans, M. M. & Horzinek, M. C. (1994) Toroviruses of animals and humans. *Advances in Virus Research* **43**, 233–273.

Lai, M. M. C. & Cavanagh, D. (1997) The molecular biology of coronaviruses. *Advances in Virus Research* **48**, 1–100.

Family: *Flaviviridae*

Enveloped 40–60 nm particles with an isometric nucleocapsid of 25–30 nm and one molecule of linear positive-sense ssRNA of 9500–12 500 nt. Differ from *Alphaviridae* by the presence of a matrix protein, the lack of subgenomic mRNAs and budding from the endoplasmic reticulum. Cytoplasmic. Spread by arthropods unless stated.

Genus: *Flavivirus*. Large group of viruses that multiply in the vertebrate host and insect or tick vector, e.g. yellow fever virus, tick-borne encephalitis virus group, dengue virus group, Japanese encephalitis virus group.

Genus: *Pestivirus*. Includes border disease virus (sheep), bovine diarrhoea virus, classic swine fever (hog cholera) virus. No vector.

Genus: *Hepacivirus*. Hepatitis C virus of humans. No vector.

Hagedorn, C. H. & Rice, C. M. (eds) (2000) The hepatitis C viruses. *Current Topics in Microbiology and Immunology* **242**, 1–375.

Ludwig, G. V. & Iaconoconnors, L. C. (1993) Insect-transmitted vertebrate viruses— *Flavividae*. *In Vitro Cellular and Developmental Biology—Animal* **29**A, 296–309.

Meyers, G. & Thiel, H.-J. (1996). Molecular characterization of pestiviruses. *Advances in Virus Research* **47**, 53–118.

Schlesinger, M. J. & Schlesinger, S. (eds) (1986) *The Togaviridae and Flaviviridae*. New York: Plenum.

Family: *Nodaviridae*

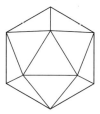

Two molecules of linear positive-sense ssRNA of 3000 and 1400 nt. in one non-enveloped 30 nm particle. Cytoplasmic multiplication with subgenomic mRNAs.

Genus: *Alphanodavirus*. Insect viruses (*Lepidoptera* and *Coleoptera*) but some grow unnaturally in suckling mice or vertebrate cells, e.g. Nodamura virus and black beetle virus.

Genus: *Betanodavirus*. Infect vertebrates, e.g. striped jack nervous necrosis virus.

Kurstak, E. (1993). *Viruses of Invertebrates*. New York: Marcel Dekker Inc.

Family: *Picornaviridae*

Non-enveloped viruses of mainly vertebrates. One molecule of linear positive-sense ssRNA of 7000–8500 nt. in 30 nm icosahedral particles. Has a 5′ VPg. Cytoplasmic.

Genus: *Enterovirus*. Acid resistant, primarily of the intestinal tract, e.g. polioviruses, most echoviruses, Coxsackie viruses of humans and various non-human enteroviruses.

Genus: *Aphthovirus*. The economically important foot-and-mouth disease viruses of cattle and other ruminants.

Genus: *Cardiovirus*, e.g. encephalomyocarditis (EMC) virus of mice.

Genus: *Hepatovirus*, e.g. human and simian hepatitis A viruses.

Genus: *Parechovirus*, e.g. human echoviruses 22 and 23.

Genus: *Rhinovirus*. Acid labile; infect the upper respiratory tract. Include about 100 common cold viruses.

Also equine rhinoviruses and various unclassified viruses of insects.

Racaniello, V. R. (ed.) (1990) Picornaviruses. *Current Topics in Microbiology and Immunology* **161**, 1–192.

Stanway, G. (1990) Structure, function and evolution of picornaviruses. *Journal of General Virology* **71**, 2483–2501.

Family: *Tetraviridae*

Non-enveloped 40 nm icosahedral particle with one molecule (6500 nt.) or two molecules (2500 and 5300 nt.) of linear positive-sense ssRNA. No infection of cultured cells. All isolated from *Lepidoptera*, e.g. Naudaurelia β virus.

Hanzlik, T. N. & Gordon, K. H. J. (2000) The *Tetraviridae*. *Advances in Virus Research* **48**, 101–168.

Kurstak, E. (1993) *Viruses of Invertebrates*. New York: Marcel Dekker Inc.

Family: *Togaviridae*

Enveloped 70 nm particles containing an icosahedral nucleocapsid and one molecule of linear positive-sense ssRNA of 9000–12 000 nt. Cytoplasmic; bud from the plasma membrane. Have an intracellular subgenomic mRNA.

Genus: *Alphavirus*. Multiply in vertebrate host and insect vector, e.g. Sindbis virus and Semliki forest virus.

Genus: *Rubivirus*. Rubella virus of humans; no vector.

Schlesinger, M. J. & Schlesinger, S. (eds) (1986) *The Togaviridae and Flaviviridae*. New York: Plenum.

Viruses with negative-sense/ambisense ssRNA genomes and a virion-associated RNA-dependent RNA polymerase (class 5)

Family: *Arenaviridae*

Enveloped isometric usually 100 nm particles with club shaped spikes. Genome is contained in two circular helical nucleocapsids, the larger with one molecule of linear negative-sense ssRNA of 7500 nt. and the smaller, also linear, of 3500 nt. Both are ambisense. May package more than two nucleocapsids per virion. Virion contains host cell ribosomes for no known function. Cytoplasmic multiplication; buds from plasma membrane. Divided into the LCM/LASV subgroup (Old World arenaviruses: lymphocytic choriomeningitis virus, lassa virus and related viruses) and the Tacaribe complex subgroup (New World arenaviruses: Tacaribe, Junin, Pichinde and related viruses).

Salvato, M. S. (ed.) (1993) *The Arenaviridae*. New York: Plenum.

Family: ***Bornaviridae***

Enveloped isometric 90 nm virions with a helical nucleocapsid and one molecule of linear negative-sense ssRNA of 9000 nt. Nuclear replication. Recently isolated from people with behavioural/neuropsychiatric problems but not known if causal. Known to infect horses since the eighteenth century and several other vertebrate species. Natural host unknown. Borna disease virus.

Nakamura, Y. *et al.* (2000) Isolation of Borna disease virus from human brain tissue. *Journal of Virology* **74**, 4601–4611.

Pringle, C. R. & Easton, A. J. (1997) Monopartite negative strand RNA genomes. *Seminars in Virology* **8**, 49–57.

de la Torre, J. C. (1994) Molecular biology of Borna disease virus: prototype of a new group of animal viruses. *Journal of Virology* **68**, 7669–7675.

Family: ***Bunyaviridae***

Enveloped isometric 100 nm particle with 10 nm spikes. Contains three helical nucleocapsids, each with one molecule of linear negative-sense ssRNA—large: 6000–12 000 nt., medium: 3500–6000 nt. and small: 1000–2000 nt. Cytoplasmic; buds from the Golgi membranes. Infect vertebrates and spread by arthropods unless stated. ***Also a genus that infects plants*** (see Section 22.2).

Genus: *Bunyavirus*. Mosquito and gnat vectors, e.g. Bunyamwera and 150 or so other viruses.

Genus: *Hantavirus*. No arthropod vector, e.g. Hantaan virus.

Genus: *Nairovirus*. Tick vector, e.g. Nairobi sheep disease virus.

Genus: *Phlebovirus*. The S RNA is ambisense. Sandfly, tick and gnat vectors, e.g. sandfly fever virus, uukuniemi virus.

Elliott, R. M. (1990) Molecular biology of the Bunyaviridae. *Journal of General Virology* **71**, 501–522.

Elliott, R. M. (ed.) (1996) *The Bunyaviridae*. New York: Plenum.

Plyusin, A. *et al.* (1996) Hantaviruses: genome structure, expression and evolution. *Journal of General Virology* **77**, 2677–2687.

Family: ***Filoviridae***

Genus: *Filovirus*. Enveloped long filamentous, sometimes branched, particles 800–900 (sometimes 14 000) nm × 80 nm with helical nucleocapsid of 50 nm diameter. One molecule of negative-sense ssRNA of 19 000 nt. Buds from plasma membrane. Includes Marburg, Ebola and Reston viruses. Highly pathogenic for humans. Natural reservoir not known.

Feldman, H. & Klenk, H.-D. (1996) Marburg and Ebola virus. *Advances in Virus Research* **47**, 1–51.

Pringle, C. R. & Easton, A. J. (1997) Monopartite negative strand RNA genomes. *Seminars in Virology* **8**, 49–57.

Family: *Orthomyxoviridae*

Enveloped pleomorphic 100 nm (sometimes filamentous) particles with a dense layer of protein spikes. Six to eight helical nucleocapsids, 9 nm in diameter with transcriptase activity. Each contains one molecule of linear negative-sense ssRNA, totalling 12 000–15 000 nt. All RNA synthesis is nuclear. Within a genus, RNA segments in a mixed infection readily assort to form genetically stable hybrid viruses. Plasma membrane budding.

Genus: *Influenzavirus A*. Genome comprises eight molecules of RNA ranging from 900 to 2350 nt. Virions have separate haemagglutinin and neuraminidase spike proteins. Undergo antigenic shift and drift. Natural reservoir is seashore birds; infect several vertebrate species including humans.

Genus: *Influenzavirus B*. Genome comprises eight molecules of RNA ranging from 900 to 2350 nt. Virions have separate haemagglutinin and neuraminidase spike proteins. Undergo antigenic drift only. Infect humans.

Genus: *Influenzavirus C*. Genome comprises seven molecules of RNA ranging from 1000 to 2350 nt. Virions have one haemagglutinin–esterase fusion spike protein. The esterase is a receptor-destroying enzyme. Undergo minor antigenic variation. Infect humans.

Genus: *Thogotovirus*. Carried by ticks and occasionally infect humans. Genome comprises six molecules of RNA. Thogoto virus and Dhori virus. Asia, Africa and Europe.

Possibly also another genus for infectious salmon anaemia virus.

Leahy, M. B. *et al.* (1997) *In vitro* polymerase activity of Thogoto virus: evidence for a new cap-snatching mechanism in a tick-borne orthomyxovirus. *Journal of Virology* **71**, 8347–8351.

Portela, A. *et al.* (1999) Replication of orthomyxoviruses. *Advances in Virus Research* **54**, 319–348.

Webster, R. G. (1998) Influenza: an emerging disease. *Emerging Infectious Diseases* **4**, 436–441.

Family: *Paramyxoviridae*

Enveloped pleomorphic particle usually 150–200 nm in diameter. Has a dense layer of fusion protein spikes and attachment protein spikes. Contains one helical nucleocapsid 13–18 nm in diameter with one molecule of linear negative-sense ssRNA of 13 000–16 000 nt. Filamentous forms of up to 400 nm common. Cytoplasmic; buds from plasma membrane. Infect vertebrates. Most, but not all, are respiratory viruses.

Subfamily: *Paramyxovirinae*

Genus: *Respirovirus*, e.g. human parainfluenzavirus type 1.

Genus: *Morbillivirus*. Measles virus, rinderpest virus and canine distemper virus.

Genus: *Rubulavirus*, e.g. mumps virus, Newcastle disease virus of poultry or avian parainfluenzavirus type 1.

Subfamily: *Pneumovirinae*

Genus: *Pneumovirus*, e.g. human and bovine respiratory syncytial viruses, pneumonia virus of mice.

Genus: *Metapneumovirus*, e.g. turkey rhinotracheitis virus.

Curran, J. & Kolakofsky, D. (1999) Replication of paramyxoviruses. *Advances in Virus Research* **54**, 403–422.

Nagai, Y. (1999) Paramyxovirus replication and pathogenesis. Reverse genetics transforms understanding. *Reviews in Medical Virology* **9**, 83–99.

Pringle, C. R. & Easton, A. J. (1997) Monopartite negative strand RNA genomes. *Seminars in Virology* **8**, 49–57.

Family: **Rhabdoviridae**

Enveloped, bullet-shaped particles 100–430 nm × 45–100 nm, containing one molecule of linear negative-sense ssRNA of 11 000–15 000 nt. Have 5–10 nm spikes. Inside is a helical nucleocapsid. Cytoplasmic and buds from plasma membrane. Infect vertebrates. *Also genera that infect plants* (see Section 22.2).

Genus: *Vesiculovirus*, e.g. vesicular stomatitis virus.

Genus: *Lyssavirus*, e.g. rabies virus, unusual as infects all mammals.

Genus: *Ephemerovirus*, e.g. bovine ephemeral fever virus.

Genus: *Novirhabdovirus*, e.g. infectious haemopoietic virus of fish.

Pringle, C. R. & Easton, A. J. (1997) Monopartite negative strand RNA genomes. *Seminars in Virology* **8**, 49–57.

Rupprecht, C. E. *et al.* (eds) (1994). Lyssaviruses. *Current Topics in Microbiology and Immunology* **187**, 1–350.

Viruses with RNA genomes that replicate through a DNA intermediate (class 6)

Family: **Retroviridae**

Enveloped 80–100 nm particles with spikes. Nucleocapsid can be isometric or a truncated cone; contains two identical copies of a linear positive-sense ssRNA of 7000–11 000 nt. Virions contain a reverse transcriptase and integrase enzymes. The DNA provirus is nuclear and integrated with host DNA. Transmission is horizontal or vertical. Associated with many different diseases. Not all viruses are oncogenic.

Genus: *'Mammalian type C retroviruses'*, e.g. murine leukaemia virus. Oncogenic.

Genus: *Alpharetrovirus*, e.g. avian leukosis virus, Rous sarcoma virus. Oncogenic.

Genus: *Betaretrovirus*, e.g. Mason–Pfizer monkey virus.

Genus: *Gammaretrovirus*, e.g. mouse mammary tumour virus. Oncogenic.

Genus: *Deltaretrovirus*, e.g. bovine leukaemia virus, human T-cell lymphotrophic virus types 1–3. Oncogenic.

Genus: *Epsilonretrovirus*, e.g. walleye dermal sarcoma virus of fish.

Genus: *Lentivirus*. Infect primates (human immunodeficiency virus types 1 and 2, simian immunodeficiency viruses), horses, sheep and goats, cattle and cats. All associated with immunodeficiency diseases.

Genus: *Spumavirus*, e.g. human spumaviruses. Named after their foamy cytopathogenic effect. No disease known.

Coffin, J. M. *et al.* (eds) (1996) *Retroviruses*. Cold Spring Harbor, NY: Cold Spring Harbor Laboratory Press.

Dalgleish, A. G. & Weiss, R. A. (1999) *HIV and the New Retroviruses* (2nd edn). New York: Academic Press.

Levy, J. A. (ed.) (1992) *The Retroviridae*, Vol. 1. New York: Plenum.

Levy, J. A. (ed.) (1993) *The Retroviridae*, Vol. 2. New York: Plenum.

Levy, J. A. (ed.) (1995) *The Retroviridae*, Vols 3 & 4. New York: Plenum.

Levy, J. A. (1998) *HIV and the Pathogenesis of AIDS* (2nd edn). Herndon, VA: ASM Press.

Viruses with a DNA genome that replicate through an RNA intermediate (class 7)

Family: *Hepadnaviridae*

A 40–48 nm enveloped particle containing an isometric nucleocapsid with DNA polymerase and protein kinase activities. One partially double-stranded circular DNA molecule that is not covalently closed. This has a complete negative-sense strand of 3000 nt. with a 5'-terminal protein, and a variable length positive-sense strand of 1700–2800 nt. The circularization overlaps the 3' and 5' termini of the negative-sense DNA. These are 'reversiviruses' that have a reverse transcriptase.

Genus: *Orthohepadnavirus*, e.g. hepatitis B (HBV) of humans, which is strongly associated with liver cancer, and woodchuck hepatitis virus.

Genus: *Avihepadnavirus*, e.g. duck hepatitis B virus.

Kurstak, E. (ed.) (1993) *Viral Hepatitis. Current status and issues*. Vienna: Springer-Verlag.

Mason, W. S. & Seeger, C. (eds) (1991) Hepadnaviruses: molecular biology and pathogenesis. *Current Topics in Microbiology and Immunology* **168**, 1–206.

22.2 Viruses that multiply in plants

Knowledge of plant virus multiplication is harder to obtain than that of animal viruses because plant cell culture systems are less manageable. Work tends to concentrate on physical properties and disease characteristics, especially as many are important in agriculture. Differences in virus proteins, translation strategy (which may not be mentioned below) and vector are important criteria in plant virus classification. Some plant viruses also multiply in their animal vector (*Bunyaviridae, Marafivirus, Reoviridae, Rhabdoviridae*), so the plant/animal virus distinction becomes uncertain.

Viruses with ssDNA genomes (class 2)

Family: *Circoviridae*
Non-enveloped small icosahedral particles with circular ss DNA. **Another genus infects animals** (see Section 22.1).

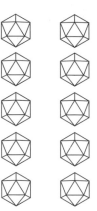

Genus: *Nanovirus*. Virions of 18 nm with 6–10 different molecules of DNA of approximately 1000 nt., each with a single open reading frame (ORF). Not known how many make up the genome. Persists but does not multiply in aphid vector, e.g. subterranean clover stunt virus.

Katul, L. *et al*. (1998) Ten distinct circular ssDNA components, four of which encode putative replication-associated proteins, are associated with the faba bean necrotic yellows virus genome. *Journal of General Virology* **79**, 3101–3109.

Family: *Geminiviridae*
Virions comprise two incomplete icosahedra, 30 nm long×18 nm, joined as Siamese twins. Non-enveloped. Has closed circular ssDNA. Nuclear replication. Persistent or not in an insect vector but does not multiply in the vector.

Genus: *Curtovirus*. Genome is one DNA molecule of 3000 nt. Does not infect monocotyledonous plants (grasses). Whitefly vector, e.g. beet curly top virus.
Genus: *Mastrevirus*. Genome is one DNA molecule of 2600 nt. Infects mainly monocotyledonous plants (grasses). Leafhopper vector, e.g. maize streak virus.

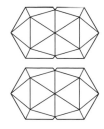

Genus: *Begomovirus*. Most genomes are two molecules of DNA each of 2500 nt. Whitefly vector, e.g. bean golden mosaic virus.

Palmer, K. E. & Rybicki, E. P. (1998). The molecular biology of the mastreviruses. *Advances in Virus Research* **50**, 183–234.

Viruses with dsRNA genomes and a virion-associated RNA-dependent RNA polymerase (class 3)

Family: *Partitiviridae*
Two molecules of linear dsRNA of 1400–3000 bp packaged in one or several non-enveloped isometric particles that lack structural detail. Have a transcriptase. Cytoplasmic. **Also genera that infect fungi** (see Section 22.3).

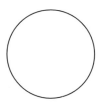

Genus: *Alphacryptovirus*. Particles 30 nm. Two RNA segments and not known if in one or two particles. Spread by seed, pollen or mechanically, e.g. white clover cryptic virus 1.
Genus: *Betacryptovirus*. Particles 38 nm. Two RNA segments. Spread by seed, pollen or mechanically, e.g. white clover cryptic virus 2.

Xie, W. S. *et al*. (1993) A third cryptic virus in beet (*Beta vulgaris*). *Plant Pathology* **42**, 464–470.

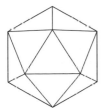

Family: *Reoviridae*

Large and widespread family. **Genera also found in vertebrates and insects** (see Section 22.1). Non-enveloped 60–80 nm icosahedral particles containing an isometric nucleocapsid with 10–12 segments of linear dsRNA of 18 000–30 000 bp. Have a transcriptase activity. Cytoplasmic. Within a genus RNA segments in a mixed infection readily assort to form genetically stable hybrid virus.

Genus: *Fijivirus*. Ten dsRNA segments. Multiply in leafhopper insect vector, e.g. Fiji disease virus.

Genus: *Oryzavirus*. Ten dsRNA segments. Multiply in insect vector, e.g. rice ragged stunt virus.

Genus: *Phytoreovirus*. Twelve dsRNA segments. Multiply in leafhopper insect vector, e.g. wound tumour virus of clover.

Nuss, D. L. & Dall, D. J. (1990) Structural and functional properties of plant reovirus genomes. *Advances in Virus Research* **38**, 249–306.

Uyeda, I. & Milne, R. G. (1995) Genomic organization, diversity and evolution of plant reoviruses. *Seminars in Virology* **6**, 85–139.

The following genus has not yet been assigned to a family

Genus: *Varicosavirus*. Two non-enveloped helical straight rods of 350 and 320 nm × 18 nm. Contain, respectively, one molecule of linear dsRNA of 7000 or 6500 bp. Cytoplasmic. Fungal vector, e.g. lettuce big-vein virus.

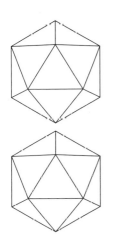

Huijberts, N. *et al.* (1990) Lettuce big-vein virus—mechanical transmission and relationship to tobacco stunt virus. *Annals of Applied Biology* **116**, 463–475.

Viruses with positive-sense ssRNA genomes (class 4)

Isometric virions

Family: *Comoviridae*

Two molecules of linear positive-sense ssRNA each encapsidated separately in a non-enveloped 30 nm icosahedral particle. Both RNAs needed for infectivity. Cytoplasmic.

Genus: *Comovirus*. RNAs of <7000 and 3981 nt. Non-persistent in beetle vector, e.g. cowpea mosaic virus.

Genus: *Fabavirus*. RNAs of <7000 and 4500 nt. Non-persistent in aphid vector, e.g. broad-bean wilt virus 1.

Genus: *Nepovirus*. RNAs of >7000 and 4000 nt. Does not multiply in nematode vector or not vectored, e.g. tobacco ringspot virus.

Chen, Z. *et al.* (1990) Capsid structure and RNA packaging in comoviruses. *Seminars in Virology* **1**, 453–466.

Harrison, B. D. & Murant, A. F. (eds) (1996) *The Plant Viruses (Polyhedral and Bipartite RNA Genomes*, Vol. 5). New York: Plenum.

Family: *Luteoviridae*

Non-enveloped icosahedral 25–30 nm particles with one molecule of linear positive-sense ssRNA of 5500–7000 nt. Does not multiply in aphid vector, but persists.

Genus: *Enamovirus*. No VPg, e.g. pea enation mosaic virus.

Genus: *Luteovirus*. No VPg, e.g. barley yellow dwarf virus.

Genus: *Polerovirus*. RNA has a 5' VPg, e.g. potato leafroll virus.

Harrison, B. D. & Murant, A. F. (eds) (1996) *The Plant Viruses (Polyhedral and Bipartite RNA Genomes*, Vol. 5). New York: Plenum.

Mayo, M. A. & Ziegler-Graff, V. (1996) Molecular biology of luteoviruses. *Advances in Virus Research* **46**, 413–460.

Family: *Sequiviridae*

Non-enveloped isometric 30 nm particle with one molecule of positive-sense linear ssRNA, 9000–12 000 nt. with a 5' VPg. Cytoplasmic. Resemble picornaviruses.

Genus: *Sequivirus*. Non-persistent in aphid vector but need a helper (Waikavirus) virus for transmission, e.g. parsnip yellow fleck virus.

Genus: *Waikavirus*. Aphid and leafhopper transmitted, e.g. rice tungro spherical virus.

Hull, R. (1996) Molecular biology of rice tungro virus. *Annual Review of Phytopathology* **34**, 275–297.

Family: *Tombusviridae*

Non-enveloped isometric 32–35 nm particle with one or two molecules of linear positive-sense ssRNA totalling 4000–5000 nt. Cytoplasmic.

Genus: *Aureusvirus*. One RNA. Soil transmitted; no vector, e.g. *Pothos* latent virus.

Genus: *Avenavirus*. One RNA. Transmitted mechanically from plant to plant and possibly also by fungi, e.g. oat chlorotic stunt virus.

Genus: *Carmovirus*. One RNA. Soilborne transmission without vector, e.g. carnation mottle virus.

Genus: *Dianthovirus*. Two RNAs (4000 and 1500 nt.) in one particle. Soilborne transmission without vector, e.g. carnation ringspot virus.

Genus: *Machlomovirus*. One RNA. Infect only grasses. Mechanical or beetle transmission, e.g. maize chlorotic mottle virus.

Genus: *Necrovirus*. One RNA. Fungal vector, e.g. tobacco necrosis virus A.

Genus: *Panicovirus*. One RNA. Transmitted mechanically, e.g. *Panicum* mosaic virus.

Genus: *Tombusvirus*. One RNA. Transmitted mechanically or by grafting, e.g. tomato bushy stunt virus.

Harrison, B. D. & Murant, A. F. (eds) (1996) *The Plant Viruses (Polyhedral and Bipartite RNA Genomes*, Vol. 5). New York: Plenum.

Russo, M. *et al.* (1994) The molecular biology of Tombusviridae. *Advances in Virus Research* **44**, 321–428.

The following genera have not yet been assigned to a family

Genus: *Idaeovirus*. One non-enveloped isometric 33 nm particle with two molecules of genomic linear ssRNA of 5500, 2200 nt. Particles also contain an mRNA encoded by RNA 2 of 1000 nt. Cytoplasmic. Transmitted in pollen (seed), e.g. raspberry bushy dwarf virus.

Harrison, B. D. & Murant, A. F. (eds) (1996). *The Plant Viruses (Polyhedral and Bipartite RNA Genomes*, Vol. 5). New York: Plenum.

Jones, A. T. *et al.* (2000) Comparisons of some properties of two laboratory strains of raspberry bushy dwarf virus (RBDV) with those of three previously published RBDV isolates. *European Journal of Plant Pathology* **106**, 623–632.

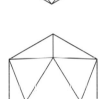

Genus: *Marafivirus*. Non-enveloped icosahedral 30 nm particle with one molecule of positive-sense linear ssRNA of 7000 nt. Cytoplasmic. Multiplies in leafhopper vector, e.g. maize rayado fino virus.

Lockhart, B. E. L. *et al.* (1985) Properties of Bermuda grass etched-line virus, a new leafhopper-transmitted virus related to maize rayado fino and oat blue dwarf viruses. *Phytopathology* **75**, 1258–1262.

Genus: *Sobemovirus*. Non-enveloped icosahedral 30 nm particle with one molecule of linear positive-sense ssRNA of 4000 nt. 5′ VPg. Cytoplasmic. Beetle vector or mechanical transmission, e.g. southern bean mosaic virus.

Makinen, K. *et al.* (1995) Characterization of cocksfoot mottle sobemovirus genomic RNA and sequence comparison with related viruses. *Journal of General Virology* **76**, 2817–2825.

Genus: *Tymovirus*

Non-enveloped icosahedral 30 nm particle with one molecule of linear ssRNA of 6000 nt. Non-persistent in beetle vectors or transmitted mechanically, e.g. turnip yellow mosaic virus.

Kadaré, G. *et al.* (1992) Comparison of the strategies of expression of five tymovirus RNAs by *in vitro* translation studies. *Journal of General Virology* **73**, 493–498.

Genus: *Umbravirus*. An unusual genus where virions are formed only with the coat protein of a coinfecting member of the luteovirus family. Linear positive-sense ssRNA of 4000 nt. with subgenomic RNAs in cells. A third component (a satellite RNA—see Section 22.5) is needed for disease and for successful transmission by aphids. Cytoplasmic. Persistent but does not multiply in aphids, e.g. groundnut rosette virus.

Naidu, R. A. *et al.* (1999) Groundnut rosette—a virus disease affecting groundnut production in sub-Saharan Africa. *Plant Disease* **83**, 700–709.

Isometric virions and virions that are short rods

Family: Bromoviridae

Non-enveloped particles. Linear positive-sense ssRNA. Tripartite genome of 2000–3000 nt. (total 8000 nt.). RNA 4 is coat protein mRNA encoded by RNA 3. Cytoplasmic.

Genus: *Alfamovirus*. Four RNAs in four bacilliform particles (short rods) 56, 43, 35 and 30 nm × 18 nm. Particles each with one molecule of RNA 1 (3644 nt.) or RNA 2 (2593 nt.) or RNA 3 (2037 nt.) or two molecules of RNA 4 (881 nt.). Infectivity requires RNAs 1–3 + coat protein or RNA 4. Cytoplasmic. Non-persistent in aphid vector, e.g. alfalfa mosaic virus.

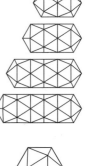

Genus: *Ilarvirus*. Four RNAs in three quasi-isometric particles of 35, 30 and 27 nm. Particles each with one molecule of RNA 1 (3491 nt.) or RNA 2 (2926 nt.) or RNA 3 (2205 nt.) + RNA 4 (900 nt.). Infectivity requires RNAs 1–3 + coat protein or RNA 4. Pollen (seed) transmitted, e.g. tobacco streak virus.

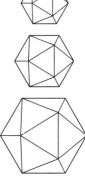

Genus: *Bromovirus*. Four RNAs in three isometric 27 nm particles. Particles each with one molecule of RNA 1 (3234 nt.) or RNA 2 (2865 nt.) or RNA 3 (2117 nt.) + RNA 4 (800 nt.). Infectivity requires RNAs 1–3 + coat protein or RNA 4. Beetle vectors, e.g. brome mosaic virus.

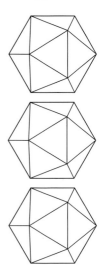

Genus: *Cucumovirus*. Three RNAs in 29 nm icosahedral particles. Particles each with one molecule of RNA 1 (3357 nt.) or RNA 2 (3050 nt.) or RNA 3 (2216 nt.) + RNA 4 (1000 nt.). Infectivity requires just the three largest RNAs. Non-persistent in aphid vector, e.g. cucumber mosaic virus.

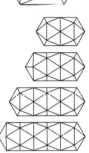

Genus: *Oleavirus*. Non-enveloped quasi-spherical to bacilliform particles of 55, 48, 43 and 37 nm × 18 nm with one RNA per virion. Genome comprises three RNAs of 3126, 2734 and 2438 nt. Also encapsidated is a fourth RNA of 2000 nt. of unknown function. Cytoplasmic. No vector known, e.g. olive latent virus 2.

Martelli, G. P. & Grieco, F. (1997) *Oleavirus*, a new genus in the family *Bromoviridae*. *Archives of Virology* **142**, 1933–1936.
Palukaitis, P. *et al.* (1992) Cucumber mosaic virus. *Advances in Virus Research* **41**, 282–348.

Virions that are rigid rods

The following genera have not yet been assigned to a family
Genus: *Benyvirus*. Non-enveloped rods of 390, 265, 100 and 85 nm × 20 nm with helical symmetry, each encapsidating one molecule of linear ssRNA of 6750, 4610, 1770 or 1470 nt. Also an RNA of 1300 nt. RNAs 1 and 2 sufficient for replication but all required for disease. Fungal vector, e.g. beet necrotic yellow vein virus.

Morales, F. J. *et al.* (1999) Emergence and partial characterization of rice stripe necrosis virus and its fungal vector in South America. *European Journal of Plant Pathology* **105**, 643–650.

Genus: *Furovirus*. Two non-enveloped rods, 300 nm × 20 nm and 100 nm × 20 nm, with helical symmetry. Contain, respectively, one molecule of linear positive-sense ssRNA of 7000 and 3500 nt. Both required for infectivity. Cytoplasmic. Fungal vector, e.g. soil-borne wheat mosaic virus.

Shirako, Y. *et al.* (2000) Similarity and divergence among viruses in the genus *Furovirus*. *Virology* **270**, 201–207.

Genus: *Hordeivirus.* Three non-enveloped rods, 148, 126, 109 nm × 20 nm, with helical symmetry. Contain, respectively, one molecule of positive-sense linear ssRNA of 3768, 3289 or 3164 nt. All required for infectivity. Cytoplasmic. Transmitted mechanically and by seed., e.g. barley stripe mosaic virus.

Edwards, M. C. *et al.* (1992) RNA recombination in the genome of barley stripe mosaic virus. *Virology* **189**, 389–392.
Jackson, A. O. *et al.* (1989) Hordeivirus relationships and genome organization. *Annual Review of Phytopathology* **27**, 95–121.

Genus: *Ourmiavirus.* Three to five non-enveloped bacilliform particles (short rods), 26–76 nm × 18–26 nm, with three molecules of linear positive-sense ssRNA of (from largest to smallest) 900–4400, 350–3300, 320–2730 or 2100 nt. RNAs have not yet been assigned to particular particles. Mechanical or seed transmission, e.g. ourmia melon virus.

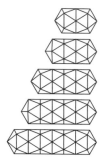

Lisa, V. *et al.* (1988). Ourmia melon virus, a virus from Iran with novel properties. *Annals of Applied Biology* **112**, 291–302.

Genus: *Pecluvirus.* Two non-enveloped rods of 245 and 190 nm × 21 nm with helical symmetry and each with a linear positive-sense ssRNA of 5900 and 4500 nt. Both required for infectivity. Fungal vector, e.g. peanut clump virus.

Herzog, E. *et al.* (1998). Identification of genes involved in replication and movement of peanut clump virus. *Virology* **248**, 312–322.

Genus: *Pomovirus.* Three non-enveloped rods of 300, 150 and 70 nm × 18 nm with helical symmetry and genome of three linear positive-sense ssRNAs of 6200, 3500 and 2500–3000 nt. Soilborne with fungal vector, e.g. potato mop-top virus.

Koenig, R. *et al.* (2000) Structure and variability of the 3′ end of RNA 3 of beet soil-borne pomovirus—a virus with uncertain pathogenic effects. *Archives of Virology* **145**, 1181–2000.

Genus: *Tobamovirus.* Non-enveloped rod, 300 nm × 18 nm, with helical symmetry. Contains one molecule of linear positive-sense ssRNA of 6395 nt. Transmitted mechanically or by seed, e.g. tobacco mosaic virus.

Dawson, W. O. (1992) Tobamovirus–plant interactions. *Virology* **186**, 359–367.
Dawson, W. O. & Lehto, K. M. (1990) Regulation of tobamovirus gene expression. *Advances in Virus Research* **38**, 307–342.

Genus: *Tobravirus.* Two non-enveloped rods, 200 nm × 22 nm and 46–115 nm × 22 nm, with helical symmetry. Contain, respectively, one molecule of linear positive-sense ssRNA of 7000 or 2000–4000 nt. The larger RNA alone is infectious and the smaller one specifies the coat protein. Both are needed for synthesis of new virions. Cytoplasmic. Nematode vector, e.g. tobacco rattle virus.

Macfarlane, S. A. (1999) Molecular biology of the tobraviruses. *Journal of General Virology* **80**, 2799–2807.

Virions that are flexuous rods (i.e. with bends but not necessarily flexible)

Family: *Closteroviridae*

Non-enveloped long highly flexuous rods 12 nm in diameter containing one or two molecules of linear positive-sense ssRNA totalling 7500–19 500 nt. Helical symmetry. Non-persistent in insect vector.

Genus: *Closterovirus*. One RNA in a 1250–2000 nm particle. Aphid, mealy bug or whitefly vectors, e.g. beet yellows virus.

Genus: *Crinivirus*. Two RNAs in separate particles of 700 and 800 nm, whitefly vector, e.g. lettuce infectious yellows virus.

Agranovsky, A. A. (1996) Principles of molecular organization, expression, and evolution of closteroviruses: over the barriers. *Advances in Virus Research* **47**, 119–158.
Coffin, R. S. & Coutts, R. H. A. (1993) The closteroviruses, capilloviruses and other similar viruses: a short review. *Journal of General Virology* **74**, 1475–1483.

Family: *Potyviridae*

Non-enveloped flexuous rod(s) with helical symmetry. There are one or two molecules of linear positive-sense ssRNA. Cytoplasmic. Genera have different vectors.

Genus: *Potyvirus*. Particle of 700–900 nm × 12 nm with one RNA molecule of 10 000 nt. A VPg at the 5′ end. Cytoplasmic. Non-persistent in aphid vector, e.g. potato virus Y.

Genus: *Ipomovirus*. One particle of 800–950 nm × 12–16 nm with one RNA molecule of 11 000 nt. Whitefly vector, e.g. sweet potato mild mottle virus.

Genus: *Macluravirus*. Non-enveloped flexuous rod, 700 nm × 13–16 nm, with one molecule of RNA of 8000 nt. Non-persistent in aphid vector, e.g. maclura mosaic virus.

Genus: *Rymovirus*. One particle of 700 nm × 11–15 nm with one molecule of RNA of 8500–10 000 nt. Mite vector, e.g. ryegrass mosaic virus.

Genus: *Tritimovirus*. One particle of 700 nm × 13 nm with one molecule of RNA of 9672 nt. Persistent in mite vector, e.g. wheat streak mosaic virus.

Genus: *Bymovirus*. Two particles of 500–600 and 250–300 nm × 13 nm. The larger with an RNA of 7632 nt. and the smaller with an RNA of 3585 nt. Fungal vector, e.g. barley yellow mosaic virus.

Chen, J. *et al.* (1999) Molecular comparisons amongst wheat bymovirus isolates from Asia, North America and Europe. *Plant Pathology* **48**, 642–647.
Gibbs, A. J. & Mackenzie, A. (1997) A primer pair for amplifying part of the genome of all potyvirids by RT-PCR. *Journal of Virological Methods* **63**, 9–16.
Salm, S. N. *et al.* (1996) Phylogenetic justification for splitting the *Rymovirus* genus of the taxonomic family *Potyviridae*. *Archives of Virology* **141**, 2237–2242.

The following genera have not yet been assigned to a family

Genus: *Allexivirus*. Non-enveloped highly flexuous rod of 800 nm × 12 nm, with one molecule of linear positive-sense ssRNA of 9000 nt. Mite vector, e.g. shallot virus X.

Sumi, S. *et al.* (1999) Complete nucleotide sequences of garlic viruses A and C, members of the newly ratified genus, *Allexivirus. Archives of Virology* **144**, 1819–1826.

Genus: *Capillovirus*. Non-enveloped flexuous rod, 600–700 nm × 12 nm, with one molecule of linear positive-sense ssRNA of 6500 nt. Transmitted through seed or by grafting, e.g. apple stem grooving virus.

Coffin, R. S. & Coutts, R. H. A. (1993) The closteroviruses, capilloviruses and other similar viruses: a short review. *Journal of General Virology* **74**, 1475–1483.

Genus: *Carlavirus*. Non-enveloped flexuous rod, 650 nm × 12 nm, with helical symmetry. Contains one molecule of linear positive-sense ssRNA of 7500 nt. Cytoplasmic. Non-persistent in aphid vector, e.g. carnation latent virus.

Foster, G. D. (1992) The structure and expression of the genome of carlaviruses. *Research in Virology* **143**, 103–112.

Genus: *Foveavirus*. Non-enveloped flexuous rod, 800 nm × 12 nm, with helical symmetry containing one molecule of linear ssRNA of 9000 nt. Particles aggregate end to end. Mechanical transmission or by grafting, e.g. apple stem pitting virus.

Martyelli, G. P. & Jelkmann, W. (1998) *Foveavirus*, a new plant virus genus. *Archives of Virology* **143**, 1245–1249.

Genus: *Potexvirus*. Non-enveloped flexuous rod, 500 nm × 13 nm, with helical symmetry. Contains one molecule of linear ssRNA of 6000–7000 nt. Cytoplasmic. Mechanical transmission, e.g. potato virus X.

Calvert, L. A. *et al.* (1996) Characterization of cassava common mosaic virus and a defective RNA species. *Journal of General Virology* **77**, 525–530.

Genus: *Trichovirus*. Non-enveloped highly flexuous rod, 500 nm × 12 nm, with one molecule of linear positive-sense ssRNA of 7500 nt. Transmitted mechanically and by grafting, e.g. apple chlorotic leaf spot virus.

Yoshikawa, N. *et al.* (1997) Grapevine berry inner necrosis virus, a new trichovirus: comparative studies with several known trichoviruses. *Archives of Virology* **142**, 1351–1363.

Genus: *Vitivirus*. Non-enveloped flexuous rod, 725–825 nm × 12 nm, with helical symmetry, crossbanding and one molecule of linear positive-sense ssRNA of 7600 nt. All transmitted mechanically, some by insects, e.g. grapevine virus A.

Martelli, G. P. *et al.* (1997) *Vitivirus*, a new genus of plant viruses. *Archives of Virology* **142**, 1929–1932.

Viruses with negative-sense/ambisense ssRNA genomes and a virion-associated RNA-dependent RNA polymerase (class 5)

Family: *Bunyaviridae*

Enveloped isometric 100 nm particle with 10 nm spikes. Each contains three helical nucleocapsids with linear negative-sense ssRNA. The largest nucleocapsid comprises an RNA of 6000–12 000 nt, the medium of 3500–6000 nt, and the smallest of 1000–2000 nt. Cytoplasmic; buds from the Golgi membranes. Spread by arthropods unless stated. **Also genera that infect animals (see Section 22.1).**

Genus: *Tospovirus*. The M and S RNAs are ambisense. Transmitted by insect (thrips) vector. Multiplies in vector, e.g. tomato spotted wilt virus.

German, T. L. *et al.* (1992) Tospoviruses: diagnosis, molecular biology, phylogeny, and vector relationships. *Annual Review of Phytopathology* **30**, 315–348.

Family: *Rhabdoviridae*

Enveloped, bullet-shaped particles 100–430 nm long and 45–100 nm in diameter, containing a helical nucleocapsid with one molecule of linear negative-sense ssRNA of 11–15 000 nt. Have 5–10 nm spikes. **Also genera that infect animals (see Section 22.1).**

Genus: *Cytorhabdovirus*. Plant viruses that mature in the cytoplasm. Spread by (and multiply in) insect vector, e.g. lettuce necrotic yellows virus.

Genus: *Nucleorhabdovirus*. Plant viruses that mature in the nucleus. Spread by (and multiply in) insect vector, e.g. potato yellow dwarf virus.

Tanno, F. *et al.* (2000) Complete nucleotide sequence of northern cereal mosaic virus and its genome organization. *Archives of Virology* **145**, 1383–1384.

The following genera have not yet been assigned to a family

Genus: *Ophiovirus*. Non-enveloped naked nucleocapsid kinked circles of 760–1300 nm × 3 nm. Three molecules of linear negative-sense ssRNA of 9000, 1700, 1500 bp. Do not infect grasses. Mechanically transmitted, e.g. citrus psorosis virus.

Alioto, D. *et al.* (1999) Improved detection of citrus psorosis virus using polyclonal and monoclonal antibodies. *Plant Pathology* **48**, 735–741.

Genus: *Tenuivirus*. Non-enveloped variable-length highly flexuous, filamentous, sometimes branched, spiral or circular narrow particles 900–1300 nm × 3–10 nm in diameter. Four molecules of linear negative-sense ssRNA of 9000, 3500, 2500 and 2000 in rice stripe virus. Segment 3 is ambisense. Particles proportional to length of their RNA. Transmission by leafhoppers, e.g. rice stripe virus.

Falk, B. W. & Tsai, J. H. (1998) Biology and molecular biology of viruses in the genus *Tenuivirus*. *Annual Review of Phytopathology* **36**, 139–163.
Ramírez, B.-C. & Haenni, A.-L. (1994) Molecular biology of tenuiviruses, a remarkable group of plant viruses. *Journal of General Virology* **75**, 467–475.

Viruses with DNA genomes that replicate through an RNA intermediate (class 7)

Family: *Caulimoviridae*

A non-enveloped particle containing one partially double-stranded circular DNA molecule that is not covalently closed (like that of hepadnaviruses), with a complete negative-sense strand of 7000–8000 nt. and an incomplete positive-sense strand with one to three discontinuities. The circularization overlaps the 3' and 5' termini of the negative-sense DNA. These are 'reversiviruses' that have a reverse transcriptase. Replication is nuclear and the genome remains episomal and does not integrate.

Genus: *Caulimovirus*. Icosahedral 50 nm particles. Non-persistent in aphid vectors, e.g. cauliflower mosaic virus.

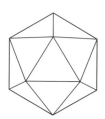

Genus: *Badnavirus*. Rod-shaped particles 130 nm × 30 nm. Non-persistent in beetle larvae, e.g. commelina yellow mottle virus.

Cheng, R. H. *et al.* (1992) Cauliflower mosaic virus: a 420 subunit (T=7) multilayer structure. *Virology* **186**, 655–668.

22.3 Viruses multiplying in algae, fungi and protozoa

Viruses with dsDNA genomes (class 1)

Family: *Adenoviridae*

Non-enveloped icosahedral particles of 60–90 nm with a fibre protein from each vertex. Contain one molecule of linear dsDNA of 25 000–42 000 bp. Replicate and assembled in the nucleus. ***Other genera infect animals*** (see Section 22.1).

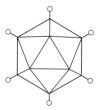

(Putative) genus: *Rhizidiovirus*. A 60 nm virion with one molecule of linear dsDNA of 25 000 bp. Infect fungi, e.g. *Rhizidiomyces* virus.

Dawe, V. H. & Kuhn, C. W. (1983). Isolation and characterization of a double-stranded DNA mycovirus infecting the aquatic fungus, *Rhizidiomyces*. *Virology* **130**, 21–28.

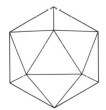

Family: *Phycodnaviridae*

Infect algae. Non-enveloped, isometric, 130–190 nm, multilayered particles containing one molecule of linear dsDNA of 160–380 000 bp. Contain internal lipid. Infect *Paramecium*, *Chlorella* and *Hydra* spp. May have a use in controlling algal blooms.

van Etton, J. L. *et al.* (1991) Viruses and virus-like particles of eukaryotic algae. *Microbiological Reviews* **55**, 586–620.

Mlot, C. (1999). Lytic virus targets harmful algal bloom. *American Society for Microbiology News* **65**, 592–594.

Viruses with dsRNA genomes and a virion-associated RNA-dependent RNA polymerase (class 3)

Family: *Hypoviridae*

Genus: *Hypovirus*. Vesicles of 50–80 nm with one molecule of linear dsRNA of 9000–13 000 bp, no structural proteins and polymerase activity. Surrounded by rough endoplasmic reticulum in infected cells. Causes hypovirulence of host chestnut blight fungus, e.g. *Cryphonectria* hypovirus 1.

Choi, G. H. & Nuss, D. L. (1992) Hypovirulence of chestnut blight fungus conferred by an infectious viral cDNA. *Science* **257**, 800–803.

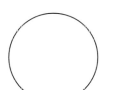

Family: *Partitiviridae*

Two molecules of linear dsRNA of 1400–3000 bp packaged in one or several non-enveloped isometric particles that lack structural detail. Have a transcriptase. Cytoplasmic. *Also genera that infect plants* (see Section 22.2).

Genus: *Partitivirus*. Particles 30–35 nm. Infect fungi, e.g. *Gaeumannomyces graminis* virus 019/6A.

Genus: *Chrysovirus*. Particles 35–40 nm. Infect fungi, e.g. *Penicillium chrysogenum* virus.

Strauss, E. E. *et al.* (2000). Molecular characterization of the genome of a partitivirus from the basidomycete *Rhizoctonia saloni*. *Journal of General Virology* **81**, 549–555.

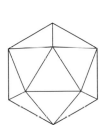

Family: *Totiviridae*

Non-enveloped particles of 30–40 nm with one molecule of linear dsRNA of 5000–7000 bp. One major capsid protein. Single shell. Cytoplasmic.

Genus: *Totivirus*. Some viruses also have other satellite RNA segments that encode killer proteins. Infect the fungus *Saccharomyces cerevisiae*.

Genus: *Giardiavirus*. Infect the protozoan *Giardia lamblia*.

Genus: *Leishmaniavirus*. One RNA. Infect protozoa: *Leishmania* spp.

Ghabrial, S. A. (1994) New developments in fungal virology. *Advances in Virus Research* **43**, 303–338.

Stuart, K. D. *et al.* (1992) Molecular organization of Leishmania RNA virus. *Proceedings of the National Academy of Sciences of the United States of America* **89**, 8596–8600.

Viruses with positive-sense ssRNA genomes (class 4)

Family: *Barnaviridae*
Non-enveloped icosahedral short rod, 50 nm×19 nm, with one molecule of linear ssRNA of 4009 nt. Infect fungi, e.g. mushroom bacilliform virus.

Ghabrial, S. A. (1994) New developments in fungal virology. *Advances in Virus Research* **43**, 303–338.

Family: *Narnaviridae*
One molecule of linear positive-sense ssRNA of 2500 nt. as a ribonucleoprotein with polymerase activity. Infect fungi.

Genus: *Narnavirus*, e.g. *Saccharomyces cerevisiae* 20S narnavirus.
Genus: *Mitovirus*, e.g. *Cryphonectria parasitica* mitovirus 1 NB631.

Ghabrial, S. A. (1994) New developments in fungal virology. *Advances in Virus Research* **43**, 303–338.

22.4 Viruses (phages) multiplying in *Archaea*, bacteria, *Mycoplasma* and *Spiroplasma*

The well-known molecular biology of phages is based on a detailed study of a few representatives, and surprisingly little is known of their comparative biology.

Viruses with dsDNA genomes (class 1)

Viruses that have some head–tail structure, here ordered by decreasing tail length

Family: *Siphoviridae*
Non-enveloped particle with a long non-contractile tail up to 570 nm×8 nm. Icosahedral head of 60 nm. Contains one molecule of linear dsDNA of 48 500 bp. Causes no host DNA breakdown. Infect bacteria including *Enterobacteria*, *Mycobacteria*, *Lactococcus*, *Methanebacteria* and *Vibrio*. Include coliphage T1, lambda (λ), chi (χ) and phi (φ) 80.

Ackermann, H.-W. (1999) Tailed bacteriophages. *Advances in Virus Research* **51**, 135–201.
Hendrix, R. W. *et al.* (eds) (1983) *Lambda II*. Cold Spring Harbor, NY: Cold Spring Harbor Laboratory Press.

Family: *Myoviridae*
Non-enveloped particle with a complex contractile tail 80–455 nm×16 nm. Contraction requires ATP. Head separated from tail by a neck. Head isometric (as shown) or elongated, 110 nm×80 nm. Has one molecule of linear dsDNA of 40 000–170 000 bp. Infect *Enterobacteria*, *Bacillus* and *Halobacteria* spp. (*Archaea*). Include the 'T-even' coliphages T2, T4 and T6, and PBS1, SP8, SP50, P1, P2, 21, 34 and Mu.

Ackermann, H.-W. (1999) Tailed bacteriophages. *Advances in Virus Research* **51**, 135–201.

Baker, T. A. (1995) Bacteriophage mu: a transposing phage that integrates like retroviruses. *Seminars in Virology* **6**, 53–63.

Karam, J. D. (ed.) (1994) *Molecular Biology of Bacteriophage T4*. Washington, DC: ASM Press.

Family: *Podoviridae*

Non-enveloped particle with a short non-contractile tail of 20 nm × 8 nm. Icosahedral head of 60 nm. One molecule of linear dsDNA of 19 000–44 000 bp. Cause host DNA breakdown. Include the coliphages T3 and T7, enterobacteria phage P22 and bacillus phage phi (φ) 29.

Ackermann, H.-W. (1999) Tailed bacteriophages. *Advances in Virus Research* **51**, 135–201.

Hausmann, R. (1988) The T7 group. In: *The Bacteriophages*, Vol. 1, pp. 259–289. Calender, R. (ed.) New York: Plenum.

Viruses that do not have a head–tail structure

Family: *Fuselloviridae*

Enveloped lemon-shaped particle 100 nm × 60 nm with short tail fibres at one pole. One molecule of circular dsDNA of 15 465 bp. Infect *Archaea* such as *Desulfolobus* and *Methanococcus*, e.g. *Sulfolobus* virus 1.

Schleper, C. *et al.* (1992) The particle SSV1 from the extremely thermophilic archaeon *Sulfolobus* is a virus. *Proceedings of the National Academy of Sciences of the United States of America* **89**, 7645–7649.

Family: *Tectiviridae*

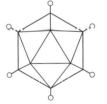

Non-enveloped icosahedral 63 nm particle. Unusual enveloped nucleoprotein surrounded by lipid. One molecule of linear dsDNA of 150 000 bp. On attachment forms a tail-like tube 60 nm × 10 nm. Infect Gram-negative bacteria, e.g. enterobacteria phage PRD1. Have similarities with adenoviruses.

Ackermann, H.-W. (1999) Tailed bacteriophages. *Advances in Virus Research* **51**, 135–201.

Bamford, D. H. *et al.* (1995) Bacteriophage PRD1: a broad host range dsDNA tectivirus with an internal membrane. *Advances in Virus Research* **45**, 281–319.

Family: *Corticoviridae*

Non-enveloped, non-tailed icosahedral 60 nm particle formed of several layers, including one of lipid. Spikes at vertices. One molecule of circular dsDNA of 9000 bp. Infect *Pseudomonas*, e.g. *Alteromonas* phage PM2.

Merino, S. *et al.* (1998) Bacteriophage PM2 nomenclature revision. *Archives of Virology* **143**, 1852–1853.

Family: *Plasmaviridae*

Enveloped pleomorphic 80 nm particle with small dense core; 50 and 125 nm particles also produced. One molecule of circular dsDNA of 12 000 bp. Infect *Mycoplasma*, e.g. *Acholeplasma* phage L2.

Maniloff, J. (1988) Mycoplasma viruses. *CRC Critical Reviews in Microbiology* 15, 339–389.
Maniloff, J. *et al.* (1994) Sequence analysis of a unique temperate phage: mycoplasma virus L2. *Gene* **141**, 1–8.

Family: *Rudiviridae*

Non-enveloped rigid rod of 500–780 nm × 23 nm with helical symmetry, with a plug and three tail fibres at each end. One molecule of linear dsDNA of 33 000 bp. Non-lytic. Infect thermophilic *Archaea*, *Sulfolobus*, e.g. *Sulfolobus* virus SIRV-1.

Prangishvili, D. *et al.* (1999) A novel virus family, the *Rudiviridae*: structure, virus–host interactions and genome variability of the *Sulfolobus* viruses SIRV1 and SIRV2. *Genetics* **152**, 1387–1396.

Family: *Lipothrixviridae*

Enveloped, rigid rod of 400 nm × 40 nm with protrusions at both ends that participate in cell attachment. One molecule of linear dsDNA of 16 000 bp. Virions stable at 100°C. Infect *Archaea*, e.g. *Thermoproteus* virus 1.

Neumann, H. & Zillig, W. (1990). Structural variability in the genome of the *Thermoproteus tenax* virus TTV1. *Molecular Genes and Genetics* **222**, 435–437.

Viruses with ssDNA genomes (class 2)

Family: *Inoviridae*

Non-enveloped rod with one molecule of circular positive-sense ssDNA of 5000–10 000 nt. Virion length depends on the length of the DNA and its conformation.

Genus: *Inovirus*. Flexuous rods of 700–2000 nm × 7 nm with DNA of 4400–8500 nt. Host bacteria not lysed. Includes Enterobacteria phage M13 and Enterobacteria phage fd.

Genus: *Plectrovirus*. Nearly straight rods of 85 or 280 nm × 10–15 nm with DNA of 4400–8500 nt. Infect *Mycoplasma* (85 nm) and *Spiroplasma* (280 nm), e.g. *Acholeplasma* phage MV-L51.

Kuo, T. T. *et al.* (1991) Complete nucleotide sequence of filamentous phage Cflt from *Xanthomonas campestris* pv citri. *Nucleic Acids Research* **19**, 2498.
Zinder, N. D. & Horiuchi, K. (1985) Multiregulatory elements of filamentous bacteriophages. *Microbiological Reviews* **49**, 101–106.

Family: *Microviridae*

Icosahedral particle with large knobs on the 12 vertices. Minimum and maximum diameters are 22 and 23 nm respectively. One molecule of circular ssDNA of 4400–5400 nt.

Genus: *Microvirus*. Infect bacteria, e.g. enterobacteria phage phi(ϕ)X174.
Genus: *Spiromicrovirus*. Infect *Spiroplasma*, e.g. *Spiroplasma* phage 4.
Genus: *Chlamydiamicrovirus*, e.g. *Chlamydia* phage 1.
Genus: *Bdellomicrovirus*, e.g. *Bdellovibrio* phage MAC 1.

McKenna, R. *et al.* (1992). Atomic structure of single-stranded DNA bacteriophage ϕX174 and its functional implications. *Nature (London)* **355**, 137–143.

The following genus has not yet been assigned to a family

Genus: 'Sulfolobus SNDV-like viruses'. Drop-shaped virions 180 nm × 80 nm containing covalently closed circular DNA of 20 000 bp that resists digestion by many restriction endonucleases suggesting extensive chemical modification. Infect *Archaea*, e.g. *Sulfolobus* virus SNDV.

Zillig, W. *et al.* (1996) Viruses, plasmids and other genetic elements of thermophilic and hyperthermophilic *Archaea*. *FEBS Microbiological Reviews* **18**, 225–236.

Viruses with dsRNA genomes and a virion-associated RNA-dependent RNA polymerase (class 3)

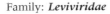

Family: *Cystoviridae*

Enveloped icosahedral 85 nm particle with 8 nm spikes. Has a 58 nm nucleocapsid and a 43 nm core. Each particle contains three molecules of linear dsRNA of 6374, 4057 and 2948 bp. Infect *Pseudomonas*, e.g. *Pseudomonas* phage phi(ϕ)6.

Mindich, L. (1988) Bacteriophage phi 6: a unique virus having a lipid-containing membrane and a genome composed of three ds RNA segments. *Advances in Virus Research* **35**, 137–176.
Mindich, L. (1995) Heterologous recombination in the segmented dsRNA genome of bacteriophage phi 6. *Seminars in Virology* **6**, 75–83.

Viruses with positive-sense ssRNA genomes (class 4)

Family: *Leviviridae*

Non-enveloped 26 nm icosahedral particles with one molecule of linear positive-sense ssRNA of 3400–4300 nt. Include enterobacteria phages R17, MS2 and Qβ.

Valegaard, K. *et al.* (1990) The three-dimensional structure of the bacterial virus MS2. *Nature (London)* **345**, 36–41.

22.5 Satellite viruses and satellite nucleic acids of animals, plants, fungi and bacteria

The replication of satellite viruses and satellite nucleic acids depends upon co-infection of a host cell with a helper virus. The satellite genome has no significant homology with that of the helper and hence differs from other types of dependent nucleic acid molecules. Here, we use a broad definition of a satellite virus—one that is incapable of independent production of progeny virus particles, yet may be able to replicate itself. *Satellite viruses* encode their own coat protein, whereas *satellite nucleic acids* do not and use the coat protein of their helper virus. Some satellite viruses/satellite nucleic acids modulate the replication of the helper virus and exacerbate or diminish disease.

Satellite nucleic acids with dsDNA genomes (class 1)

Non-enveloped isometric virions with a *Myoviridae* helper (enterobacteria phage P2). One molecule of linear dsDNA of 11 627 bp. Interesting that the head–tail helper virion protein can form an isometric virion, e.g. enterobacteria P4 satellite.

Bertani, L. E. & Six, E. W. (1988) The P2-like phages and their parasite, P4. *Annual Review of Genetics* **24**, 465–490.

Satellite viruses with ssDNA genomes (class 2)

Family: ***Parvoviridae***
Non-enveloped icosahedral particle of 18–26 nm which contains no enzyme. One 5000-nt. linear molecule of *either* negative- *or* positive-sense ssDNA per particle. On extraction these form an artefactual double strand. Nuclear replication. **There are also independently replicating (autonomous) parvoviruses of animals** (see Section 22.1).

Subfamily: *Parvovirinae*
Genus: *Dependovirus*. Dependent on helper adenovirus or herpesvirus for efficient replication, but in some cell cultures replicate independently. Particles package 90% negative-sense DNA. Encode their own coat protein. Adeno-associated viruses. Infect vertebrates.

Berns, K. I. & Giraud, C. (1996). Biology of adeno-associated virus. *Current Topics in Microbiology and Immunology* **218**, 1–23.

Satellite nucleic acids with ssDNA genomes (class 2)

A genome of 682 nt. with no ORF. Geminiviruses helpers, e.g. tomato leaf curl virus satellite DNA. Infect plants. Particle morphology not yet determined.

Dry, I. B. *et al*. (1997). A novel subviral agent associated with a geminivirus: the first report of a DNA satellite. *Proceedings of the National Academy of Sciences of the United States of America*, **94**, 7088–7093.

Satellite nucleic acids with dsRNA genomes (class 3)

Non-enveloped isometric particles with a *Totiviridae* helper virus. Linear dsRNA of 500–1800 bp inside a virion made of helper virus coat protein, e.g. satellite of *Saccharomyces cerevisiae* M virus. Infect fungi.

Satellite viruses with positive-sense ssRNA genomes (class 4)

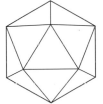

Isometric particles with one linear ssRNA.

Tobacco necrosis virus satellite virus subgroup
Particles of 17 nm containing one molecule of RNA of 1239 nt. Fungal vector. Infect plants.

Chronic bee-paralysis virus-associated satellite virus subgroup
Particles of 12 nm containing three molecules of RNA totalling 1000 nt. Infect animals.

Satellite nucleic acids with positive-sense ssRNA genomes (class 4)

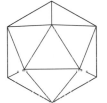

Large linear mRNA satellites
Encode a non-structural protein. Rarely modify disease. Linear RNA of 800–1500 nt., e.g. arabis mosaic virus large satellite RNA. Infect plants.

Satellite nucleic acids with negative-sense ssRNA genomes (class 5)

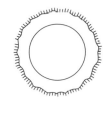

Deltavirus. Hepatitis delta virus. Enveloped isometric particle of 36 nm with one circular RNA molecule of 1700 nt. Dependent on hepatitis B virus which contributes the envelope, but expresses 2 delta antigen proteins which form the core. RNA has self-cleaving (ribozyme) activity. May exacerbate hepatitis B virus infection. Infects humans.

Karayiannis, P. (1998) Hepatitis D virus. *Reviews in Medical Virology* **8**, 13–24.
Taylor, J. M. (1999) Replication of human hepatitis delta virus: influence of studies on subviral plant pathogens. *Advances in Virus Research* **54**, 45–60.

Satellite nucleic acids with ssRNA genomes (unclassified as make no mRNA)

Have no functional ORF.

Small linear RNA satellites

Linear RNA of 340 nt., e.g. cucumber mosaic virus satellite RNA. Infect plants.

Circular RNA satellites

Covalently closed circular RNA of usually 350 nt. Smallest is 220 nt. All form a double-stranded rod-like viroid RNA. This is encapsidated. Also called virusoids. Have self-cleaving (ribozyme) activity. Five types all with sobemovirus helpers, e.g. velvet tobacco mottle virus satellite RNA. Infect plants.

Collmer, C. W. & Howell, S. H. (1992) Role of satellite RNA in the expression of symptoms caused by plant viruses. *Annual Review of Phytopathology* **30**, 419–442.

Maramarosch, K. (ed.) (1991) *Viroids and Satellites: Molecular parasites at the frontier of life.* Boca Raton, FL: CRC Press.

Roossinck, M. J. *et al.* (1992) Satellite RNAs of plant viruses: structures and biological effects. *Microbiological Reviews* **56**, 265–279.

Tien, P. & Wu, G. (1991) Satellite RNA for the biocontrol of plant diseases. *Advances in Virus Research* **39**, 321–339.

Vogt, P. K. & Jackson, A. O. (eds) (1999) Satellites and defective viral RNAs. *Current Topics in Microbiology and Immunology* **239**, 1–180.

22.6 Viroids (unclassified as make no mRNA)

Viroids are small, circular, single-stranded, infectious RNAs (246–370 nt.) which are never encapsidated and have no helper virus (and so are distinguished from circular RNA satellites), and many are serious plant pathogens. They have extensive internal base pairing, so that the RNAs resemble double-stranded rods rather than circles. There is no ORF in either sense and hence they encode no protein and are not classified under the Baltimore scheme. Some viroids are replicated in the nucleus by the host's RNA polymerase II. Others are found in chloroplasts. Transmitted through vegetative propagation of the host, by seed, by aphids or through mechanical damage, and so overcome the problem of there being no receptor for naked RNA. Infect plants.

Family: *Pospiviroidae*

Have a central conserved sequence, e.g. potato spindle tuber viroid, coconut cadang cadang viroid.

Family: *Avsunviroidae*

No central conserved sequence but undergo self-cleavage, e.g. avocado sunblotch viroid.

Maramarosch, K. (ed.) (1991) *Viroids and Satellites: Molecular parasites at the frontier of life.* Boca Raton, FL: CRC Press.

Symons, R. H. (1990) Viroids and related pathogenic RNAs. *Seminars in Virology* **1**, 75–162.

22.7 Further reading and references

Further information and references can be obtained from many sources but a good starting place is:

http://life.anu.edu.au/viruses/ICTVdB/index.htm (though somewhat dated now)

http://www.mirror.ac.uk/sites/www.virology.net/index.html

Fields, B. N., Knipe, D. M. & Howley, P. M. (eds) (1996) *Virology* (3rd edn), Vols 1 and 2. Hagerstown, MD: Lippincott-Raven.

Francki, R. (1985) *Atlas of Plant Viruses*, Vols 1 and 2. Boca Raton, FL: CRC Press.

Granoff, A. & Webster, R. G. (eds) (1999) *Encyclopedia of Virology* (2nd edn), Vols 1–3. New York: Academic Press.

Matthews, R. E. F. (1992) *Fundamentals of Plant Virology*. New York: Academic Press (a more manageable text).

Matthews, R. E. F. (1992) *Plant Virology* (3rd edn). New York: Academic Press (for advanced students).

van Regenmortel, M. H. V., Fauquet, C. M., Bishop, D. H. L., Carstens, J. M., Estes, M. K., Lemon, S. M., Maniloff, J., Mayo, M. A., McGeoch, D. J., Pringle, C. R. & Wickner, R. B., (eds) (2000) *Virus Taxonomy: Seventh Report of the International Committee on Virus Taxonomy of Viruses*. San Diego, CA: Academic Press.

Rothman-Denes, L. B. & Weisberg, R. (1995) Recent developments in bacteriophage virology. *Seminars in Virology* **6**, 1–83.

Index

Page numbers in *italics* refer to figures; pages in **bold** refer to tables.
Since the major subjects of this book are viruses, few entries are listed under this
and 'viral' keywords. Readers are advised to seek more specific references.
Abbreviations used in subentries: EBV, Epstein–Barr virus; LCMV, lymphocytic
choriomeningitis virus; TMV, tobacco mosaic virus; VZV, varicella-zoster virus.